Knowledge, Innovation and Internationalisation

As firms increasingly rely upon knowledge as a key factor for innovation, the ability to innovate is increasingly perceived as a key asset for being competitive in international markets. This new volume argues that innovation, knowledge and internationalisation should be viewed as tightly related concepts. It provides a stimulating and comprehensive framework for understanding key tendencies in modern economics, as well as an overview of the state of the art in the three fields covered.

The first section explores in detail the relationship between knowledge and the innovative capability of firms, focusing on key topics such as social capital, intentional knowledge diffusion and unintentional knowledge spillovers. Section two examines the drivers and the impact of innovation strategies, assessing the role of technological advantage, networking and R&D investments in innovation, as well as the impact of innovation on the labour market. The third and final section examines the ongoing internationalisation process faced by 'global' economies. The topics explored in each section are tightly linked, ensuring that a strong thematic thread runs through the collection.

Piergiuseppe Morone is associate professor at the Sapienza – University of Rome, Italy.

Routledge Studies in Global Competition
Edited by *John Cantwell*, Rutgers,
the State University of New Jersey, USA
and *David Mowery*,
University of California, Berkeley, USA

1 **Japanese Firms in Europe**
Edited by Frédérique Sachwald

2 **Technological Innovation, Multinational Corporations and New International Competitiveness**
The case of intermediate countries
Edited by José Molero

3 **Global Competition and the Labour Market**
Nigel Driffield

4 **The Source of Capital Goods Innovation**
The role of user firms in Japan and Korea
Kong-Rae Lee

5 **Climates of Global Competition**
Maria Bengtsson

6 **Multinational Enterprises and Technological Spillovers**
Tommaso Perez

7 **Governance of International Strategic Alliances**
Technology and transaction costs
Joanne E. Oxley

8 **Strategy in Emerging Markets**
Telecommunications establishments in Europe
Anders Pehrsson

9 **Going Multinational**
The Korean experience of direct investment
Edited by Frédérique Sachwald

10 **Multinational Firms and Impacts on Employment, Trade and Technology**
New perspectives for a new century
Edited by Robert E. Lipsey and Jean-Louis Mucchielli

11 **Multinational Firms**
The global-local dilemma
Edited by John H. Dunning and Jean-Louis Mucchielli

12 **MIT and the Rise of Entrepreneurial Science**
Henry Etzkowitz

13 **Technological Resources and the Logic of Corporate Diversification**
Brian Silverman

14 **The Economics of Innovation, New Technologies and Structural Change**
Cristiano Antonelli

15 **European Union Direct Investment in China**
Characteristics, challenges and perspectives
Daniel Van Den Bulcke, Haiyan Zhang and Maria do Céu Esteves

16 **Biotechnology in Comparative Perspective**
Edited by Gerhard Fuchs

17 **Technological Change and Economic Performance**
Albert L. Link and Donald S. Siegel

18 **Multinational Corporations and European Regional Systems of Innovation**
John Cantwell and Simona Iammarino

19 **Knowledge and Innovation in Regional Industry**
An entrepreneurial coalition
Roel Rutten

20 **Local Industrial Clusters**
Existence, emergence and evolution
Thomas Brenner

21 **The Emerging Industrial Structure of the Wider Europe**
Edited by Francis McGowen, Slavo Radosevic and Nick Von Tunzelmann

22 Entrepreneurship
A new perspective
Thomas Grebel

23 Evaluating Public Research Institutions
The U.S. advanced technology program's intramural research initiative
Albert N. Link and John T. Scott

24 Location and Competition
Edited by Steven Brakman and Harry Garretsen

25 Entrepreneurship and Dynamics in the Knowledge Economy
Edited by Charlie Karlsson, Börje Johansson and Roger R. Stough

26 Evolution and Design of Institutions
Edited by Christian Schubert and Georg von Wangenheim

27 The Changing Economic Geography of Globalization
Reinventing space
Edited by Giovanna Vertova

28 Economics of the Firm
Analysis, evolution and history
Edited by Michael Dietrich,

29 Innovation, Technology and Hypercompetition
Hans Gottinger

30 Mergers and Acquisitions in Asia
A global perspective
Roger Y. W. Tang and Ali M. Metwalli

31 Competitiveness of New Industries
Institutional framework and learning in information technology in Japan, the U.S and Germany
Edited Cornelia Storz and Andreas Moerke

32 Entry and Post-Entry Performance of Newborn Firms
Marco Vivarelli

33 Changes in Regional Firm Founding Activities
A theoretical explanation and empirical evidence
Dirk Fornahl

34 Risk Appraisal and Venture Capital in High Technology
New ventures
Gavin C. Reid and Julia A. Smith

35 Competing for Knowledge
Creating, connecting and growing
Robert Huggins and Hiro Izushi

36 Corporate Governance, Finance and the Technological
Advantage of nations
Andrew Tylecote and Francesca Visintin

37 Dynamic Capabilities Between Firm Organisation and Local Systems of Production
Edited by Riccardo Leoncini and Sandro Montresor

38 Localised Technological Change
Towards the economics of complexity
Cristiano Antonelli

39 Knowledge Economies
Innovation, organization and location
Wilfred Dolfsma

40 Governance and Innovation
Maria Brouwer

41 Public Policy for Regional Development
Edited by Jorge Martinez-Vazquez and François Vaillancourt

42 Evolutionary Economic Geography
Location of production and the European Union
Miroslav Jovanovic

43 Broadband Economics
Lessons from Japan
Takanori Ida

44 Targeting Regional Economic Development
Edited by Stephan J. Goetz, Steven C. Deller and Thomas R. Harris

45 Innovation, Knowledge and Power
Theodora Asimakou

46 **Creativity, Innovation and the Cultural Economy**
Edited by Andy C. Pratt and Paul Jeffcutt

47 **Co-opetition Strategy**
Giovanni Battista Dagnino and Elena Rocco

48 **Knowledge Intensive Entrepreneurship and Innovation Systems**
Evidence from Europe
Edited by Franco Malerba

49 **Innovation in Complex Social Systems**
Edited by Petra Ahrweiler

50 **Internationalization, Technological Change and the Theory of the firm**
Edited by Nicola De Liso and Riccardo Leoncini

51 **Territory, Specialization and Globalization in European Manufacturing**
By Helena Marques and Francisco Puig

52 **Institutional Diversity and Innovation**
Continuing and emerging patterns in Japan and China
Cornelia Storz and Sebastian Schäfer

53 **Innovation and Economic Crisis**
Daniele Archibugi and Andrea Filippetti

54 **The Communications Industries in the Era of Convergence**
Catherine Mulligan

55 **Innovation, Technology and Knowledge**
Charlie Karlsson, Borje Johansson and Roger R. Stough

56 **Evolution of Competition Laws and their Enforcement**
Pradeep S. Mehta

57 **The Economics of Structural Change in Knowledge**
Francesco Quatraro

58 **Economic Geography and the Unequal Development of Regions**
Jean-Claude Prager and Jacques-François Thisse

59 **Social Networks, Innovation and the Knowledge Economy**
Edited by Isabel Salavisa and Margarida Fontes

60 The Economics of Creativity
Ideas, firms and markets
Edited by Thierry Burger-Helmchen

61 Epistemic Economics and Organization
Forms of rationality and governance for a discovery oriented economy
Anna Grandori

62 Universities, Cities and Regions
Loci for knowledge and innovation creation
Edited by Roberta Capello, Agnieszka Olechnicka and Grzegorz Gorzelak

63 Knowledge, Innovation and Internationalisation
Essays in honour of Cesare Imbriani
Edited by Piergiuseppe Morone

Knowledge, Innovation and Internationalisation

Essays in honour of Cesare Imbriani

Edited by Piergiuseppe Morone

Routledge
Taylor & Francis Group
LONDON AND NEW YORK

First published 2013 by Routledge
2 Park Square, Milton Park, Abingdon, Oxon OX14 4RN

Simultaneously published in the USA and Canada
by Routledge
605 Third Avenue, New York, NY 10017

Routledge is an imprint of the Taylor & Francis Group, an informa business

British Library Cataloguing in Publication Data
A catalogue record for this book is available from the British Library

Library of Congress Cataloging in Publication Data
A catalog record for this book has been requested

ISBN: 978-0-415-69311-0 (hbk)
ISBN: 978-0-203-38549-4 (ebk)

Typeset in Times New Roman
by RefineCatch Limited, Bungay, Suffolk

Contents

List of figures xiii
List of tables xiv
Notes on contributors xvi

Knowledge, innovation and internationalisation: a roadmap 1
PERGIUSEPPE MORONE

PART I
Knowledge 15

1 **Knowledge and innovation in social networks: an assessment of industrial clusters' literature** 17
ELISA GIULIANI AND F. XAVIER MOLINA-MORALES

2 **Friends and rivals: modelling the social relations of inventors** 38
LORENZO CASSI AND LORENZO ZIRULIA

3 **Knowledge and innovative entrepreneurship: social capital and individual capacities** 59
UWE CANTNER AND MICHAEL STUETZER

4 **Agglomeration economies, forward linkages and regional specialisation in business services in the EU27** 91
VALENTINA MELICIANI AND MARIA SAVONA

PART II
Innovation 109

5 **Technological competences and regional innovation networks: measurement and visualisation with patent data** 111
ANDREAS PYKA, BERND EBERSBERGER AND DOMINIK HARTMANN

6 **Implementation cycles and the new economy in retrospect** **130**
PASQUALE SCARAMOZZINO, JONATHAN TEMPLE AND NIR VULKAN

7 **The impact of technological change on employment and skills: theory and empirical evidence** **155**
MARIACRISTINA PIVA AND MARCO VIVARELLI

8 **Design and evaluation of fiscal incentives for business innovation in Latin America: lessons learned and future developments** **171**
GUSTAVO A. CRESPI

PART III
Internationalisation **195**

9 **Trade policy and internationalisation** **197**
DOMINICK SALVATORE

10 **Models of speculative attacks and crashes in international capital markets** **216**
GIANCARLO MARINI AND GIOVANNI PIERSANTI

11 **Firms' internationalisation, innovation strategies and product quality** **243**
PIERGIUSEPPE MORONE AND FRANCESCO RENNA

12 **Do foreign ownership and technological environment affect firm survival? Evidence on Italian manufacturing and services firms** **258**
FILIPPO REGANATI, ANNA M. FERRAGINA AND ROSANNA PITTIGLIO

Index 275

Figures

0.1	The book structure at a glance	3
2.1	Social distance and citations	41
2.2	An example of knowledge set	44
2.3	The three social structures	48
2.4	Number of patents over time	52
2.5	Social distance and knowledge diffusion (normalised to average citations)	54
3.1	The moderating effect of teams' variety of functional background	80
4.1	Specialisation in business services in EU regions – comparative advantage in 2003	97
4.2	Moran scatterplot of specialisation in business services	97
5.1	Regional municipalities and their contribution to the overall patent score	119
5.2	Revealed technological advantages in Bremen	121
5.3	Revealed technological disadvantages in Bremen	122
5.4	International comparison of Bremen's RTAs	123
5.5	Patent-based innovation network of Bremen	124
5.6	Internationalisation of the innovation profile in Bremen	126
6.1	Annual MFP growth rate, non-farm private business	133
6.2	Annual data on Initial Public Offerings	134
6.3	The co-movement of IPOs and MFP growth	135
6.4	Annual MFP growth (manufacturing, non-durables) and IPOs	136
6.5	Annual MFP growth (manufacturing, durables) and IPOs	137
6.6	The declining volatility of MFP growth	140
6.7	The declining volatility of MFP growth in manufacturing	141
6.8	The declining volatility of IPOs	142
8.1	Latin America and the Caribbean innovation at a glance	173
8.2	Percentage of countries with fiscal incentives for business innovation	177
9.1	Growth of world trade and GDP, 2000–2009 (annual percentage changes)	209
10.1	(a) Unique equilibrium; (b) multiple equilibria	225

Tables

1.1 Geography for the diffusion of knowledge 20
1.2 The structure of knowledge networks and differences of knowledge networks and other local networks 22
1.3 The dynamics of knowledge networks 24
1.4 Empirical evidence on cluster internal differences 27
1.5 Resource-Based View's explanation of the cluster internal differences 29
1.6 Relational or Social Capital view's explanation of the cluster internal differences 31
2.1 Patent per inventor distribution. Different simulation runs 50
3.1 Social capital variables for predicting venture success 68
3.2 Independent variables for predicting venture success 70
3.3 Descriptive statistics and intercorrelation matrix for solo entrepreneurs 72
3.4 Descriptive statistics and intercorrelation matrix for entrepreneurial teams 73
3.5 Use of social capital between solo entrepreneurs and entrepreneurial teams 74
3.6 The effect of social capital use 75
3.7 The differential use of social capital 78
3.8 Summary of results 81
4.1 The determinants of specialisation in business services 1999–2003 – Spatial Durbin Model estimates 100
6.1 Correlations between MFP growth and lagged IPOs 137
6.2 Do IPOs help to predict future MFP growth? 138
8.1 Input additionality – testing for crowding in/crowding out effects 186
8.2 Output additionality – testing for productivity impacts 187
9.1 Factors responsible for deindustrialisation 199
9.2 US job loss rates by industry (per cent) 200
9.3 Tariffs on non-agricultural products in the United States, the European Union, Japan and Canada 2009 (percentages) 204
9.4 Tariffs on non-agricultural products in China, India, Russia, Brazil, Korea, and Mexico in 2009 (percentages) 205

9.5 Anti-dumping investigations initiated in 2009 and 2010 by G20 members 205
9.6 Agricultural subsidies and producer-subsidy equivalent in various OECD countries in 2005 and 2009 206
9.7 Nations and products most affected by protectionist measures, September 2008 and September 2009 207
9.8 Trade openness in the G20 since 1980 210
9.9 Movement of people 211
9.10 Major world net exporters and importers of capital in 2009 212
11.1 Summary statistics 249
11.2 Mean export turnover by business strategy 250
11.3 Probit analysis 251
11.4 Linear model (restricted to exporting firms) 253
12.1 Distribution of Italian firms by size and ownership status (percentages, sample average) 260
12.2 Distribution of Italian manufacturing firms by size and ownership status (percentages, sample average) 261
12.3 Distribution of Italian service firms by size and ownership status (percentages, sample average) 261
12.4 Average exit rate by ownership status and sector (percentages) 262
12.5 Estimation results: Cox Proportional Hazard Model 266
12.6 Estimation results by technological intensity in manufacturing: Cox Proportional Hazard Model 268
12.7 Estimation results by knowledge intensity in the services: Cox Proportional Hazard Model 269

Contributors

Elisa Giuliani is Associate Professor of Management at the University of Pisa. Her research focuses on the impact of the private sector upon processes of socio-economic development. Her research has been published in several international journals such as the *Journal of Economic Geography*, *Research Policy*, *Regional Studies*, and *World Development*.

F. Xavier Molina-Morales is full Professor at the Business Administration Department Universitat Jaume I Castelló (Spain); his research interests include industrial clusters, inter-organisational relationships, social capital and innovation. His research has been published in international journals such as *Research Policy*, *European Urban and Regional Studies*, and *European Planning Studies*.

Lorenzo Cassi is Maître de Conférences at the Centre d'Economie de la Sorbonne-University Paris 1and research fellow at Observatoire des Sciences et des Techniques of Paris. His research interests include economics of innovation and technical change, technology policy and cooperative R&D, knowledge transfer and social network analysis.

Lorenzo Zirulia is Assistant Professor of Applied Economics at the University of Bologna, and fellow at KITeS Bocconi University and at the Rimini Centre for Economic Analysis. His main research interests are the economics of innovation, science and technology, industrial economics, and social and economic networks.

Uwe Cantner is Full Professor of Economics at the Friedrich Schiller University Jena (Germany) and Pofessor of Economics at the University of Southern Denmark in Odense. His research interests are economics of innovation, evolutionary economics, entrepreneurship, and productivity measurement.

Michael Stuetzer is a post-doctoral research fellow at the Australian Centre for Entrepreneurship Research at the Queensland University of Technology. His main research interests include human capital and start-up success factors.

Valentina Meliciani is Full Professor of Economics at the Faculty of Political Science of the University of Teramo. She has worked in EU-funded research projects on the role of services for innovation and economic growth. Her research focuses on the impact of technology on international competitiveness and on regional disparities.

Maria Savona is Senior Lecturer at SPRU, Science and Technology Policy Research at the University of Sussex, UK. She has published widely in top innovation journals such as *Research Policy*, *Cambridge Journal of Economics*, *Journal of Evolutionary Economics*, *Journal of Economic Survey*, and *The Service Industries Journal*.

Andreas Pyka has held the chair for innovation economics at the University of Hohenheim, Stuttgart, since April 2009. His fields of research are Neo-Schumpeterian Economics and Evolutionary Economics with a special emphasis on numerical techniques for analysing dynamic processes of qualitative change and structural development.

Bernd Ebersberger currently holds a Professorship for Economics and Management of Innovation with the Management Center Innsbruck. His research interests are innovation systems and policy, open and interactive innovation, and the internationalisation of innovation activities.

Dominik Hartmann is an associate researcher at the Department for Innovation Economics, University of Hohenheim. His research focuses on entrepreneurship and innovation networks as well as the relations between structural change and human development.

Pasquale Scaramozzino is Professor of Economics at SOAS, University of London and at Università di Roma Tor Vergata. He has taught at the University of Bristol and at University College London.

Jonathan Temple is an economist at the University of Bristol. His current research interests include foreign aid, economic growth and structural transformation, the interactions between growth and labour markets, and regional growth.

Nir Vulkan is Associate Professor of Economics at the Said Business School, and a Fellow of Worcester College, both at Oxford University. He is the author of two books and dozens of articles, mostly in the area of e-commerce and market design.

Mariacristina Piva is Associate Professor in Economic Policy at the Università Cattolica del Sacro Cuore, Italy. Her research interests are the economics of innovation and labour economics fields, with particular attention on the role of human capital in the innovative process. She has published in various refereed journals.

Marco Vivarelli is Full Professor at the Università Cattolica del Sacro Cuore, Italy. He is Research Fellow at IZA-Bonn and Honorary Professor at SPRU-Sussex University. He has published extensively in the fields of industrial organisation, labour economics, and the economics of innovation.

Gustavo A. Crespi is a Lead Science and Technology Specialist at IADB. He has published numerous articles in journals such as Word Development, Research Policy, Journal of Technology Transfer, Oxford Review of Economic Policy, Technovation and Small Business Economics. He is also a member of the editorial board of Research Policy and the International Journal of Technology Learning, Innovation and Development.

Dominick Salvatore is Distinguished Professor and Director of the PhD Program in Economics at Fordham University in New York, and Honorary Professor at Shanghai Finance University and Hunan University. He has published many books and articles in leading economics journals, and has given many lectures around the world.

Giancarlo Marini is Professor of Economics at the University of Rome 'Tor Vergata'. His research interests are economic theory and stabilisation policy, monetary economics, environmental economics, fiscal policy, public debt and intergenerational equity, international economics, theory of growth and innovation, and the economics of European integration.

Giovanni Piersanti is Professor of Economics at the University of Teramo and the University of Rome 'Tor Vergata'. His research interests are economic theory and stabilisation policy, monetary economics, macroeconomic models of intertemporal optimisation, macroeconometric models with rational expectations, currency, and financial crises.

Piergiuseppe Morone is Associate Professor of Economics at the Sapienza University of Rome. His research interests range from knowledge economics and economics of innovation to trade and development. He has published in several international journals such as *The Journal of Evolutionary Economic*, *JASSS*, *Futures*, and *Journal of International Development*.

Francesco Renna is Associate Professor of Economics at the University of Akron (USA). He received his PhD from the University of Arizona in 2002. His research interests are labour and health economics. His research has been published in several professional journals and presented at major international conferences.

Filippo Reganati is Professor of Economics at the Sapienza University of Rome. He received his PhD in Economics from Reading University (UK). His research focuses on foreign direct investment and productivity spillovers, international trade in imperfect competitive markets, and applied industrial organisation.

Anna Maria Ferragina is Assistant Professor of Economics at the University of Salerno, research fellow at DISES and CELPE (University of Salerno), and a member of the Steering Committee of FEMISE. Her interests are trade, regional integration, FDI, and firm dynamics.

Rosanna Pittiglio is Assistant Professor of Economics at the Second University of Naples. At present she is working on her PhD in Economics (Dundee, UK), carrying out research on intra-industry trade. Her research interests focus on the economics of multinational firms, intra-industry trade, international fragmentation, industrial location, foreign direct investment and international trade.

Knowledge, innovation and internationalisation

A roadmap

Piergiuseppe Morone

1 Knowledge, innovation and long-term competitiveness

The idea that firms compete through innovating activities and knowledge creation is as old as Economic Theory itself. Its importance has been at the core of economic thinking from Adam Smith onwards. Smith's emphasis on the division of labour, in fact, could be understood not only as a way to increase productivity via specialisation but, more importantly, as a way of organising the development of skills, and also of different kinds of knowledge. The division of labour originates via a process of learning by doing, which contributes to an increase in skills and expertise, and thus to enhanced accumulation of specialised knowledge, thereby generating the possibility for novel production (Loasby, 1999, p. 32). This, in turn, translates into economic wealth.

This idea was echoed in Schumpeter's view, which considered innovation as the only reliable basis for long-term technological competitiveness, as opposed to short-term price competitiveness that might be based on monopolistic practices or market power (Cantwell, 2002).

As observed by Porter (1990), a long-lasting competitive advantage (higher-order advantages, in the words of the author) is rooted in: (a) the acquisition of advanced skills and capabilities; (b) specialised and risky learning activities; and (c) large investments in physical facilities, as well as research and development and marketing. By combining these elements, firms succeed in creating those dynamic economic process which, in a Schumpeterian perspective, are solely capable of overcoming price competition and initiating a process of creative destruction. Firms' competitiveness, therefore, depends upon their ability to create stable advantages, which runs parallel with their ability to create new specialised knowledge.

Here, the competitiveness of a firm should be understood as the possession of the capabilities needed for sustained economic growth in an internationally competitive selection environment, in which there are others (countries, clusters, or individual firms, depending upon the level of analysis) that have an equivalent but differentiated set of capabilities of their own (Cantwell, 2005, p. 544). Hence, being competitive implies having a leading position in a global environment, a position which is acquired by gaining access to strategic assets. At the same time, the internationalisation process gradually converts most formerly important localised factors into ubiquities. In the author's words, *ubiquification* is the process through which a previously valuable locational capability loses its economic relevance once it becomes available to all firms in virtually any location (Maskell, 1999, p. 58).

This process shifts the competitiveness of internationalised firms onto one of the remaining localised factors upgraded by the process of internationalisation: the labour costs. Firms in high-cost countries, however, are at disadvantage in this respect and cope with this difficulty in various ways. Some raise their capital/labour ratio through massive investments, while others outsource or relocate part or all of their activities to low-cost areas.

Many entrepreneurs, however, meet the challenge in a different way, which does not involve cost reductions. They attempt to regain a long-lasting competitive advantage through enhanced knowledge creation. This idea is at the heart of the so-called knowledge-based economy, where the competitive edge of many firms operating in high-cost countries has ultimately shifted from static price competition towards dynamic improvement, favouring those who can create knowledge faster than their competitors (Maskell, 1999, p. 113).

With this in mind, the questions that should be asked are: how do firms create such knowledge in order to enhance their competitiveness? How does knowledge relate to innovation? And how does this lead to global competitiveness? As we shall see, knowledge is created through learning, which occurs at various intertwined and often overlapping levels: firm level, cluster level, and global level.

This book aims to address these very questions, first understanding the knowledge creation process which occurs through learning activities, and subsequently linking it to innovation and internationalisation. As this book attempts to show, these three concepts (knowledge, innovation and internationalisation) add up to one, and, in this sense, should be conceived as a virtuous path leading firms (and countries) to higher competitive positions. Accordingly, the book is structured into three interlinked sections (precisely, Knowledge, Innovation, and Internationalisation), and the chapters in each section address the core issue of competitiveness from different (but complementary) perspectives. Figure 0.1 provides an at-a-glance illustration of the way these concepts, their links, and a set of other side elements have been explored in this book.

The left part of the figure provides a simplified 'broad structure' of the core concepts and links leading to competitiveness. The circled labels presented on the right-hand side of the figure provide some examples of the many elements that fall within the realm of the analysis developed through the 12 chapters which comprise this book. This shows how, within an apparently simple structure, the system (like any other social system) displays an inherent degree of complexity. This book tries to give an account of such complexity whilst remaining consistent with the proposed broad structure.

2 Knowledge creation and learning processes

In recent years, the debate on knowledge (creation, diffusion, appropriability, etc.) has gained momentum, which has led to the emergence of a brand new field of

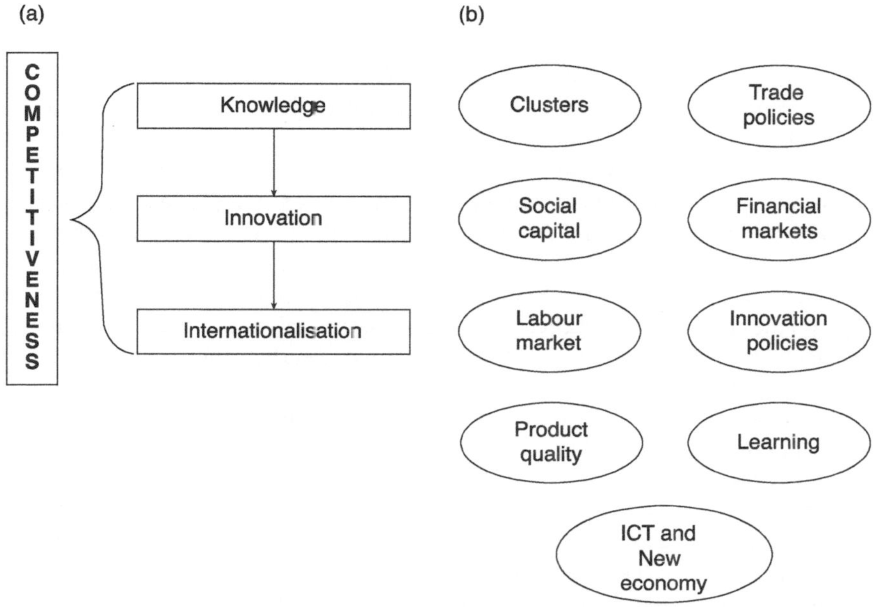

Figure 0.1 The book structure at a glance. (a) Pathway to higher competitiveness; (b) elements along the way to higher competitiveness

economics: knowledge economics. This field of study emerged as an autonomous area of enquiry, distinct from information economics, precisely because of the semantic difference between the economic meanings of 'information' and 'knowledge'. More specifically, and as stated by numerous authors, the acquisition of *knowledge* enhances individual capabilities by changing cognitive abilities, while *information* takes the form of structured data that requires knowledge to be articulated by means of processes, in order to be useful. To put it differently, information can be considered as a part of knowledge that can easily be partitioned and transmitted, either through computer networks or in a written form. Schematically, knowledge can be defined as 'organised information', and information as 'organised data'.

To further clarify the distinction between these two fundamental categories of analysis, Foray (2004) suggests analysing the conditions under which knowledge and information can be reproduced and shared. The reproduction of knowledge entails a complex process of learning, while the reproduction of information is based on the mere action of duplication. The mobilisation of cognitive resources is therefore a necessary condition for the reproduction of knowledge, while the reproduction of information is done exclusively through the use of physical resources (a photocopier, for example).

A more subtle distinction is the one between tacit and codified knowledge. In two celebrated papers of 1958 and 1967, Polanyi highlighted the existence of a tacit dimension of knowledge, to be distinguished from explicit (or codified) knowledge, yet bound to it by a strong complementary relationship. The nature of tacit knowledge may be understood by the fact that, as Polanyi himself suggests, we all know more than we can tell. What we can tell is, in fact, codified (or codifiable) knowledge; residually, what we know but cannot tell is tacit knowledge.

In this sense, tacit knowledge cannot be diffused through structured communication processes. It is a type of knowledge borne out of experience and, as such, is linked to direct actions and can be transferred only through direct interactions. Tacit knowledge, therefore, is a disembodied form of knowledge, often acquired without awareness, simply by means of face-to-face interactions. Codified knowledge, on the contrary, can be encoded into books, manuals and any digital support, and can easily spread geographically, a phenomenon earlier referred to as *ubiquification* of strategic resources.

This distinction is, in the view of many, at the heart of the regional dimension of firms' local clusters. In fact, firms' clusters are primarily perceived as knowledge-exchange-oriented activities. Knowledge is exchanged through interaction, and the structure of the interaction has a strong influence upon the amount of knowledge exchanged (Steiner, 2011). We shall continue our analysis by briefly looking at the reasons why firms show a growing tendency to cluster, exchanging knowledge either formally or informally. This behaviour reflects a tendency to mobilise external (to the firm) sources of innovation rather than internal ones. Logically (and methodologically), this is an antecedent to the *knowledge clusters* and *innovation* dyad, which will be investigated further in the following section.

In order to address this preliminary point, we shall look at the kind of knowledge needed in order to innovate, who possesses such knowledge, and how it can be acquired by innovating firms. As pointed out (Morone and Taylor, 2010), the simultaneous on-going processes of knowledge deepening and knowledge widening – which lead to a growing specialisation of competences, as well as to a general expansion of the range of available technologies – calls for new learning efforts from firms. Innovative firms need specialised knowledge, as well as additional types of knowledge, which increasingly lie outside the firm itself. However, because of its tacit component, knowledge – and especially new knowledge – can be difficult to acquire in the market, so firms seek some form of collaboration with other firms and/or institutions that possess the required knowledge and are keen to share it on a reciprocal basis,. This trend is reflected in the growing number of strategic research and development alliances which have been struck since the early 1980s (Hagedoorn, 2002). Having a portfolio of alliances acts as insurance for a firm not wishing to be taken by surprise by new technological developments, and increases their innovative capability (Cowan, 2004).

Hence, firms act to create links through which to access the disparate and specialised knowledge needed to innovate. Such bilateral links, when considered together, constitute a web of connections which can be configured as an innovation cluster. This brings us back to the *knowledge clusters* and *innovation* debate, which was briefly introduced earlier.

3 Knowledge and innovation

The discussion so far leads us directly to one of the two central themes of this book: the linkage between knowledge and innovation (the second theme being the linkage between innovation and internationalisation). As observed by **Giuliani and Molina-Morales** (Chapter 1), knowledge clusters have been rightly perceived as the locus of innovation. A large body of empirical research has found that clustered firms show a higher innovative capacity than isolated firms. In this regard, it is relevant to question whether such clusters are inherently local in nature. The cumulative and tacit dimension of knowledge and the associated need for proximity provide one possible positive answer to this question. However, **Pyka *et al.*** (Chapter 5) observe how the higher transferability of knowledge (due to the rapid diffusion of information and communication technologies, multinational enterprises, the global market place, and open innovations) and the importance of accumulation of knowledge lead to both centripetal and centrifugal forces.

Indeed, empirical assessments provide examples of cooperation and of knowledge exchange where proximity is not at all necessary. Epistemic communities (groups of individuals whose communication rests entirely on ICT) do not need face-to-face contact; similarly the ubiquity of single actors via portable phones, the Internet and email is certainly a possible way to cooperate (Torre, 2008).

In this vein, Breschi and Lissoni (2009) questioned the need for proximity, stressing how the distinction between tacit and codified knowledge is blurry, and instead pointed to the importance of inventor mobility as a way of transferring knowledge across geographical boundaries. However, as they themselves observe, the most fundamental reason why geography matters in constraining the diffusion of knowledge is that mobile researchers are not likely to relocate in space, and so their co-invention network is also localised.

Cassi and Zirulia (Chapter 2) provide a modelisation of such co-invention networks, developing an agent-based theoretical model where firms (agents) are embedded in a social network and the actual use of the network is an economic choice. Firms typically have other options for innovating, other than interacting socially – i.e. they can learn individually and, in this sense, social learning bears an opportunity cost. It emerges that inventors' relationships are affected by the structure of the network in which the inventors are embedded. This finding calls for further empirical efforts in assessing the core features of regional systems, such as the network structures, the sectorial technological competences, and the processes through which they evolve.

This call has been taken up by various authors in this book, as well as in other recent scientific works. An effort in this direction was made by Cantner and Graf (2006), who studied knowledge diffusion patterns in Jena (Germany). They focused on the structure of networks, investigating the relation between the embeddedness of firms within networks and individual innovative performance.

Their empirical investigation was conducted using patent data and the tools provided by social network analysis, which allowed an examination of the evolution of two types of innovative networks (the first one linking innovators by the kind of technological knowledge they created, and the second one relating to the notion of knowledge transfer through personal relationships). The authors found evidence in support of the idea that the dynamics of the system are directed towards an increasing focus on the core competencies of the local innovation system: i.e. innovators on the periphery of the network exit and new entrants position themselves closer to the core of the network. Additionally, they found hints of the importance of short-term interpersonal linkages in shaping networks of cooperation and even entire local innovation systems.

Few other authors have used social network analysis in order to assess the impact of network structure upon knowledge flows. Among these studies, Reid *et al.* (2008) analysed the North-Western Ohio greenhouse cluster, showing that suppliers communicate with a wide variety of growers, thus playing a central role in the transmission of information among themselves. Steiner and Ploder (2008) combined social network analysis with parametric analysis and in-depth interviews in order to analyse various forms of network relations and the possible contents of the interactions of firms exchanging knowledge. The authors argued that these interactions are distinctively structured, and that the diffusion of knowledge is highly selective and strongly dependent upon the position of firms within networks and on their absorptive capacity.

Cantner and Stuetzer's contribution to this book (Chapter 3) takes a further step along this line. In their chapter, the authors look closely at the use of social capital among solo entrepreneurs and entrepreneurial teams in the venture creation process, and assess the relevance of strong and weak ties for both types of start-up. Using appropriate data obtained for the German state of Thuringia, the authors provide a fine-grained definition of social capital, which goes beyond *potential* access to network resources and information, pointing at *actual* use of the network contacts. Their empirical investigation hints at the existence of structural differences between solo entrepreneurs and entrepreneurial teams, which adds up to more than just the number of people involved in the firm's management and refers to the non-trivial interactions between human and social capital.

Meliciani and Savona (Chapter 4) also conducted a parametric investigation into structural and spatial determinants of specialisation in business services at the regional level. Along with a traditional explanation of the spatial distribution of industries, the authors consider two further elements: first, they account for the presence of Hirschmann's forward linkages, and second, they explicitly consider the spatial dependence of sources of agglomeration economies in a spatial econometric framework. Along with social network analysis, spatial econometrics provides a valuable tool for empiric studies. The underlying idea is that *space matters*; more precisely, what happens in one region is related to what happens in neighbouring regions. This is a key concept which was emphatically underlined by Tobler's first law of geography, which states that: 'Everything is related to everything else, but near things are more related than distant things'.

It appears that the literature on regional networks of firms and knowledge spillover is very important and attracts the interest of many scholars. As discussed above, this is also due to the growing number of tools available for an empirical understanding of the black-box of knowledge networks (Steiner, 2011). In this context, the contribution of **Giuliani and Molina-Morales** (which opens the first section of this book) is most welcome as it provides a comprehensive assessment of the recent literature, and aims at unpacking the concept of localised knowledge spillovers. They do this by addressing three core and intertwined issues: (1) the importance of geography for the diffusion of knowledge; (2) the structure of knowledge networks and differences of knowledge networks and other local networks; and (3) the dynamics/stability of knowledge networks over time. This analysis is followed by a discussion on the relations between such knowledge clusters and their ability to innovate; this is done by pointing at firms' heterogeneity within clusters, a concept which, from a different stand, is also central in the so-called 'new' new trade theory, which aims at explaining the different performances of firms in terms of their ability to penetrate international markets (we will come back to this point later on in this chapter).

In addition to this, **Pyka *et al.*** explore the possibility of mapping technological competences and innovation networks for regions by using patent data. The importance of this study lies in the possibility of measuring the technological advantages and disadvantages of regions with respect to other geographical entities. Looking at the Bremen region, the authors assess the regional technological advantages in respect of a nationwide context (i.e. Germany), a supranational context (i.e. OECD countries) and an international context. This investigation paves the way for the second core theme of this book, i.e. the linkage between innovation patterns and internationalisation trends. However, before moving on to this topic, we shall consider several other aspects associated with innovative activities, which are central in the literature.

4 Assessing the impact of innovation activities and innovation policies

Chapters 6, 7 and 8 of this book give accounts of some relevant aspects associated with innovative activities. These are: (1) the macroeconomic effect of the new economy boom on investment volatility and cyclical productivity growth; (2) the labour market effect of innovation, and specifically the impact of technological change on employment; and (3) the impact of innovation policies on firm-level technological upgrading in Latin American countries.

These are all notable aspects of the broad concept of innovation as they add to the discussion developed above (i.e. the source and the locus of innovation), focusing on the effects of innovation activities at both the macro and micro levels.

Scaramozzino *et al.* (Chapter 6) explore the sustained output growth experienced in the USA in the 1990s as a result of the New Economy boom. As observed by the authors, despite the stability of output growth, there was uncertainty about whether this growth could be sustained, which was linked to the uncertainty about the productivity of new technologies. Hence, this decade was characterised by an unusually stable output growth combined with a high degree of uncertainty surrounding the trend growth rate. As shown by the authors, these stylised facts fit well with the classic Shleifer (1986) model of implementation cycles, when extended to incorporate the possibility of uncertainty about the underlying growth rate. In fact, when incorporating uncertainty, the equilibria with delayed implementation (which are associated with volatility) are eliminated from the model outcomes, because delay becomes risky. Instead, immediate implementation equilibria emerge and the output growth stabilises. This contribution provides readers with a better understanding of the macro effects of the New Economy boom, and adds to our understanding of the sustained growth period experienced in the USA during the 1990s.

In Chapter 7 **Piva and Vivarelli** tackle another important effect of innovation – i.e. its impact upon the labour market which, as the authors state, has regained momentum in the literature precisely due to the New Economy boom. This chapter provides a critical reassessment of the theoretical and empirical literature on the quantitative and qualitative impact of technological change upon employment, comparing the comparative explanatory power of the competing theories from both macro and micro perspectives.

A key finding of this investigation is the heterogeneous effect that can be observed when decomposing technological change into product and process innovation. In fact, process innovation tends to be labour-saving, while product innovation shows a labour-friendly nature. As for the 'compensation mechanism', which in principle should assure a complete and automatic compensation of the initial labour-saving impact of process innovations, the authors observe that it might be partially impeded due to institutional settings, market structures, macroeconomic cycles, and the expectations of economic players. Finally, large and statistically robust empirical evidence in support of the so-called skill-bias technological change hypothesis is observed among North American and European countries, both in manufacturing and in service sectors.

Chapter 8 concludes the second section of the book by providing an assessment of innovation policies in Latin America, a region where several market failures have long impeded private companies from adopting modern technologies and business practices that would help them to improve their productivity and competitiveness. **Crespi** provides a qualitative 'meta analysis' of these programmes (and their respective impact evaluations) in order to take stock of the learning achieved so far, and proposes specific recommendations to maximise their impact upon firms' productivity. A preliminary question posed by the author concerns whether public investments have a 'crowding-out' effect on private investments or whether, on the contrary, they have a 'crowding-in' impact, prompting input additionality. It appeas that there is clear evidence that fiscal incentives have been effective in increasing firms' investments in innovative projects, and they have been also effective in leveraging private resources for this investment. Subsequently, Crespi looks at output additionality and productivity – i.e. did these investments result in rise a rise in productivity and increased innovativeness?

Also, in this case the impact of fiscal policies seems to be positive once enough time has elapsed since the approval of the grant. Empirical evidence suggests, in fact, that output additionality and labour productivity start to pick up only three to five years after the start of an innovation project.

5 Innovation and internationalisation

In this section we shall address the second central theme of this book – the linkage between innovation and internationalisation. As stated above, growing world competitiveness puts firms under the threat of being marginalised if they are unable to cope with it. Thus, many scholars have advocated enhanced innovation as one of the most effective strategies, especially for firms operating in rich countries and facing high input costs. From this perspective, the first-order effect of internationalisation would be to reallocate resources towards the most productive (knowledge-intensive) firms, which self-select to operate in international markets, whereas less innovative firms are unable to penetrate foreign markets. This *self-selection* hypothesis typically goes through the productivity gaps and sunk-costs theory: only those firms who are efficient enough to bear the entry costs and intense competition of the export market will start exporting.

Hence, firms that engage in international trade do so because they already possess competitive advantages in knowledge-intensive production processes and have adopted world-class technologies in order to compete abroad (Martijn and Hein, 2012). However, there is a second hypothesis, which rests on the idea that intended and unintended international knowledge flows stimulate the post-entry innovative performance of firms. In other words, firms operating in foreign markets gain access to technical expertise from international buyers and competitors (World Bank, 1993), which fuels their innovative performance. Hence, this *learning-by-exporting* hypothesis points at the potential spill-back effects upon productivity of internationalisation of the firm: it is expected that firms opening to international markets would gain the opportunity to upgrade their technology through the import of knowledge-intensive capital goods, as well as learning from their partner's business practices and capabilities (Martijn and Hein, 2012).

It is most likely that both hypotheses are at work, which suggests the existence of a tight and circular link between knowledge, innovation and internationalisation. Indeed, this is the core idea of the whole book, which finds the logical conclusion of the loop in its third section.

With this general framework in mind, the third section opens up with a preliminary (and much needed) investigation of the real degree of internationalisation observed across world economies over recent decades. To this end **Salvatore** (Chapter 9) provides an accurate analysis of trends in current thinking on trade policy, as well as the real degree of internationalisation in the world economy today. As shown by the author, the rapid internationalisation of the world economy during the past three decades has increased the international competitiveness of those countries that have opened themselves more widely to the world economy. At a macro level, this has made a significant contribution to the growth of large nations, whereas small open economies captured the benefits of openness before the current period of rapid internationalisation and so their growth now depends more on internal factors. Salvatore shows also that, in spite of the on-going trend toward internationalisation, free trade has recently been put under unprecedented attack, both theoretically and empirically. Protectionism has been advocated in order to reverse the process of deindustrialisation in advanced countries, protect jobs and wages from globalisation and outsourcing, and mitigate the effects of the world economic crisis.

In this respect, the speculative attacks and crashes in international markets played a key role in prompting arguments in favour of restrictions upon free trade and capital mobility. Indeed, the present economic crisis poses serious questions about the abilities of national governments to effectively counter critical situations, which spread rapidly across the world. Analysts need to reflect upon the degree of complexity imposed by internationalisation trends, cross-border capital mobility, and contagion mechanisms.

To this end, **Marini and Piersanti** (Chapter 10) provide a comprehensive assessment of the theoretical models of speculative attacks, which point toward inconsistent macroeconomic policies and self-fulfilling expectations as the two principle explanations of financial crises. These two approaches are often referred to as first or second generation models of currency crisis. As the authors suggest, the financial sector's fragility and contagion mechanism, which typically emerges at the onset of financial and currency turmoils, prompted the development of new models; however, these new models still fit in the two-approaches-based classification and should be conceived as an extension of such setups that better evaluate the financial side of the economy.

Chapters 11 and 12 go back to a microanalysis of the phenomenon. Specifically, building on innovation and new trade theory, **Morone and Renna** empirically test the self-selection hypothesis, considering, along with innovation, the effect of quality upgrading on the exporting performance of Italian SMEs operating in the manufacturing sector. The novelty of their investigation rests on their measurement of quality, which is captured by both a dummy variable for investments in product quality improvements and by a dummy for compliance with ISO9000 standard as a measure of quality upgrading. The findings are in line with the authors' expectations: the best performing firms meet the new challenges posed by internationalisation not by aspiring to acquire competitiveness through cost reduction, but by generating entrepreneurial rents through enhanced knowledge creation, which is reflected in product innovation and quality upgrading.

The link between knowledge, innovation and internationalisation is taken one step further in the last chapter of this collection, where **Reganati *et al.*** investigate how foreign ownership affects the survival prospects of Italian firms according to the technological intensity of the sector in which they operate. Their investigation departs from some stylised facts broadly accepted in the literature – i.e. foreign multinational enterprises are typically more footloose than domestic non-multinational firms. This finding has been explained as being due to the fact that multinational firms can more easily relocate production between countries in response to adverse shocks in the host country. However, as pointed out by the authors, in this literature one issue that has often been overlooked concerns the factors which may moderate (or strengthen) the footloose behaviour of foreign multinational firms. Specifically, the authors suggest that the technological intensity of an industry might moderate the footloose behaviour of multinationals. There might be many reasons for this. As suggested by De Propris and Driffield (2006), the existence of location-specific factors might encourage multinational enterprises to commit themselves to a particular locality. These factors include local tacit or uncodified knowledge, often employed in technologically intense sectors, which cannot be disseminated outside a geographical cluster and which constitutes the intangible assets of a cluster. As the latter are immobile and embedded, they can then be a key factor in preventing the easy exit of multinationals from local markets, and hence reduce their propensity to be 'footloose'.

6 Summing up: a brief note on methodologies

As the 12 chapters of this collection show, the circular link between knowledge, innovation and internationalisation is a complex one, and involves many channels through which this loop's effects are exerted.

The contributions in this book capture some aspects of this complexity, addressing the issue from several perspectives and employing several methodologies. The book is, in fact, a collection of empirical and theoretical investigations, which make use of various heterogeneous tools. The theoretical contributions, for instance, span from agent-based models to close analytical models, whereas the empirical studies span from econometric studies to network analysis and field-based case studies.

Such a multitude of approaches, far from being detrimental to the overall consistency of the book, should be understood (at least, in the hope of the editor!) as an added value to our knowledge and understanding of the phenomena under investigation.

In fact, the methods used to conduct scientific investigations influence the selection of the phenomena under study, as well as the conclusions drawn. In this regard, increasing the variety of methods enriches our understanding of the phenomenon under investigation on the one hand, while, on the other hand, it could provide conflicting conclusions.

However, any scientific discipline (and in this regard Economics is no different from any other body of knowledge) advances precisely because someone questions an existing wisdom. Both completely new ideas and the extensions to existing ideas are created as a reaction to a prevailing wisdom.

For those who own the truth, their only task is to make sure that others accept that truth. This, however, does not move scientific knowledge forward. As suggested by Karl Popper (1934), scientific knowledge grows through a process of making hypotheses about the nature of problems and the subsequent falsification or testing of these hypotheses. According to Popper, it is the duty of every scientist to try to disprove or reject her/his hypotheses. This process of hypotheses creation and disproval provides researchers with a different idea of truth, a truth that is historically contingent, and 'probably true' at best.

Having understood this is my major intellectual debt towards my academic mentor, Cesare Imbriani.

References

Breschi, S. and Lissoni, F. (2009) 'Mobility of skilled workers and co-invention networks: an anatomy of localised knowledge flows', *Journal of Economic Geography*, 9(4): 439–68.

Cantner, U. and Graf, H. (2006) 'The network of innovators in Jena: an application of social network analysis', *Research Policy*, 35(4): 463–80.

Cantwell, J. (2002) 'Innovation, profits and growth: Schumpeter and Penrose', in C. Pitelis (ed.) *The Growth of the Firm – the Legacy of Edit Penrose*, Oxford University Press, Oxford.

Cantwell, J. (2005) 'Innovation and competitiveness', in J. Fagerberg, D. Mowery and R. Nelson (eds) *The Oxford Handbook of Innovation*, Oxford University Press, Oxford.

Cowan, R. (2004) 'Network models of innovation and knowledge diffusion', MERIT-Infonomics Research Memorandum Series, N. 2004-016.

De Propris, L. and Driffield, N. (2006) 'The importance of clusters for spillovers from foreign direct investment and technology sourcing', *Cambridge Journal of Economics*, 30(2): 277–91.

Foray, D. (2004) *Economics of Knowledge*, Cambridge, MA: MIT Press.

Hagedoorn, J. (2002) 'Inter-firm partnerships: an overview of major trends and patterns since 1960', *Research Policy*, 31(4): 477–92.

Loasby, B.J. (1999) 'The significance of Penrose's theory for the development of economics', *Contributions to Political Economy*, 18(1): 31–45.

Martijn, A.B. and Hein, R. (2012) 'The effects of internationalization on innovation: firm-level evidence for transition economies', Tjalling C. Koopmans Research Institute Discussion Paper Series N 12-04, Utrecht University.

Maskell, P. (1999) 'Competitiveness, localised learning and regional development' in R. Treptow, T. Hastings, P. Maskell and J. Musil (eds) *Economic and Cultural Transitions Towards a Learning City. The Case of Jena*. OECD, Friedrich-Schiller-Universität. OECD Publishing.

Morone, P. and Taylor, R. (2010) *Knowledge Diffusion and Innovation*, Cheltenham: Edward Elgar Publishing.

Polanyi, M. (1958) *Personal Knowledge*, London: Routledge & Kegan Paul.

Polanyi, M. (1967) *The Tacit Dimension*, London: Routledge.

Popper, K. (1934) *The Logic of Scientific Discovery*, (the revised English translation, with new appendices and footnotes, was published in 1959.) London: Hutchinson.

Porter, M.E. (1990) *The Competitive Advantage of Nations*, Macmillan, London.

Reid, N., Smith, B.W. and Caroll, M.C. (2008) 'Cluster Regions – a Social Network Perspective', *Economic Development Quarterly*, 22(4): 345–52.

Shleifer, A. (1986) 'Implementation Cycles', *Journal of Political Economy*, 94(6): 1163–90.

Steiner, M. (2011) 'Regional knowledge networks', in P. Cooke, B. Asheim, R. Boschma, D. Schwarz and F. Toedtling (eds) *Handbook of Regional Innovation and Growth*, Cheltenham: Edward Elgar Publishing.

Steiner, M. and Ploder, M. (2008) 'Structure and strategy within heterogeneity: multiple dimensions of regional networking', *Regional Studies*, 42(6): 793–815.

Torre, A. (2008) 'The role played in temporary geographical proximity in knowledge transmission', *Regional Studies*, 42(6): 869–89.

World Bank (1993) *The East Asian Miracle: Economic Growth and Public Policy*, New York: Oxford University Press.

Part I

Knowledge

1 Knowledge and innovation in social networks

An assessment of industrial clusters' literature

Elisa Giuliani and F. Xavier Molina-Morales

1.1 Introduction

Spatially clustered business activities have existed in economic studies for a long time (Marshall, 1920). Geographic agglomerations of firms specialised in one or more connected industries are here defined as industrial clusters. Introduced by Alfred Marshall at the beginning of the last century, the concept was revamped at the end of the 1970s by Italian scholars interested in understanding the similarities between Marshallian districts and Italian industrial districts (Bagnasco, 1977; Becattini, 1979). Next, the interest in districts or clusters[1] spread internationally, becoming a widely studied phenomenon. Cases of agglomerations can be found both in advanced countries (Aydalot and Keeble, 1988; Sabel, 1989; Schmitz and Musyck, 1994; Saxenian, 1994) and developing countries (Schmitz, 1995; Rabellotti, 1995; Cassiolato and Lastres, 1999), as well as in knowledge-intensive manufacturing sectors (Aydalot and Keeble, 1988; Breschi and Lissoni, 2009; Ter Wal, 2009) and more traditional manufacturing sectors (Rabellotti, 1995; Staber, 2011). In spite of the existence of some significant contextual differences between clusters across the world, they are generally considered to share some key characteristics such as local intense formal and informal relationships, which favour inter-organisational trust and external economies.

Since the 1980s, industrial clusters have come to play an increasingly central role in the analysis of the economic growth and competitiveness of countries (Krugman, 1991; Porter, 1998). In particular, a wealth of empirical research has found that clustered firms show a higher innovative capacity than isolated firms (Porter, 1990; Baptista and Swann, 1998; Baptista, 2000). Building upon the original contribution of Marshall (1920), for many years scholars have attributed this success to the presence of three types of external economies: the presence, at the local level, of a pool of skilled human resources; the local availability of inputs; and the presence of technological spillovers. On these grounds, subsequent scholars have developed a wealth of conceptual frameworks and terminologies to explain the power of 'meso-level' forces for cluster competitiveness.

Among these meso-level forces, scholars have pointed at the importance of relational aspects. For instance, Italian scholars of industrial clusters have long emphasised the importance of localised social and productive 'thickening', an expression that points to the fact that economic actors are embedded within dense local social and productive networks (Becattini, 1989). American economic geographers have used the concept of *untraded* interdependencies (Storper, 1997) to refer to labour markets, public institutions, and locally or nationally derived rules of action, customs, understanding and values, which were unique within regional and local contexts and which boosted their development potential. Finally, economists have referred to the concept of *localised knowledge spillovers*, and have interpreted the high innovativeness of clustered firms by conceiving local knowledge as a public good, which spreads pervasively within spatially bounded areas (Audretsch and Feldman, 1996; Baptista and Swann, 1998).

While these concepts were mainstreamed during most of the 1980s and 1990s, the profound transformation of clusters in light of the new global distribution of production and innovation has prompted scholars to call for new interpretative frameworks of cluster innovation and competitiveness. In particular, scholars have started to question 'meso-level' interpretations of the success of clusters and have proposed alternative analyses. Among the critical views, we consider the following to be of key relevance for the wealth of studies that have followed thereafter.[2] First, scholars questioned the over-emphasis on local knowledge spillovers (LKS) as a concept that explains most of cluster innovativeness, and became dissatisfied with the way that it was measured and understood. In their seminal paper, Breschi and Lissoni (2001) argued that:

> The concept of LKS is no more than a 'black box', whose contents remain ambiguous. On the one hand, its frequent citation serves well an evocative purpose, i.e. it helps to signal a strong interest in coupling 'geography' with 'innovation' as research themes. On the other, by contrast, it provides the researcher with an escape route to avoid studying the specific mechanisms through which the two phenomena are linked.
>
> (Breschi and Lissoni, 2001, p. 976)

This critical view sparked the subsequent generation of studies aimed at opening up the localised knowledge spillover black box, as discussed in section 2.

Second, most of the early studies on clusters treated firms as homogeneous actors and generally overlooked the importance of individual agency and characteristics in explaining cluster innovativeness. This means that the extant literature has assumed that there is a high degree of internal homogeneity within clusters. This view still predominates today, to the point that, in a recent contribution, Ter Wal and Boschma (2011, p. 921) argued that research on clusters '[does] not pay close attention to the fact that firms are highly heterogeneous in terms of capabilities, strategies, and routines [. . .] in that literature, clusters matter and not so much firms.'

In this chapter we attempt a review of the most salient of the papers which address the limitations mentioned above. In particular, section 2 discusses the advancements in the measurement of localised knowledge spillovers, taking as its reference the recent literature on local knowledge and innovation networks in industrial clusters. Section 3 reviews some of the critical studies that have taken into account the heterogeneity of cluster firms (and other actors) in order to explain their performance. Section 4 concludes by tracing some promising lines of research in this area of investigation.

1.2 Unpacking localised knowledge spillovers

In the concluding section of their article, Breschi and Lissoni (2001) call for a new research agenda in the study of localised knowledge spillovers. The core of their argument is that, within clusters, there are different kinds of knowledge flows, which should not all be placed within a unique bundle. They criticise the fact that the extant literature pays insufficient attention to disentangling the various determinants of spillovers, or to studying whether and how all firms benefit from their presence purely as a result of their co-location. Indeed, the generation of local spillovers – often referred to as the 'local buzz' – is only in part a random process of knowledge flowing from one firm to the other. In fact, the overall phenomenon cannot be entirely due to pure chance. Furthermore, knowledge flows through different channels: labour mobility, trade of goods, informal 'chit chat', communities of practice, formal meetings, and so on. All of these channels differ in their potential to transfer valuable knowledge and respond to different logics and incentives. More importantly, not all actors in the cluster get exposed to the same amount of local knowledge, as conventional localised knowledge spillover stories would presume. To understand how knowledge is diffused within a cluster, and also the degree to which firms benefit from localised knowledge spillovers, it is necessary to study the networks through which knowledge gets transferred within the cluster. However, the conventional literature is also rather superficial in terms of understanding the characteristics of networks that transport knowledge across organisations. As remarked by Staber:

> all economic action in industrial districts is said to be embedded in a dense web of network ties [. . .] But beyond the widespread reference to dense networks as a characteristic of successful industrial districts, many investigators are surprisingly silent about the structure of networks.
>
> (Staber, 2001, p. 537)

The need for a new research agenda oriented toward unpacking the concept of localised knowledge spillovers has sparked, among others, a new wave of cluster studies, which has introduced methodological novelties in the measurement of localised knowledge spillovers and, more generally, in that of local networks. Through methods of Social Network Analysis (SNA) (Wasserman and Faust, 1994),[3] scholars have started to look at the way flows of knowledge between cluster firms were distributed within local networks, allowing for a potentially neater understanding of the drivers and consequences of knowledge networks in clusters. In this section we review a selection of influential papers on this subject, a summary of which is presented in Tables 1.1–1.3. Papers in this area of research aim at answering three types of questions.

Table 1.1 Geography for the diffusion of knowledge

Author(s), Year	*Research Questions/ Objective*	*Context*	*Industry*	*Methodology*	*Main results*
Breschi and Lissoni (2009)	Contribution of inventors' mobility and network of inventors to the diffusion of localised knowledge	Europe	Drugs, biotechnology and organic chemistry	Social Network Analysis + Econometrics on EU Patent Data	The key motivation why geography matters in constraining the diffusion of knowledge is that mobile researchers are not likely to relocate in space. Informal social ties are not significant drivers of knowledge diffusion at the local level.
Parra-Requena *et al.* (2010)	Analysis on whether knowledge access depends on firms' capacity to share goals and culture with other members of the district	Spain	Footwear	Survey data on 224 firms + econometric analysis	Knowledge sharing is not explained solely by territorial commonalities, but rather by the existence of a shared cultural vision of firms (cognitive dimension of social capital).
Balland (2011)*	Drivers of collaborative networks	Europe	Global navigation satellite systems	Social Network Analysis (SIENA) on FP7 Projects	Geographical, institutional and organisational proximity matters in initiating collaborations. Cognitive and social proximity not relevant for starting new collaborations.
Scherngell and Hu (2011)*	How spatial, economic and technological characteristics affect collaborations across regions	China	Several fields	Social Network Analysis + econometrics on co-publication data	Geography matters (scientific collaborations decrease by geographical distance), but geographic proximity per se is not sufficient to ensure that a collaboration takes place.

Note: (*) This paper does not analyse cluster networks, but it looks at how and whether geographic proximity matters for network evolution. For this reason it has been included in the literature review.

The first group of studies (Table 1.1) focuses on whether geography matters for the generation of knowledge linkages, and on the significant drivers of localised knowledge spillovers. While most studies seem to confirm that geography matters (Balland, 2011; Scherngell and Hu, 2011; see also Maggioni and Uberti, 2009), the predominant view is nevertheless sceptical about attributing too much importance to geography. Parra-Requena *et al.* (2010) found that cognitive proximity acts as a mediator between geographical proximity and knowledge acquisition in clustered firms. They suggest that firms acquire knowledge from other culturally similar firms, rather than from those that are simply geographically close.

Breschi and Lissoni (2009) question the overall significance of localised knowledge spillovers, and query the real contribution of inventors' mobility vis a vis inventors informal networks in explaining the diffusion of localised knowledge. They find that labour mobility explains a great part of the local diffusion of knowledge, and argue that the key motivation for why knowledge is locally bounded is that inventors are resilient to relocating in space. By contrast, they find that informal ties play a minor role in the diffusion of local knowledge, a result that is at odds with much of the conventional knowledge about clusters and innovation. Hence, being able to unpack the concept of localised knowledge spillovers seem to have downplayed geography in favour of other explanations.

The second strand of studies (Table 1.2) takes a detailed look at localised knowledge spillovers. It focuses more specifically on measuring what nobody had measured in the past: the structure of knowledge networks – that is, the way in which knowledge linkages are distributed across co-located organisations. The first result, found consistently across different study contexts, is that knowledge networks are distributed in space in a very uneven and fragmented way (Giuliani, 2007; Boschma and Ter Wal, 2007; Sammarra and Biggiero, 2008; Morrison and Rabellotti, 2009; Ter Wal, 2012). In turn, this means that not all firms benefit from or are capable of generating flows of local knowledge – in contrast to most of the conventional wisdom and to the Marshallian 'knowledge in the air' metaphor. These studies have also analysed why knowledge networks are unevenly distributed. For instance, Giuliani and Bell (2005) looked at firm-level characteristics and found that firms which have accumulated more skills and experience tend to be part of a local elite of knowledge-intensive firms, exchanging knowledge much more frequently than firms with weak internal skills and capabilities, which instead remain peripheral within the local knowledge network. Their message, however, is not that firms with more advanced capabilities are likely to be central in *all* cluster knowledge networks, but that intra-firm capabilities influence the ways in which knowledge is transferred and received. Other studies corroborate this view, showing that technically distant firms are unlikely to communicate with each other (on this, see also Morrison and Rabellotti, 2009).

Broekel and Boschma (2011) push the argument of inter-organisational proximity further, showing that in the context of the Dutch aviation industry organisational, technological, social and geographical proximities are all associated with the formation of inter-organisational knowledge linkages. A parallel strand of studies has looked at whether local knowledge networks overlap with other types

Table 1.2 The structure of knowledge networks and differences of knowledge networks and other local networks

Author(s), Year	*Research Questions/Objective*	*Context*	*Industry*	*Methodology*	*Main results*
Giuliani and Bell (2005)	Structure of the knowledge network and connection between intra- and extra-cluster knowledge systems	Chile	Wine	Case Study + Social Network Analysis	Firms with stronger knowledge bases more central in the local network and more open to external knowledge.
Giuliani (2007)	Differences between knowledge and business networks	Chile and Italy	Wine	Case Study + Social Network Analysis + Econometrics	Firms with stronger knowledge bases more central in the network. Knowledge diffusion is selective. Business networks differ from knowledge networks.
Boschma and Ter Wal (2007)	Structure of knowledge network in clusters and impact on firm-level performance	Italy	Footwear	Case Study + Social Network Analysis	The local knowledge network is quite weak and unevenly distributed among the local firms. Local ties not enough for innovative performance.
Sammarra and Biggiero (2008)	Analysis of the technological, market and managerial knowledge	Italy	Aerospace	Case Study + Social Network Analysis	In most collaborative relationships, partners exchange technological knowledge together with market and managerial knowledge.
Giuliani (2008)	Differences between knowledge and trade networks	Costa Rica	High-tech	Case Study + Social Network Analysis	Trade and knowledge networks do not overlap significantly. The formation of knowledge ties depends on the strategy of firms.
Giuliani (2010b)	Analysis of the factors that underpin the formation of knowledge networks in clusters	Chile	Wine	Case Study + Social Network Analysis (QAP)	Trade networks do not carry knowledge. Partial importance of friendship networks in the transfer of local knowledge.
Morrison and Rabellotti (2009)	Analysis of the nature and overlap between knowledge and information networks	Italy	Wine	Case Study + Social Network Analysis	Information and knowledge networks differ.
Broekel and Boschma (2011)	Analysis of the factors that underpin the formation of knowledge networks	The Netherlands	Aviation industry	Case Study + Social Network Analysis (QAP)	Organisational, technological, social and geographical proximity all favour knowledge linkages between organisations.

of networks in the cluster. For instance, Giuliani (2007) finds that business networks – defined as being formed by any type of business relationship, including information exchange, trade, meeting at business association events, and the like – are structured in a very different way as compared to the local knowledge network. While the latter is selective, the former is highly pervasive – consistent with the well-known Marshallian metaphor of 'industrial atmosphere'. In other cases scholars have explored the significance of trade linkages[4] as channels of knowledge in clusters and argue that the association between these should not be taken for granted. For instance, in a study on the local linkages of high tech multinational (MNC) subsidiaries in Costa Rican Export Processing Zones, Giuliani (2008) observes that multinationals are unwilling to provide their local suppliers with significant training and technical advice. Rather, they tend only to select suppliers which already possess the required skills, limiting the need for technology transfers as far as possible. More research is needed to understand whether trade linkages are still a channel of knowledge within clusters, particularly in light of the growing global disintegration of production and the subsequent reduction of local value chains.

Giuliani (2010a) and Broekel and Boschma (2011) have tried to empirically validate the relevance of friendship ties as a means of inter-organisational knowledge transfer – a channel often advocated by cluster scholars, but never really analysed. Their results show that friendship is an important driver of knowledge transfer, but they also highlight that not all friends exchange knowledge, leaving open other possible complementary interpretations. Finally, other studies have explored the extent to which networks of technical knowledge overlap with other types of knowledge networks. Here, the results are contrasting: while some find significant differences between technical and market knowledge networks (Boschma and Ter Wal, 2007) and between knowledge and innovation networks (Morrison and Rabellotti, 2009), Sammarra and Biggiero (2008) find that local partners exchange technological knowledge together with market and managerial knowledge. They conclude that collaboration for innovation in competitive high-tech clusters occurs through the exchange of heterogeneous (technological, market and managerial) knowledge. In fact, they postulate that 'increasing reliance on innovation collaborations to transfer specialised technological capabilities among partner firms will increase firms' need/opportunity to exchange managerial knowledge through network ties.' (2008, p. 820).

The third and final strand of studies (Table 1.3) asks a fundamental question: what happens to local knowledge networks over time? This question is rooted in the idea that, if knowledge spillovers flow within cluster boundaries in a structured and selective way, it becomes interesting to understand whether any structural characteristics of knowledge networks change over time, and in what direction. Do knowledge networks in clusters become denser and more cohesive over time? And if so, why? Scholars have attempted to explore different dimensions of these questions, and given the paucity of studies investigating these broad research questions, it is hard to find a coherent synthesis of the findings.[5]

Table 1.3 The dynamics of knowledge networks

Author(s), Year	*Research Questions/ Objective*	*Context*	*Industry*	*Methodology*	*Main results*
Cantner and Graf (2006)	Dynamics of local innovation ties	Germany	Several sectors	Social Network Analysis on Patent Data	Job mobility drives future connections. Technological overlap increases over time.
Giuliani (2011) (Giuliani and Bell 2008)	Drivers of knowledge network dynamics	Chile	Wine	Social Network Analysis (SIENA)	Knowledge networks are structurally stable over time. Social drivers (cohesion) and firm-level characteristics responsible for network stability.
Wilhelmsson (2009)	Drivers of inventors network dynamics and role of geography	Sweden	Several sectors	Social Network Analysis on Patent Data + Econometrics	Geographical agglomerations and industry diversity play a role in explaining networking.
Ter Wal (2009)	How network dynamics change along cluster life cycle	Germany	Biotechnology	Social Network Analysis (SIENA) on Patent Data + Econometrics	Geographical proximity much more important in the early stages of cluster development processes. Social drivers (closure) gain importance as the cluster becomes mature.
Ter Wal (2012)	How differences in cluster life cycle affects the evolution of local networks of collective learning	France	High-tech	Social Ego-Network Analysis on Patent Data	Inventors networks are fragmented also in high tech clusters. Spin-off and start-ups important drivers for network evolution in IT. Geography does not matter for the formation of innovation ties in Life Sciences (most ties are outward oriented).
Staber (2011)	What drives the duration of dyadic linkages among clusters	Germany	Textile and clothing	Social Network Analysis + econometrics	Ties are more durable in uncertain environments. Ties were formed through triadic closure, but those formed through an indirect broker lasted less than those formed directly. Competitors play a crucial role as information brokers but relationships do not last.

The focus of research has essentially been on the micro-level factors that contribute to the evolution of knowledge networks in clusters. Cantner and Graf (2006) and Ter Wal (2012) find that job mobility and spin-offs stimulate changes in local knowledge networks. Others borrow concepts from economic sociology and empirically test whether social mechanisms of new tie formation – among which are reciprocity, transitivity, and homophily[6] – explain the evolution of knowledge networks in clusters (Giuliani, 2010b; Staber, 2011). Besides relational aspects, scholars have advanced the idea that firm-level characteristics also play a role in the evolution of knowledge networks in clusters. In her analysis of network dynamics in a Chilean wine cluster over a four-year time period, Giuliani (2010b) shows that it is not correct to attribute the evolution of a knowledge network to one single factor. Network dynamics depend upon a set of concurrent – and sometimes contrasting – micro-level connectivity choices, and do not respond to a unique logic. In fact, she finds that the change of the knowledge network is due not only to a set of mechanisms of social cohesion (i.e. reciprocity and transitivity), but also to the heterogeneity of firms in terms of their knowledge bases, as firms that are particularly resource-poor display poor socialisation dynamics, and do not take part in or benefit from the presence of socially cohesive behaviours at the local level. It is hard to generalise from one single case-study, but nonetheless it serves to show that: first, geographical proximity plays a limited role, as firms belonging to the same cluster show widely different socialisation patterns; and second, social mechanisms of new tie formation explain only part of the story and, consequently, that firm-level heterogeneity is an important piece of the knowledge dynamics puzzle.

Finally, Ter Wal (2009) and Ter Wal and Boschma (2011) have elaborated a conceptual framework to explain the evolution of innovation networks in clusters, according to which the geographical proximity of firms plays a role in the formation of knowledge ties only at the early stages of the cluster development process, with it becoming less relevant as the cluster enters the maturity phase. In the latter phase, they argue that socialisation processes – such as transitivity – become more prominent in explaining innovative ties. Their studies do not look at actor level characteristics, but they open up a new research agenda about the role of geographical, social, and firm-level drivers of network change along the cluster life cycle.

1.3. Inter-firm heterogeneity and innovative performance in cluster

As discussed in section 1, one of the critiques of *meso-level* interpretations of cluster performance is that studies insist upon considering the cluster as a whole and do not give sufficient attention to the characteristics, behaviours, and performances of the individual firms within the cluster. This view is probably due to the primary purpose of cluster studies in emphasising the importance of the meso-level factors – e.g. geography, social embeddedness, institutional thickness, etc. – in explaining economic behaviour and performance. For instance, in his seminal work, Becattini (1979), questioned the interpretation of firm performance as being dependent upon the industry structure (Scherer and Ross, 1990; Caves, 1992) and advanced the idea that a firm's performance was enhanced by their belonging to an industrial cluster. Within clusters, local resources are considered to be shared collectively and to be neither exclusive to any individual enterprise, nor available outside the cluster. In other words, this view conceives the success of firms to come primarily from the fact that they are located within a cluster, as cluster firms perform better than firms operating in isolation and outside clusters, even if they belong to the same industry and to the same national economic system (Signorini, 1994; Paniccia, 1999).

In spite of the fact that this view became dominant in the traditional cluster literature, a growing number of studies have discussed the need for a focus on inter-firm heterogeneity within clusters, as well as on the role of the individual firm. As an illustrative example of these studies, Lazerson and Lorenzoni (1999) published a critical review of the idea of internal homogeneity in clusters. They argued that most of the literature assumes that all firms in clusters are relatively homogenous and that firms do not deserve individual attention. Even accepting the fact that local institutions and a broader socio-economic characterisation of clusters shape and condition the economic behaviour of firms, the authors emphasise individual entrepreneurs as the leading players in the district. Although these studies do not constitute a coherent strand of research, they more or less explicitly share an interest in the analysis of how inter-firm heterogeneity affects the performance of firms within clusters.

In this section we review a selection of papers on this subject, which are presented in Tables 1.4–1.6. Within this literature we envisage three main areas of interest. First, some studies question the homogeneity view of clusters and seek empirical evidence of their internal differences. Second, other authors explain firm heterogeneity on the basis of their different endowment of internal resources. Finally, further studies focus on the inherent heterogeneity of networks of relationships of firms in order to explain how the performance of the firms varies.

The first group of authors comes to question homogeneity by finding empirical evidence of internal differences (see Table 1.4). On this front, the earlier works were undertaken by Rabellotti and Schmitz (1999), who compare different cases of clusters and find evidence of a clear heterogeneity within clusters in terms of firm size and performance. Using the cluster analysis technique, they have classified a large sample of firms from different clusters into groups which are internally homogeneous in terms of performance, size, collective efficiency, and other structural features. Within each analysed cluster there are several of these groups, which are internally homogeneous but differ among themselves, implying the existence of a clear internal heterogeneity within the clusters (Rabellotti and Schmitz, 1999). Similarly, taking the idea of core and periphery networks, Molina-Morales and Martínez-Fernández (2009) evidenced the heterogeneity among cluster firms and distinguished two basic groups of firms with significant differences in terms of relationships, resources and performance. The group of

Table 1.4 Empirical evidence on cluster internal differences

Author(s), Year	*Research Questions/ Objective*	*Context*	*Industry*	*Methodology*	*Main results*
Rabellotti and Schmitz (1999)	Investigate firms' heterogeneity within clusters	Italy, Brazil, Italy	Shoes	Sample surveys. Two explanatory multivariate techniques: factor and cluster analysis	There is a clear heterogeneity within clusters in terms of size and performance.
Molina-Morales and Martínez-Fernández (2004)	Investigate firms' heterogeneity within clusters	Spain	Manufacturing firms	Survey (288 firms) + econometric analysis	Individual firms develop their own particular resources, relationships and networks and, as a result, the exploitation of the shared resources is irregular. These differences explain the differences in value creation among firms.
Molina-Morales and Martínez-Fernández (2009)	Examine the degree of heterogeneity within industrial districts	Spain	Ceramic tile	Survey (118 firms) econometric analysis	The existence of heterogeneity among the member firms within industrial districts. In fact, two distinct groups of firms can be distinguished.
Molina-Morales *et al.* (2011)	Explore the existence of significant differences in the efficiency levels of two groups of firms in the cluster	Spain	Ceramic tile	Survey (118 firms) econometric analysis, Data Envelopment Analysis (DEA) technique	Findings confirming internal heterogeneity in terms of efficiency.

firms which occupy peripheral positions in the cluster showed significantly less density of interactions and strength in ties and displayed higher values in the indicators of innovation. More specifically, core firms were more efficient than peripheral enterprises in the improvement of existing technologies, whereas the latter focused more on radical innovations. In the same vein, Molina-Morales *et al.* (2011), using the *Data Envelopment Analysis* (DEA) technique, found that peripheral firms presented higher productive efficiency than firms occupying core positions in the network. Finally, more recently, Tomas-Miquel *et al.* (2011) analysed the evolution of firm size, as an indicator of inter-firm heterogeneity, taking the cases of several Spanish clusters during the last two decades. The authors find an increasing dispersion of cluster firms' size and show that firm size tends to move away from average values, in both directions: large firms become larger and small ones smaller. This longitudinal picture suggests an asymmetric cluster with relevant, growing internal differences.

More broadly, studies which look at cluster firm heterogeneity and performance have taken two theoretical perspectives: the Resource-Based View (RBV) of the firm (Wernerfelt, 1984; Barney, 1991), which explores cluster firms' differences on the basis of their knowledge endowments; and the Relational or Social Capital view (Granovetter, 1985).

With regard to studies taking the Resource-Based View (see Table 1.5), Molina-Morales and Martinez-Fernandez (2004) specifically addressed the question of internal heterogeneity and performance. This research applied the RBV to clusters using the notion of *shared resources* in order to define those strategic resources which firms can obtain in clusters. The study was based on a sample of 288 Spanish manufacturing firms belonging to industrial districts; the results concluded that individual firms develop their own particular resources, relationships and networks and, as a result of this, internal resources vary depending upon the degree to which firms participate or exploit the shared competitive resources in the district. This in turn affects the performance of individual firms.

The internal resources also generate combinative effects with external resources (Hervas-Oliver and Albors-Garrigos, 2009) or with managerial ties (Zhang and Li, 2008) enhancing the performance of clustered firms. First, internal resources are drivers for accessing, exploiting and combining external knowledge in clusters, and the observed synergic effect upon the interaction between internal and relational resources is related to innovation performance (Hervas-Oliver and Albors-Garrigos, 2009). On the other hand, Zhang and Li (2008) documented the relationships between managerial ties, firm resources and performance within Chinese clusters. Their study used data from 163 firms in two different clusters. The authors concluded that the heterogeneity of firms' internal resources is a critical source of differences in cluster firm performance. These studies are in line with McCann and Folta (2011), who drew upon the Knowledge-Based View and investigated which firms benefited most from agglomeration. They showed that firms benefit asymmetrically from clusters; in particular, firms with higher knowledge stocks and younger firms demonstrate stronger performance in terms of number of patents.

Table 1.5 Resource-Based View's explanation of the cluster internal differences

Author(s), Year	*Research Questions/ Objective*	*Context*	*Industry*	*Methodology*	*Main results*
Zhang and Li (2008)	Document the relationships between managerial ties, firm resources and performance within a cluster	Guangdong Province (China)	Textile, Electronics	Survey on a sample of 163 firms and econometric analysis	The heterogeneity in managerial ties and internal resources are critical sources of differences in cluster firm performance.
Hervas-Oliver and Albors-Garrigos (2009)	Explore the role of firms' internal and relational resources in explaining innovation in clusters	Spain	Ceramic tile	Survey on a sample of 148 firms + semi-structured interviews and questionnaires + econometric analysis	Internal resources determine how external resources are accessed, exploited and combined in clusters. Internal and relational resources are important, and influence performance.
McCann and Folta (2011)	Analyse what kind of firms benefit most from agglomeration	United States	Biotechnology	Bioscan data + econometric analysis	Firms may benefit asymmetrically from clusters; in particular, firms with higher knowledge stocks and younger firms demonstrate stronger clustering-performance relationships. Patenting is the dependent variable.

Studies taking the Relational or Social Capital view consider that firm behaviour and outcomes are affected by the pattern of relationships maintained with other firms and organisations. The idea of heterogeneity is inherent to social relations. In fact, networks of social relations penetrate irregularly and to differing degrees (Granovetter, 1985). Studies which apply this perspective to cluster studies (see Table 1.6) have analysed the role of the networks of relationships of firms, and how they affect the performance of firms. In clusters, it can be suggested that firms are interrelated in highly differentiated ways and therefore their networks present distinct opportunities and constraints (see also section 2). In this respect, McEvily and Zaheer (1999) show that firms' distinctive patterns of network ties are related to their competitive capabilities. Their study largely supports the idea that firms that take advantage of services available from regional (cluster) institutions achieve stronger competitive capabilities.

In an extensive study, Cooke *et al.* (2005) address the question of whether firms that make greater use of social capital display differences in business performance, as compared with other firms. They found that relational embeddedness is an important indicator for SME performance at the firm-level, but not at the regional level. Finally, in the specific context of biotechnology, Whittington *et al.* (2009) found that geographic and network positions both have independent and contingent effects upon organisational innovation. Both geographic and structural explanations for differential firm-level innovation depend upon the specific context. Some geographic and network locations convey considerable benefits, while others do not; in other words, firms vary significantly in terms of innovation performances depending upon the specific cluster to which they belong. Giuliani's (2012) study on wine clusters shows that firms that embed in local dense and close networks perform better than firms that bridge structural holes, but her analysis also reveals that connections external to the clusters are more valuable than local ones in promoting the transfer of the diverse knowledge that is needed for innovation and product success.

The different types of ties matter in different ways for cluster firms. Mesquita and Lazzarini (2008) found that vertical ties enhance productivity along the supply chain, while horizontal ties enable collective resource use as well as joint product innovation. Furthermore, McDermot *et al.* (2009) suggest that access to a variety of knowledge resources explains how firms upgrade products, and show that upgrading depends upon maintaining ties to firms whose centrality and bridging qualities exceed the average values in the cluster. In the same vein, Bell (2005) studied the differential effects of managerial and institutional ties on firm innovativeness, concluding that it was the centrality in the managerial tie network that enhanced firm innovativeness.

In spite of differences of perspective and methods used, all of the selected research presents a common conclusion regarding the internal heterogeneity and performance issues of clusters. Individual firms within clusters have distinct and different attributes, stock of internal resources, and network positioning. As a consequence of that, individual firms show significant differences in performance.

Table 1.6 Relational or Social Capital view's explanation of the cluster internal differences

Author(s), Year	*Research Questions/ Objective*	*Context*	*Industry*	*Methodology*	*Main results*
McEvily and Zaheer (1999)	What explains the differences in firms' abilities to acquire competitive capabilities?	United States	Job shop manufacturers operating in the metalworking sector	Survey of 309 executives and econometric analysis	One source of important firm heterogeneity is the idiosyncratic and unique manner in which firms are embedded in networks. Participation in regional institutions mediates the link between bridging ties and the acquisition of competitive capabilities.
Cooke *et al.* (2005)	The social capital effect on the performance of SMEs in 12 UK regions	UK	High-technology; medium-technology; manufacturing, low-technology manufacturing and knowledge-based services and other business services	Survey, questionnaires, interviews, econometrics (correlation analysis)	Relational embeddedness is an important indicator for SMEs performance at the individual level, but not at the regional level.
Bell (2005)	What is the relationship between industry clusters, network centrality, and firm innovativeness?	Canada	Mutual fund companies	Survey and econometric analysis	The study shows the differential effects of managerial and institutional ties on firm innovativeness. Centrality in the managerial tie network enhances firm innovativeness.

(Continued)

Table 1.6 (Continued)

Author(s), Year	*Research Questions/ Objective*	*Context*	*Industry*	*Methodology*	*Main results*
Mesquita and Lazzarini (2008)	Model how collaboration among SMEs in environments of weak infrastructure and institutions help them achieve greater collective efficiencies and access to global markets	Argentina	Furniture	Survey, questionnaire (232 responses) and econometric analysis	Different types of ties matter in different ways for SMEs' collective efficiencies. For instance, vertical ties yield manufacturing productivity along the supply chain, while horizontal ties enable collective resource use as well as joint product innovation.
McDermott *et al.* (2009)	Which ties influence on product upgrading?	Argentina	Wine	Ethnographic comparative analysis and deductive analysis 115 firms	To explain how firms upgrade products, a firm's access to a variety of knowledge resources depends on its ties not simply to any or many organisations or institutions, but particularly on its ties to those that excel in their centrality and bridging qualities.
Giuliani (2012)	What type of connections influence product success?	Chile, Italy	Wine	Case study and econometric analysis on 105 firms	Local network closure explains product success better than positions in which structural holes are locally bridged. External connections are more important than local ones for product success.

1.4 Conclusions

This chapter reviews the recent advancements in the literature on industrial clusters, knowledge spillovers, and innovative performance. It focuses on recent literature which questions the traditional conception of clusters as homogeneous communities of actors, which benefit – more or less equally – from the presence of external economies and other meso-level forces. It shows, in light of the recent empirical evidence, first that firms do not benefit equally from the presence of localised spillovers, as some are more prepared and more skilled in both generatin and taking advantage of spillovers – thus questioning the Marshallian popular metaphor that knowledge is 'in the air'. It also discusses the factors that explain knowledge spillovers in a more powerful way than pure geographic proximity (see section 2). Second, it shows that firms in clusters behave and perform differently – a result that questions the overall significance of clustering in explaining firms' performances. However, this does not mean that clusters play no role in fostering the performance of firms. Rather, scholars propose to integrate cluster-centred perspectives with others – such as the RBV and the relational view (section 3) – which emphasise the internal resources of firms and the nature and characteristics of their local ties. Taking into account firms' internal heterogeneity, as well as the diversity of networks formed by firms within clusters, proves to be a fruitful approach for the study and understanding of firm-level performance in clusters. The main contribution of this chapter is to integrate all of these critical approaches and to provide a comprehensive and updated overview of the research on this subject.

Nevertheless, our literature review leaves a number of new research opportunities open. Future research seems likely to head in two directions. One direction is that of investigating the impact of extra-cluster connections on local learning processes – a subject that we have not discussed in this chapter. Scholars have already produced a large and fairly consolidated body of knowledge on this topic, which shows the importance of connecting local learning processes in clusters with extra-cluster sources of technologies and knowledge (i.e. from suppliers or clients, universities and other organisations) (among many others, see: Bell and Albu, 1999; Humphrey and Schmitz, 2002; Bathelt *et al.*, 2004). In order to understand how local actors tap into extra-cluster knowledge, some scholars have recently explored the role of *gatekeepers* – that is, firms with strong connections outside the cluster which also contribute to the diffusion and recombination of external knowledge within the local context (among others see: Giuliani and Bell, 2005; Morrison, 2008; Giuliani, 2011; Graf, 2011). In spite of these recent contributions, there seems to be scope for further research on this subject. A second strand of promising research concerns the evolution of clusters and of their knowledge networks. In this chapter we have reviewed the most recent studies on this topic, and we consider that this to be an area of great interest and with significant opportunities for novel contributions in the future.

Notes

1 Of the different definitions of geographical agglomerations, probably the most popular and well-accepted are those of industrial districts and clusters. For the purpose of our research we use the concept of 'cluster' defined as geographical agglomerations of firms specialised in one or more connected industries. For a discussion on the differences across definitions and conceptualisations of industrial clusters, see Lazzeretti (2006); Martin and Sunley (2003).

2 In this chapter we do not consider a potential third critical view, which looks at the importance of linkages external to the cluster for competitiveness (see e.g. Bell and Albu, 1999; Bathelt *et al.*, 2004; Wolfe and Gertler, 2004).

3 Social Network Analysis is a distinct research perspective within the social sciences, and it is based on the assumption that relationships among interacting actors are important to explain their nature, behaviour, and outputs. To rigorously study relationships, SNA uses graph theory.

4 The relevance of trade linkages for the performance of cluster firms will also be discussed in Section 3.

5 Probably, the lack of studies on this subject is due to the inherent difficulty of collecting network data over time. Recently, scholars have approached this difficulty by using co–patent data to track knowledge flows in regions.

6 Reciprocity refers to the tendency of pairs of actors to form mutual connections between each other; transitivity refers to the tendency of actor *a* and actor *b* to form a tie if *a* and *b* are both connected to a third actor *c*. Homophily is the tendency of actors to form ties with similar others.

References

Audretsch, D. and Feldman, M.P. (1996) 'R&D spillovers and the geography of innovation and production', *American Economic Review*, 86(3): 630–40.

Aydalot, P. and Keeble, D. (1988) *High technology industry and innovative environments: The European experience*, London and New York: Routledge.

Bagnasco, A. (1977) *Tre Italie: la problematica territoriale dello sviluppo italiano*, Bologna: Il Mulino.

Balland, P.A. (2011) 'Proximity and the evolution of collaboration networks: Evidence from research and development projects within the Global Navigation Satellite System (GNSS) industry', *Regional Studies*, 45(8): 1059–78.

Baptista, R. (2000) 'Do innovations diffuse faster within geographical clusters?', *International Journal of Industrial Organization* 18(3): 515–35.

Baptista, R. and Swann, P. (1998) 'Do firms in clusters innovate more?', *Research Policy* 27(5): 525–40.

Barney, J.B. (1991) 'Firm resources and sustained competitive advantage', *Journal of Management*, 17(1): 99–120.

Bathelt, H., Malmberg, A. and Maskell, P. (2004) 'Clusters and knowledge: Local buzz, global pipelines and the process of knowledge creation', *Progress in Human Geography*, 28(1): 31–56.

Becattini, G. (1979) 'Dal Settore Industriale al Distretto Industriale. Alcune considerazioni Sull'unità di Indagine in Economia Industriale', *Rivista di Economia e Politica Industriale*, 1: 1–8.

Becattini, G. (1989) 'Sectors and/or districts: some remarks on the conceptual foundations of industrial economics', in E. Goodman and J. Bamford (eds) *Small firms and industrial districts in Italy*, Routledge.

Bell, G.G. (2005) 'Clusters, networks, and firm innovativeness', *Strategic Management Journal*, 26(3): 287–95.

Bell, M. and Albu, M. (1999) 'Knowledge systems and technological dynamism in industrial clusters in developing countries', *World Development*, 27(9): 1715–34.

Boschma, R. and Ter Wal, A. (2007) 'Knowledge networks and innovative performance in an industrial district: The case of a footwear district in the south of Italy', *Industry and Innovation*, 14(2): 177–99.

Breschi, S. and Lissoni, F. (2001) 'Knowledge spillovers and local innovation systems: A critical survey', *Industrial and Corporate Change* 10(4): 975–1005.

Breschi, S. and Lissoni, F. (2009) 'Mobility of skilled workers and co-invention networks: an anatomy of localized knowledge flows', *Journal of Economic Geography* 9(4): 439–68.

Broekel, T. and Boschma, R. (2011) 'Knowledge networks in the Dutch aviation industry: the proximity paradox', *Journal of Economic Geography*, Advance Access, May 26 2011, 1–25.

Cantner, U. and Graf, H. (2006) 'The network of innovators in Jena: an application of social network analysis', *Research Policy*, 35(4): 463–80.

Cassiolato, J. and Lastres, H.M.M. (1999) *Globalização and Inovação Localizada: Experiencias de Sistemas Locais no Mercosul*, Edição: IBICT.

Caves, E.R. (1992) *American industry: Structure, conduct, performance*, 7th edn, Prentice Hall.

Cooke, P., Clifton, N. and Oleaga, M. (2005) 'Social capital, firm embeddedness and regional development', *Regional Studies*, 39(8): 1065–77.

Giuliani, E. (2007) 'The selective nature of knowledge networks in clusters: Evidence from the wine industry', *Journal of Economic Geography*, 7(2): 139–68.

Giuliani, E. (2008) 'Multinational corporations and patterns of local knowledge transfer in Costa Rican high-tech industries', *Development and Change*, 39(3): 385–407.

Giuliani, E. (2010a) 'Distretti industriali, reti e comunità di knowledge worker: un'analisi empirica nel settore vitivinicolo cileno', *Finanza Marketing e Produzione*, 28(1): 77–115.

Giuliani, E. (2010b) 'Network dynamics in regional clusters: The perspective of an emerging economy', *Papers in Evolutionary Economic Geography (PEEG)*, 1014, Utrecht University, Section of Economic Geography.

Giuliani, E. (2011) The role of technological gatekeepers in the growth of industrial clusters: Evidence from Chile, *Regional Studies*, 45 (10): 1329–48.

Giuliani, E. (2012) 'Clusters, networks and firms' product success. An empirical study', *Management Decision*, forthcoming.

Giuliani, E. and Bell, M. (2005) 'The micro-determinants of meso-level learning and innovation: evidence from a Chilean wine cluster', *Research Policy*, 34(1): 47–68.

Graf, H. (2011) 'Gatekeepers in regional networks of innovators', *Cambridge Journal of Economics*, 35(1): 173–98.

Granovetter, M. (1985) 'Economic action and social structure: The problem of embeddedness', *American Journal of Sociology*, 91(3): 481–510.

Hervas-Oliver, J.L. and Albors-Garrigos, J. (2009) 'The role of the firm's internal and relational capabilities in clusters: when distance and embeddedness are not enough to explain innovation', *Journal of Economic Geography*, 9(2): 263–83.

Humphrey, J. and Schmitz, H. (2002) 'How does insertion in Global Value Chains affect upgrading in industrial clusters?', *Regional Studies*, 36(9): 1017–27.

Krugman, P. (1991) *Geography and trade*, Cambridge: MIT Press.

Lazerson, M. and Lorenzoni, G. (1999) 'The firms that feed industrial districts: A return to the Italian source', *Industrial and Corporate Change*, 8(2): 235–66.

Lazzeretti, L. (2006) 'Distritos industriales, clusters y otros: un análisis trespassing entre la economía industrial y la gestión estratégica', *Economía Industrial*, 359: 59–72.

McCann, B.T. and Folta, T.B. (2011) 'Performance differentials within geographics clusters', *Journal of Business Venturing*, 26(1): 104–23.

McDermott, G.A., Corredoira, R.A. and Kruse, G. (2009) 'Public-private institutions as catalysts of upgrading in emerging market societies', *Academy of Management Journal*, 52(6): 1270–96.

McEvily, B. and Zaheer, A. (1999) 'Bridging ties: A source of firm heterogeneity in competitive capabilities', *Strategic Management Journal*, 20(12): 1133–56.

Maggioni, M.A. and Uberti, T.E. (2009) 'Knowledge networks across Europe: which distance matters?', *Annals Regional Science*, 43(3): 691–720.

Marshall, A. (1920) *Principles of economics*, London: MacMillan.

Martin, R. and Sunley, P. (2003) 'Deconstructing clusters: chaotic concept or policy panacea?', *Journal of Economic Geography*, 3(1): 5–35.

Mesquita, L.F. and Lazzarini, S.G. (2008) 'Horizontal and vertical relationships in developing economies: implications for SMEs' access to global markets', *Academy of Management Journal*, 51(2): 359–80.

Molina-Morales; F.X. and Martínez-Fernández, M.T. (2004) 'How much difference is there between industrial district firms? A net value creation approach', *Research Policy*, 33(3): 473–86.

Molina-Morales, F.X. and Martínez-Fernández, M.T. (2009) 'Does homogeneity exist within industrial districts? A social capital-based approach', *Papers in Regional Science*, 88(1): 209–29.

Molina-Morales, F.X., Martínez-Fernández, M.T. and Coll Serrano, V. (2011) 'Eficiencia de las empresas del distrito industrial cerámico de Castellón: Un análisis comparativo mediante medidas no radiales', *Revista de Estudios Regionales*, 1: 155–77.

Morrison, A. (2008) 'Gatekeepers of knowledge within industrial districts: who they are, how they interact', *Regional Studies*, 42(6): 817–35.

Morrison, A. and Rabellotti, R. (2009) 'Knowledge and information networks in an Italian wine cluster', *European and Planning Studies*, 17(7): 983–1006.

Paniccia, I. (1999) 'The performance of IDs. Some insights from the Italian case', *Human Systems Management*, 18(2): 141–59.

Parra-Requena, G., Molina-Morales, F.X. and García-Villaverde, P. (2010) 'The mediating effect of cognitive social capital on knowledge acquisition in clustered firms', *Growth and Change*, 41(1): 59–84.

Porter, M.E. (1990) *The competitive advantage of nations*, London: Macmillan.

Porter, M.E. (1998) 'Clusters and the new economics of competition', *Harvard Business Review*, November–December: 77–90.

Rabellotti, R. (1995) 'Is there an 'Industrial District Model'? Footwear districts in Italy and Mexico compared', *World Development*, 23(1): 29–41.

Rabellotti, R. and Schmitz, H. (1999) 'The internal heterogeneity of industrial districts in Italy, Brazil and Mexico', *Regional Studies*, 33(2): 97–108.

Sabel, C. (1989) 'Flexible specialization and the re-emergence of regional economies', in P. Hirst and J. Zeitlin (eds) *Reversing industrial decline? Industrial structure and policy in Britain and her competitors*, Oxford: Berg.

Sammarra, A. and Biggiero, L. (2008) 'Heterogeneity and specificity of inter-firm knowledge flows in innovation networks', *Journal of Management Studies*, 45(4): 800–29.

Saxenian, A. (1994) *Regional advantage: Culture and competition in Silicon Valley and Route 128*, Cambridge, Mass: Harvard University Press.

Scherer, F.M. and Ross, R. (1990) *Industrial market structure and economic performance*, 3rd edn, Houghton Mifflin.

Scherngell, T. and Hu, Y. (2011) 'Collaborative knowledge production in China: Regional evidence from a Gravity Model Approach', *Regional Studies*, 45(6): 755–72.

Schmitz, H. (1995) 'Collective Efficiency: growth path for small-scale industry', *Journal of Development Studies*, 31(4): 529–66.

Schmitz, H. and Musyck, B. (1994) 'Industrial districts in Europe: Policy lessons for developing countries', *World Development*, 22(6): 889–910.

Signorini, L.F. (1994) 'The price of Prato, or measuring the ID effect', *Papers in Regional Science*, 73(4): 369–92.

Staber, U. (2001) 'The structure of networks in industrial districts', *International Journal of Urban and Regional Studies*, 25(3): 537–52.

Staber, U. (2011) 'Imitation without interaction: How firms identify with clusters', *Organization Studies*, 31(2): 153–74.

Storper, M. (1997) *The regional world territorial development in a global economy*, New York: Guilford Press.

Ter Wal, A.L.J. (2009) 'The dynamics of the inventor network in German biotechnology: Geographical proximity versus triadic closure', Paper to be presented at the DRUID Summer Conference 2009; Copenhagen Business School, Denmark.

Ter Wal, A.L.J. (2012) 'Cluster emergence and network evolution: A longitudinal analysis of the inventor network in Sophia-Antipolis', *Regional Studies*, in press.

Ter Wal, A.L.J. and Boschma, R. (2011) 'Co-evolution of Firms, Industries and Networks in Space', *Regional Studies*, 45(7): 919–33.

Tomas-Miquel, J., Molina-Morales, F.X. and Exposito-Langa, M. (2011) 'Evolution of Spanish industrial districts: How are they evolving and adapting in the face of globalization?', in F. Belussi and U. Staber (eds) *Managing networks of creativity* (forthcoming).

Wasserman, S. and Faust, K. (1994) *Social network analysis. Methods and applications*, Cambridge: Cambridge University Press.

Wernerfelt, B. (1984) 'A resource-based view of the firm', *Strategic Management Journal*, 5(2): 171–80.

Whittington, K.B, Owen-Smith, J. and Powell, W.W. (2009) 'Networks, propinquity, and innovation in knowledge-intensive industries', *Administrative Science Quarterly*, 54(1): 90–122.

Wilhelmsson, M. (2009). 'The spatial distribution of inventor networks', *The Annals of Regional Science*, Springer, 43(3): 645–668.

Wolfe, D.A. and Gertler, M.S. (2004) 'Clusters from the inside and out', *Urban Studies*, 41(5-6): 1071–93.

Zhang, S. and Li, X. (2008) 'Managerial ties, firm resources, and performance of cluster firms', *Asia Pacific Journal of Management*, 25(4): 615–33.

2 Friends and rivals

Modelling the social relations of inventors

Lorenzo Cassi and Lorenzo Zirulia

2.1 Introduction

Recently, the role of social networks in the creation and diffusion of knowledge has been intensively studied, both empirically (Giuliani and Bell, 2005; Fleming *et al.*, 2007) and theoretically (Cowan and Jonard, 2004; Cassi and Zirulia, 2008; Morone and Taylor, 2010). The assumption that a large part of scientific and technical knowledge is tacit and only partially codifiable, and that face-to-face interactions are required for its transmission, leads to the conclusion that social connections typically matter (Powell and Grodal, 2005). In this chapter, we make a contribution to the theoretical literature on this topic. We develop a model where a population of inventors are rivals in the production of patents. Inventors are embedded in the social space, and this affects the process of knowledge creation and diffusion. The main element of novelty in our work consists in extending the literature in favour of better comparability of our results with those of empirical analysis. This is done by explicitly introducing into the model the counterpart of proxies for knowledge and knowledge diffusion used in these works: in the case of inventors, patents and patent's citations.

Of the different streams of empirical literature on networks and knowledge, our attention in this chapter is focused on studies that investigate the relationship between knowledge flows, as measured by patent citations, and social distance. These works (e.g. Breschi and Lissoni, 2004) find that the probability of observing a citation tie between two patents decreases as the social distance between the two patents increases, which is suggestive of the idea that knowledge flows (with decay) through the social network. Our model replicates this evidence and, more importantly, provides new theoretical results on the impact of network structure upon knowledge flows, in a form that is potentially testable by empirical researchers.

The structure of the chapter is as follows. Section 2 positions our work within the perspective of the existing theoretical literature and creates a link with empirical analysis, thus motivating the extensions we propose. Section 3 describes the model in detail, while Section 4 reports the results from numerical simulations. Section 5 concludes.

2.2 The theoretical and empirical literature on social networks and knowledge

In its formal structure and focus of analysis, this chapter belongs to the theoretical literature on networks and knowledge, initiated by the work of Cowan and Jonard (Cowan and Jonard, 2004). This literature builds on two basic elements. First, it represents the economy as a *heterogeneous* population of agents involved in the creation and diffusion of knowledge. Second, these agents are located on a network, which implies that each agent can 'interact' (both in the process of knowledge creation and diffusion) only with a subset of other agents in the society (their 'neighbourhood'). Given these commonalities, the existing models differ on several dimensions: the network can be exogenous (Cowan and Jonard, 2003a) or endogenous (Cowan *et al.*, 2003); knowledge has been represented in different ways, as a stock (Morone and Taylor, 2004), as vector of real positive scalars (Cowan and Jonard, 2004), or as a pair constituted by a scalar and an angle (Cowan *et al.*, 2003); different diffusion processes (pairwise barter, as in Cowan and Jonard, 2004; broadcasting, as in Cowan and Jonard, 2003a) and knowledge creation processes (recombination by the receiver, as in Cowan and Jonard, 2003a; individual learning, as in Cowan *et al.*, 2004). Further details have been added to characterise the process of knowledge creation and diffusion in specific institutional contests, such as the scientific community (Cowan and Jonard, 2003b). The most recent contributions (e.g. Baum *et al.*, 2010; Morone and Taylor, 2010) focused on the endogenous network formation of 'agents' that are interpreted as firms, thus relating to, and trying to rationalise, the results of the vast literature on inter-firm technological networks (Zirulia, 2009). In this chapter we take a different route, by making a number of contributions to the current state of the literature, which are empirically motivated by the specific context we have in mind, i.e. a population of inventors engaged in the production of patents. We believe that the modifications we propose are both theoretically interesting *per se*, and useful in bridging the gap with the empirical literature on this issue. First, we consider the role of social networks as an option. This has already been done in the model of Cassi and Zirulia (2008). The idea is that, while agents are embedded in a social network, the actual use of the network is an (economic) choice. Agents will have typically other options for reaching their goal (in this case, learning and consequently producing new patents), rather than interacting socially. Quite obviously, agents will have the possibility of learning individually: thus, it could be possible that the structure of the network affects its *use*.

Second, we relax the assumptions of non-rivalry between agents. If agents seek knowledge not for its own sake, but in order to produce 'objects' that embody such knowledge, agents will typically care about the knowledge of other individuals, since collectively they participate in a number of 'races'. This is true for inventors, but also for scientists engaged in the production of papers. These considerations imply, as described in detail in the next section, that we will model the problem of choice between learning modes, and we will do that by endowing agents with a simple rule of thumb that implicitly takes into consideration the rivalry between them.

A third, more significant extension concerns the representation of knowledge. Under this aspect, our attempt to create connections with the empirical literature is explicit. There is now a significant amount of empirical work on the structure and role of social networks in 'knowledge-intensive' communities, both in the realm of *Open Science* (Newman, 2000), and in the realm of *Proprietary Technology* (Balconi *et al.*, 2004; Breschi and Lissoni, 2004; Agrawal *et al.*, 2006). The assumptions that motivate such an empirical program of research are indeed similar to the assumptions that lie at the core of the theoretical literature. Since technical and scientific knowledge is largely 'tacit' and 'non-codifiable', face-to-face interactions, and consequently social networks, are important for knowledge diffusion and creation. In that respect, the empirical and theoretical literatures are largely complementary: at both levels, for instance, it is possible to study how the specific characteristics of a technical or scientific field impact upon the emergent properties of a social network structure, or the relationship which occurs between distances in the social and 'cognitive' space. However, the empirical approach needs to tackle the issue of the 'non-observability' of knowledge, and, especially in large groups, the difficulties of measuring social networks between inventors and scientists. This leads to the question of proxies.

The natural solution to these problems has been to use the 'objects' in which knowledge is embodied. In the case of inventors, these are given by patents (Griliches, 1990). Patents are clearly correlated with knowledge possessed by inventors; there is a large availability of data; patent citations can be used to track 'knowledge' flows (Jaffe *et al.*, 1993) among individuals and consequently to locate an individual in the 'cognitive' space. Furthermore, patent data can also be used to map the social network of inventors, using co-patenting to infer the existence of a social link between the inventors in the research team (e.g. Breschi and Lissoni, 2004).

Within the different streams of empirical literature on the broad theme of knowledge and networks, we focus on a recent set of empirical analyses which has used patent and patent citations in order to investigate the relationship between social relations and knowledge flows.

The starting point of these works is to better understand the mechanisms underlying knowledge flows and why knowledge spillovers are geographically bounded. All of the articles use a refinement of the methodology originally proposed by Jaffe *et al.* (1993). In this seminal article, the three authors considered patent citations as possible paper trails left by knowledge spillovers. Accordingly, they set up an original test to show that citations to prior art tend to originate disproportionally from the same geographic area as the cited patents. In order to separate the spillover effect from other agglomeration effects, they defined a control sample of patents *similar* to the citing ones in terms of technological content and filed in the same period, but not reporting the same reference to prior art. In other words, the control sample represents the potential citations. Comparing the localisations of the pairs citing-cited patents with the control-cited ones allows, in principle, the isolation of the knowledge spillover effect from other effects, such as, for instance, agglomeration.

While Jaffe *et al.* (1993) applied this methodology in considering the geographic localisation of inventors, Breschi and Lissoni (2005) applied it in considering the social space defined by the network of inventors, and found the association between patent citations and geographic co-location to be greater for socially connected inventors than for others. In other words, Breschi and Lissoni show that geographical proximity matters for knowledge spillovers because knowledge flows through social networks, and social networks are, for the most part, geographically bounded. In addition to that, Agrawal *et al.* (2006) show that patents from inventors who move from one region to another continue to be cited by their previous colleagues, reflecting the fact that ties from past collaborations facilitate knowledge flows across regions.

Extending this line of reasoning, Breschi and Lissoni (2004), Cassi (2004), and Singh (2005) have investigated the relationship between knowledge flows and social distance. In particular, these authors have shown that the probability of observing a citation tie between two patents decreases as the social distance between them increases, where the social distance between two patents has been defined as the minimum distance between all of the distances between the pair of inventors belonging to the two teams responsible for the patents.[1] Figure 2.1 (Cassi, 2004) reports the social distance calculated between inventors participating in patents linked by an *actual* citation (full line) compared to the distance between patents linked by a *potential* citation (dot line); more specifically, it compares the frequencies of a citation between two patents (vertical axe – in per cent value) relative to the social distance separating the two (horizontal axe – geodesic distance).

Citations seem to occur between inventors socially close to each other. The frequencies initially drop very rapidly: direct connection or being at distance 2 seems to be important, and a distance greater than 4 seems not to have any

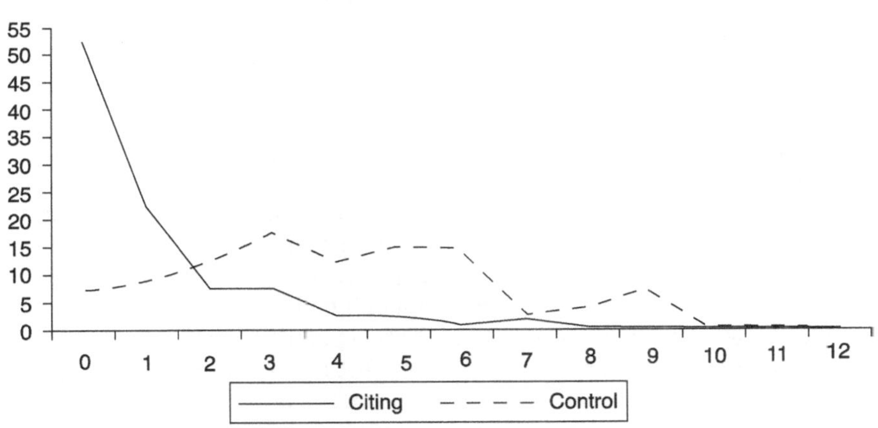

Figure 2.1 Social distance and citations

Source: Cassi (2004)

role. This suggests that knowledge flows through networks are characterised by fast decay.

In order to create a connection with the empirical literature as previously described, in our model we will represent knowledge in a way that makes possible, in the same space, the representation of patents and patent citations. This makes it possible to produce theoretical results that are directly comparable with empirical ones. In particular, we aim at investigating the determinants of knowledge flows as measured by patent citations. In that respect, not only we will look at the effect of social distance on the probability of citation, but we will also study how different network structures impact upon the overall intensity of knowledge flows and how the effect of social distance is mediated by the network structure and other characteristics of the system. As the description of the model will hopefully make clear, in representing patents and citations we try to capture two distinctive features of these objects in reality. First of all, knowledge flows can be measured only if both the 'sender' and the 'receiver' have obtained an innovation, and then a patent. Second, a citation between two patents does not necessarily entail knowledge flows between the two inventors, since the related patents may be the outcome of autonomous learning.

2.3 The model

The model considers a population of N economic agents who create and diffuse knowledge over time. We will refer to these economic agents as individual inventors, but one could easily consider them as a team of researchers or firms. In this case, some of the assumptions require a different interpretation, which we will discuss. The goal of inventors is to transform their knowledge into new patentable ideas, and for that reason the relationships between them are characterised by a certain degree of rivalry. Furthermore, inventors are located on a lattice, which is exogenously given and time-invariant, and it identifies the social space. In the process of knowledge creation, inventors can learn in two ways: individually, through introspection or by accessing a piece of codified knowledge embodied in existing patents; and socially, by interacting with other inventors in their social neighbourhood.

Clarification is necessary as to what we mean by individual and social learning when agents are organisations instead of individual inventors. 'All learning takes place inside individual human heads; an organisation learns in only two ways: (a) by the learning of its members, or (b) by ingesting new members who have knowledge the organisation didn't previously have' (Simon, 1991, p.125). Since in our model the social structure is given, Simon's option *b* (e.g. hiring someone) is excluded by assumption. Therefore, the learning of the organisation is reduced to the learning of its members (i.e. Simon's option *a*). According to this, an organisation learns *individually* if some of its members learn, and this could likely occur via the interaction of members. This could be thought of as an intra-firm learning phenomena. Otherwise, organisations learn *socially* when (at least) two individuals belonging to two different organisations interact with each other.

This section is structured as follows: first, we describe how knowledge is represented in our model and, given this representation, how we characterise patents and patent citations. Then, we discuss the options agents have for learning, and the rule of thumb we assume they adopt in choosing between them. Finally, we describe the structure of the social networks we will consider in our numerical exercises.

2.3.1 Knowledge representation, patents and patent's citations

A primary contribution of this chapter lies in the representation of knowledge. The rationale for developing such a representation is twofold. First, we intend to go beyond a scalar or vector representation of knowledge. These formal representations are simple, and are often convenient in a number of applications: however, they fail in capturing some of the notions put forth appreciatively in the economics of knowledge, and then it is meaningful, per se, to try to improve upon them. Second, the representation we propose is functional in order to explicitly introduce the role of patents and patent citations.

The key notion in our model is that of *knowledge space*. The knowledge space is given by a regular bi-dimensional lattice (i.e. a rectangular grid of cells). Formally, it is denoted as $K \subset N^2$. Consequently, it is possible to identify the single cell with a pair of natural numbers $(k_1, k_2) \in N^2$.

In each period t, each agent is characterised by a knowledge set $K_i^t \subseteq K$. Graphically, the knowledge possessed by each individual can be represented by cells that are 'turned on' in the grid of knowledge space. In Figure 2.2, the collections of dark cells are an example of a knowledge set for an individual.

Patents are defined as a subset of the knowledge space as well. A patentable idea is defined as a *squared* subset of cells in the knowledge set, whose dimension exceeds a lower bound l. We denote a patent as $P_j \subseteq K$, with P as the set of all conceivable patents. An inventor 'produces' a patentable idea at time t whenever such a subset is entirely included in K_i^t. For a patentable idea to become a patent, a novelty criterion must be satisfied. Define K_p^t as the set of existing patents at time t. The novelty criterion requires that $\frac{|K_p^t \cap P_j|}{|P_j|} < x$ – that is, that the number of cells belonging to the intersection of K_p^t and P_j – cannot exceed a certain percentage x of the cardinality of P_j.

In this set-up, we can quite naturally define *relations* within the set of patents (i.e. patents citations). A patent P_1 must cite a *pre-existing* patent P_2 whenever $P_1 \cap P_2 \neq \{0\}$. Citations, as we will see, can be used to introduce the notion of distance in the patent space. In turn, distance in the space of patents can be used to proxy distance among agents in the knowledge space, and to compare it with distance in the 'social' space.

To summarise, in each period t, we need to keep track of $N + 1$ knowledge sets: one for each agent, representing their private knowledge, and one set of 'social

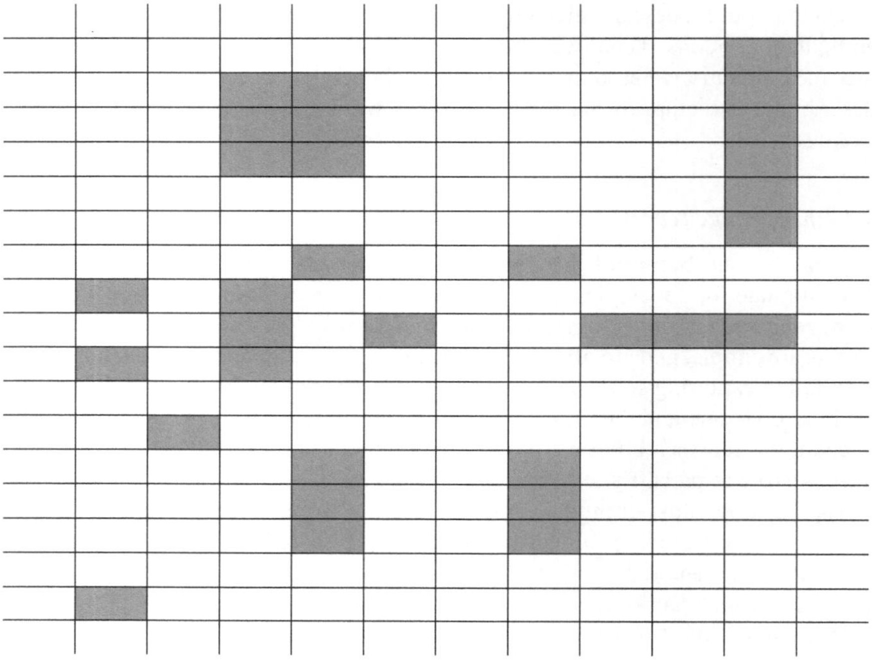

Figure 2.2 An example of knowledge set

knowledge' corresponding to the set of patents K_p^t. We note how 'social' knowledge (Machlup, 1980) will not generally coincide with the sum of knowledge for all individuals. Thus, it can be studied as the ability of the system to 'transform' private knowledge into social knowledge.

2.3.2 Learning

Learning occurs at time t whenever $K_i^{t-1} \subset K_i^t$. Learning is modelled as the outcome of a random mechanism which gives the probability that a cell (k_1, k_2) that does not belong to K_i^{t-1} will belong to K_i^t We assume that learning can occur in two ways: individually, through introspection or by accessing a piece of codified knowledge embodied in existing patents; and socially, by interacting with other inventors in their social neighbourhood. Furthermore, we assume no 'reduction' in knowledge, in the sense that $(k_1, k_2) \in K_i^{t-1} \Rightarrow (k_1, k_2) \in K_i^t$.

The forms of learning we have in mind share the property of being *local*. In our context, learning is local in the sense that the probability that an agent learns in a specific area depends solely upon interaction with a knowledge set that include areas close to that area (according to the metric implicitly defined in the knowledge space); the knowledge set can be the knowledge set of that agent, or of the agent they are interacting with, or a set of social knowledge.

If learning occurs individually for agent i, there are two inputs in the learning process: the existing knowledge set K_i^{t-1} and the set of 'social knowledge' K_p^{t-1}. Define as $n_i(k_1, k_2)$ the number of cells in the Moore neighbourhood of (k_1, k_2) that belong to K_i^{t-1}. Similarly, define as $n_p(k_1, k_2)$ the number of cells in the Moore neighbourhood of (k_1, k_2) that belong to the intersection of K_p^{t-1} and K/K_i^{t-1}, plus (k_1, k_2) if it belongs to K_p^{t-1}.

Thus, the probabilistic mechanism of learning is specified as follows:

$$Pr_{ind}\left((k_1, k_2) \in K_t^t \middle| (k_1, k_2) \notin K_i^{t-1}\right) = q^*(1 - e^{-\lambda[n_i(k_1, k_2) + rn_p(k_1, k_2)]})$$

$$\times \forall (k_1, k_2) \notin K_i^{t-1} \tag{2.1}$$

with $0 < q < 1$, $0 < r < 1$ and $\lambda > 0$. The assumption $r < 1$ captures the idea that agents have only an imperfect ability for absorbing the piece of codified knowledge that is embodied in patents.

Now, with regard to social learning, assume that agents i and j interact. Define as $n_j^{abs}(k_1, k_2)$ the number of cells of the Moore neighbourhood of (k_1, k_2) that belong to the intersection of K_i^{t-1} and K_j^{t-1}, and as $n_j(k_1, k_2)$ the number of cells in the intersection of the Moore neighbourhood of (k_1, k_2) belonging to the intersection of K/K_i^{t-1} and K_j^{t-1}, plus (k_1, k_2) if it belongs to K_j^{t-1}.

Formally, the probabilistic mechanism of learning is specified as follows:

$$Pr_{soc}\left((k_1, k_2) \in K_t^t \middle| K_i^{t-1}\right) = q^*\left(1 - e^{-\lambda[n_j^{abs}(k_1, k_2) + an_j(k_1, k_2)]}\right)$$

$$\times \forall (k_1, k_2) \notin K_i^{t-1} \tag{2.2}$$

with $0 < \alpha < 1$. $\alpha < 1$ captures the role of absorptive capacity (Cohen and Levinthal, 1989): cells that do not belong to the knowledge set of agent i have a smaller impact upon increasing the probability of learning.

We need to comment upon the nature of the social learning mechanism we have in mind. The crucial assumption is that any social interaction creates learning opportunities for both agents: the process is bi-directional in nature. In general, the process will be asymmetric, both *ex ante* (in terms of probability of learning) and *ex post* (in terms of actual learning). We believe such an assumption is not, in fact, restrictive. In abstract terms, it is consistent with the view that any form of interaction between two 'knowledge-carrying' agents can potentially lead to a process of knowledge creation for each of them.

2.3.3 Choosing the learning mechanisms and the dynamics of the system

Up to now, we have described the learning options that are available to individuals. We now need to specify the criterion agents use to choose between these options. In order to do that, we start by describing how the system evolves over time.

In each period, two agents that are connected in the social network are drawn with uniform probability. In other words, we extract one link among those existing in the social network. These two agents are the only two that are allowed to learn at period t.

We assume that individual learning and social learning are mutually exclusive: agents must choose between individual learning and social learning. Furthermore, we assume that social learning occurs only when both agents are willing to interact socially.

In principle, agents should compare the gain from each form of learning and select the best option. This is what is done in the simplified framework of Cassi and Zirulia (2008). In this model, however, things are more complex. Assuming that agents have perfect foresight of the benefits from each of the two forms of learning is quite a high demand. It implies perfect information about the knowledge set of the involved agents (in particular, about the knowledge set of other agents) and a perfect understanding of the learning mechanisms. Furthermore, we need to take into account the fact that agents are rivals in the production of patents, and that they should consider the effects of their decisions on the probability of learning (and patenting) by other agents, for the negative effect on their probability of patenting. Taking rationality to the limit, agents should not act myopically, but should evaluate the long-term effect of their choices on the more general processes of knowledge creation and diffusion.

For these reasons, we take a different route. We assume that agents follow a simple rule of thumb, which embodies some assumptions about the information they possess and capture (albeit roughly), and the incentive agents have in choosing one form of learning or the other. We assume that other agents' knowledge is unobservable, and is not taken into account in the choice of mode of learning. Instead, we assume that an agent has a fairly good perception of their own knowledge. In particular, we assume that an agent knows how far they are from a new patent.

Individual learning is a more focused form of learning: agents learn around their existing area of expertise. Social learning, instead, is modelled as a strategy of 'exploration': through interaction, inventors can acquire knowledge far from their current strengths. When an agent is close to patenting, this means they have a strong expertise in a specific area of the knowledge space: if they engage in individual learning, they will have, in general, a high probability of learning the pieces of information that are missing to produce the new patent; otherwise, if they engage in social learning, it may happen that they will learn information far from their current focus.

Notice also that if the potential partner possesses a piece of information that is missing in order for agent i to produce the patent, the probability that such a partner is also close to patenting will, in general, be significant. Then, the advantage of obtaining from the partner the key piece of missing information must be compared with the risk of losing a new patent, when the missing piece could be acquired though individual learning. Our assumption is consistent with the argument put forth by von Hippel (1987) in his case study of informal know-how exchange in the American mini-mills steel industry.[2]

> [. . .] when 'Engineer A' encounters a product or process development problem he finds difficult, he activates his network by calling 'Engineer B', an appropriately knowledgeable contact who works for a directly competing (or non-competing) firm, for advice. B makes a judgment as to the competitive value of the information A is requesting. If it seems to him vital to his own firm's competitive position, he will not provide it. However, if it seems useful but not crucial [. . .] B will answer his request as well he can and/or refer him to other experts of his acquaintance.
>
> (von Hippel, 1987, p. 292)

For these reasons, we assume a simple rule of thumb to drive agents' decisions regarding the use of social networks: agents will use the network only if they are both more distant from patenting than a given threshold w. Formally, agents will use the network if:

$$\forall P_r \in P/K_p^{t-1}: \frac{\left|P_r \cap K_p^{t-1}\right|}{\left|P_r\right|} \leq x, \left|P_r/K_i^{t-1}\right| \leq w \tag{2.3}$$

and they will use individual learning otherwise.

As a last remark, notice that the fact that agents use the social network only if both agree implies that it can happen that an agent would learn sometimes individually even if they are far from patenting (but the other agent is close). Then, the rule we propose in fact implies that agents will learn individually *for sure* if close to patenting, while, if not, the choice of mode is the outcome of a random event (from the point of view of the agent).

After the agents have chosen the mode of learning, learning occurs. If a patent is produced, it is registered at the 'patent office' and citations are assigned. In a case where the agent can patent a number of mutually exclusive patents, the patent with the highest dimension is chosen. If both agents can file for the patent, the patent is assigned randomly to one of them. Then, the economy moves to the next period.

2.3.4 The network structure

As previously stated, we assume the social network to be exogenous. Therefore, following Cassi and Zirulia (2008), we assume that inventors are 'embedded' in a pre-existing web of social relations. While these relations have not been formed for the purpose of exchanging knowledge, they constitute opportunities for knowledge exchange if the rule of thumb described in Section 3.3 suggests that it is convenient to do so.

Following the current literature which focuses on exogenous networks (Cowan and Jonard, 2003a), we keep network density (i.e. the number of total exiting links) fixed, while we vary its structure along two dimensions: average path length and average cliquishness. Average path length is a global concept, defined as the average number of steps separating two randomly chosen agents. Average

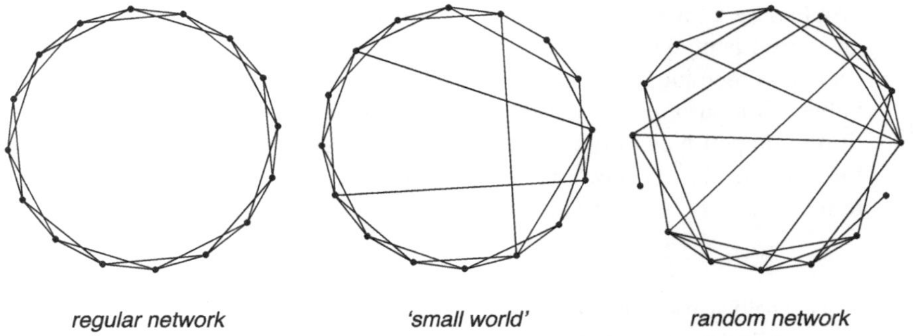

Figure 2.3 The three social structures

cliquishness is a local concept, and measures the degree of link redundancy in an agent's neighbourhood.

We are going to compare three kinds of structure: 'regular networks', 'random graphs' and 'small world'. Figure 2.3 gives a graphical example for each class of network, for graphs with 16 nodes and 32 edges.

A regular network is characterised by high cliquishness and high average path length, a random network by low cliquishness and low average path length, and the 'small world' by a distance which is close to the average distance of a random network, and a cliquishness that is instead closer to that of regular networks.

In their important contribution, Watts and Strogatz (1998) proposed a model that is able to generate these different structures as a function of a single parameter.

Consider a set of agents $I = \{1, \ldots N\}$. For any i, j define the variable $\chi(i, j)$, which takes value 1 if the two agents are connected, and 0 otherwise. Then, the network $G = \{\chi(i, j);\ i, j \in I\}$ corresponds to the list of all pairwise relationships between agents. Then the neighbourhood of agent i is formally defined as $\Gamma_i = \{j \in I/\{i\}: \chi(i, j) = 1\}$. A path connecting two agents i and j is defined as a set of pairwise relationships $\{(i, i_l), \ldots; (i_k, j)\}$ such that $\chi(i, i_l) = \cdots = \chi(i_k, j) = 1$; the distance between i and j $d(i, j)$ is given by the number of steps in the shortest path between i and j. Given this notation we can define the average path length as $L = \sum_{i,j \in I} \frac{d(i,j)}{(N(N-1)/2)}$. The cliquishness of a set $S \subseteq I$ is defined as the proportion of pairwise relationships in S over the total number of possible relationships, formally $cl(S) = \frac{\sum_{i,j \in S} \chi(i,j)}{\#S(\#S-1)/2}$. In the model, average cliquishness is given by the average value of $cl(S)$ for $S = \Gamma_i$, $i \in I$, that is $C = \sum_{i \in I} \frac{cl(\Gamma_i)}{N}$.

Watts and Strogatz consider a family of graphs with a given number of nodes N (the agents) and total number of links among them, $\frac{n * N}{2}$. They start from a ring of n vertices, i.e. a regular periodic lattice with n nearest neighbours (n even).

Loosely speaking, they consider N agents located on a circle, each of them connected (by undirected edges) with the n nearest neighbours.

The next step is to implement the following algorithm: choose a vertex and the edge that connects it to its nearest neighbour in a clockwise sense. With probability p, re-connect this edge to a vertex chosen uniformally at random over the entire ring, with duplicate edges forbidden; with probability $1 - p$, leave the edge in place. The process is repeated by moving clockwise around the ring, considering each vertex in turn until one lap is completed. Next, consider the edges that connect the vertices to their second nearest clockwise neighbours. As before, randomly rewire each of these edges with probability p, and continue the process, circulating around the ring and proceeding outward to more distant neighbours after each lap, until each edge in the original lattice has been considered once.

Varying p, it is possible to build graphs having different values for average path length and average cliquishness, $L(p)$ and $C(p)$. Both $L(p)$ and $C(p)$ are monotonically decreasing in p. At one extreme, for $p = 0$, we have a regular network, with high average cliquishness and high average path length. At the other extreme, for $p = 1$, we have a random network, where both average path length and average cliquishness are low. Indeed, the important contribution by Watts and Strogatz is to show that trade-off between the local concept of cliquishness and the global notion of average distance is not as severe as it could appear at first sight. For a non-negligible region of parameter p's space (approximately, the interval [0.01; 0.1]), they show that $L(p) \approx L(1)$, but $C(p) >> C(1)$. When in a regular network some 'shortcuts' are created, connecting distant parts of the graphs, average path length is dramatically reduced, but cliquishness is substantially preserved. Watts and Strogatz call this network structure 'small world'.

This 'small world' model is a convenient representation for comparing different sociological theories on the structural properties that are conducive to the formation of social capital (Verspagen and Duysters, 2004). On one hand, regular networks, characterised by close triads and dense and cohesive subgroups, may favour the emergence of common rules and language, reduce ex ante informational asymmetries, and increase ex post cooperation (Coleman, 1988); on the other hand, random networks, characterised by low levels of links redundancy, may favour access to diverse sources of information (Burt, 1992). Small world networks, being in some sense intermediate between the two, may guarantee the benefits of both.

In our model, we will consider network structures in line with Watts and Strogatz. Assuming that inventors are located on a circle, and starting from a regular network, we apply Watts and Strogatz's algorithm for three different values of p, generating the three different typologies of networks.

2.4 Numerical simulations

In this paragraph we discuss the results of the numerical simulations. In our exercise, we vary three parameters.[3] First of all, we consider three values for the parameter p in the rewiring procedure, so that we compare three network structures: regular networks ($p = 0.007$), 'small worlds' ($p = 0.07$), and random networks ($p = 0.7$). Then, we modify two other parameters in order to compare the three network structures under different scenarios. The first parameter is r in Eq. (2.1), which captures the role of patents in individual learning, since it measures the ability to decode the information represented in a patent. The second parameter we are going to consider is w, which determines the use of social networks: agents will use the network only if they are both more distant from patenting than a given threshold w. This can be interpreted as a degree of competition of the patent race: the higher the parameter value, the harsher the competition between inventors. In the baseline scenario, r is fixed equal to 0.25 and w to 8. Then, we modify just one parameter at a time, to allow us to appreciate its effect relative to what is obtained under the baseline simulation set of parameters, as reported in the appendix. In particular, we will consider $r = 0.1$ (i.e. the use of social networks is a more suitable way to learn than individual learning compared to the baseline scenario), and $w = 12$ (i.e. the degree of competition among inventors is higher in this case compared to the baseline scenario).[4] The following table displays the different simulations we ran, reporting some basic statistics about the number of patents per inventor distribution at the steady state, i.e. when the last patent is assigned and therefore the patents produced 'occupied' all of the knowledge space available (this occurs around period 42,000 in our runs).

Table 2.1 shows that the mean number of patents per inventor, and their total number, is similar across runs. In particular, the three network structures guarantee an almost equally efficient transformation of the knowledge space into patents. Furthermore, all of the distributions are highly skewed: most inventors have few patents, while a few inventors have a large number of patents. This is exactly what we have observed in real data about patents. Given the structure of the model, the result is not surprising: the learning mechanisms we imposed are highly cumulative and naturally generate an unequal distribution of patents.

However, the distribution varies, as measured by the median, the maximum number of patents and by the degree of skewness, independently from the values

Table 2.1 Patent per inventor distribution. Different simulation runs

Simulation run	*Network*	*Average*	*Median*	*Std dev*	*Skewness*	*Max*	*Min*
baseline	REGULAR	7,67	4,5	13,00	6,72	119	0
	SMALL WORLD	7,72	5,5	10,68	6,56	99	0
	RANDOM	7,54	2	19,61	5,85	162	0
$r=1$	REGULAR	7,64	6	7,01	2,46	46	0
	SMALL WORLD	7,61	5	10,55	4,31	81	0
	RANDOM	7,68	2	21,81	5,95	167	0
$w=12$	REGULAR	7,51	5	8,57	2,83	57	0
	SMALL WORLD	7,54	5	8,05	2,34	52	0
	RANDOM	7,54	2	23,42	7,32	213	0

of other parameters. The distribution is more skewed for the random network than for the small world and the regular network (and related to that, the median is lower and the maximum is higher). The latter have a more uniform distribution of patents in the left part of the distribution, and a smaller number of patents in the right extreme.

Note that there are reasons why one should be concerned with equity issues: if inventors are firms, unequal distribution of patents can negatively affect concentration in industries, and through it, social welfare.

Figure 2.4 reports the number of patents granted over time as a function of different network structures, for each scenario. For the sake of readability, the random network has been considered as a benchmark. For instance, the cumulative number of patents produced under the small world structure has been divided by the number produced by the random network and then rescaled to zero, i.e. the level of production of random networks corresponds to the abscissa axe. A negative value means that the production of small world is smaller than that of the random network, and *vice versa*.

While the behaviour of the total number of patents is very similar across the three network structures, some divergence seem to emerge in the process towards a steady state (especially between period 20,000 and 32,000), where the random network produces a greater number of patents. This seems particularly true when the capacity of decoding patents is smaller (i.e. $r = 0.1$, the second plot reported in Figure 2.4).

As we discussed in Section 2, the primary aim of our paper is to compare our theoretical results with the empirical analysis of the relationship between social distance and the probability of citations. The first remark to make is that we cannot replicate the same exercise exactly. Indeed, the empirical procedure, now standard in the literature, requires building a set of control patents. These are patents in the same technological class and with the same application date of the citing patent, but which did not cite the cited patent (refer to the same prior art). Informally, these are defined as patents that could have cited a patent, but which did not. Control patents cannot exist in our model: the probability that a patent cites another patent if it should is 1. For this reason we have to take a different route. In the following figures, in order to assess the role of social distance in affecting citations (and subsequently knowledge diffusion) we will report the average number of citations per pair of agents, according to the social distance between inventors. In order to make the different network structure independently comparable in terms of the number of citations effectively reported, we report a relative measure: average number of citations for a given social distance by the overall average number of citations.

In our model, assessing the effect of network structures upon the process of knowledge creation and diffusion, as captured by relations of citations across patents, is equivalent to assessing the role of two properties of networks, namely cliquishness and average distance.

Ceteris paribus, both an increase in cliquishness and a decrease in average distance increases knowledge diffusion. Low average distance has a positive effect

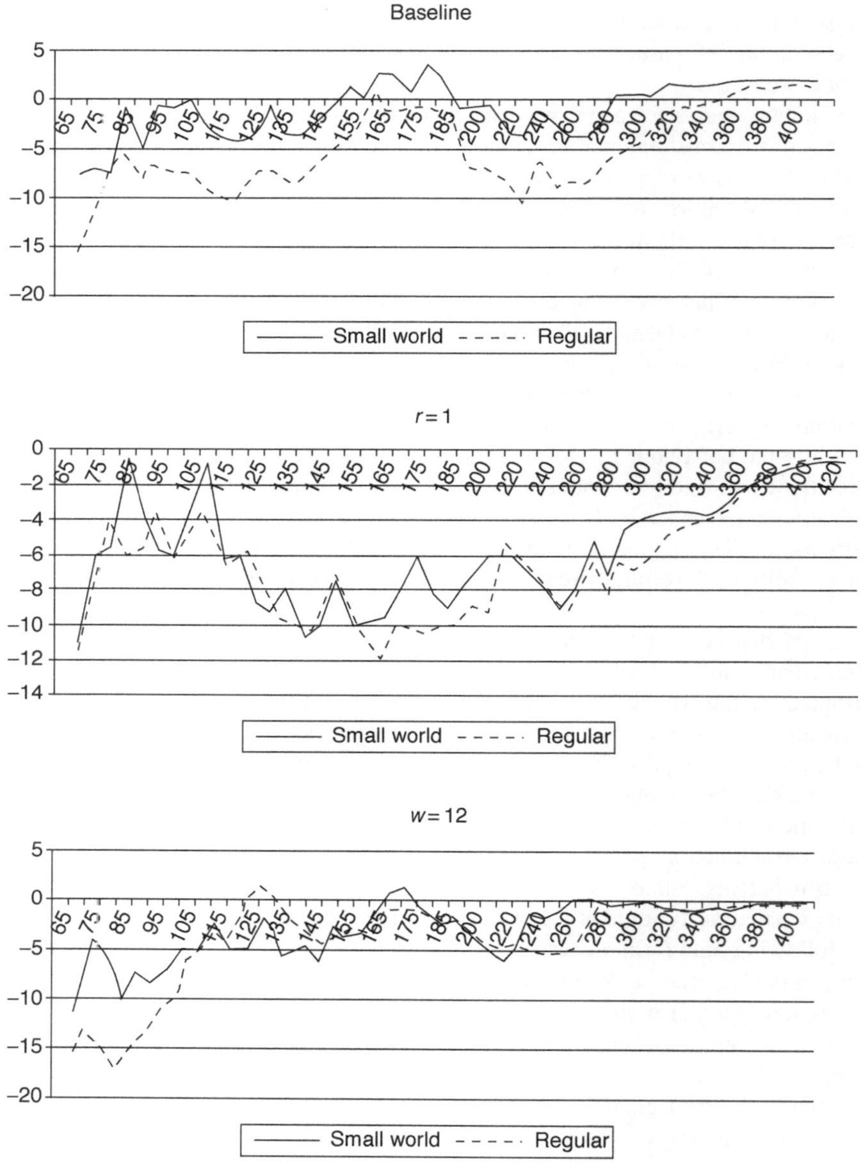

Figure 2.4 Number of patents over time

upon knowledge diffusion because it gives access, in a few steps, to knowledge created elsewhere in the network. An important note is needed, however: since social learning occurs stochastically, but with a positive probability, the expected amount of time for a piece of knowledge to flow from one agent to another decreases with social distance, but the probability of learning tends to 1 as time elapses.[5] Nevertheless, expected time matters when knowledge flows are captured by patent citations. If knowledge flows slowly from *i* to *j*, then there is a low probability that agent *j* can 'invent around' this piece of knowledge and file a patent (thus making knowledge flows visible) before some other agent, in particular agent *i*, anticipates it in the 'patent race'.

High cliquishness has a positive effect upon knowledge diffusion, too. Suppose we take two agents, *i* and *j*, such that *j* possesses the piece of knowledge that *i* does not have. Note that a direct link between *i* and *j* does not guarantee that the piece of knowledge will flow from *i* to *j*. First, we modelled learning as a stochastic process: when agents *i* and *j* interact, *i* learns only with some probability. Second, there is the 'strategic effect', which is related to the requirement of a double coincidence of wants: when the pair *ij* is selected, agent *i* could refuse to interact with *j*.

A cliquish network has the property that there are multiple, short paths linking two connected agents *i* and *j*. This increases the probability that the piece of knowledge possessed by agent *j* flows to agent *i*, for two reasons. First, there is the probability that knowledge flows from *i* to *j* *indirectly* through some alternative short path linking *i* and *j*. Second, the probability that *i* learns directly from *j* increases. This is due to the local nature of learning. For the sake of concreteness, consider an agent *k*, who is a common acquaintance of *i* and *j*. *j*'s piece of knowledge can lead *k* to learn another piece of knowledge which is close to the previous one in the knowledge space. If this piece of knowledge is 'transmitted' to *i*, this increases the probability that *i* learns from *j*, due to the role of absorptive capacity (Eq. 2.2). Notice that the mechanisms we have just described do not involve the sources of benefits (common language, reputation enhancement, information diffusion) normally attributed to 'cliquish' networks.

Figure 2.5 reports the average number of citations as a function of social distance, in relative terms. The value of the average citation is reported in the legend of the graph.

The figures are noteworthy for two reasons: first, independently of the network structure, the average number of citations decreases with social distance, consistent with the empirical evidence; second, the structure of the network affects such a relationship.

The negative relationship between citations and social distance derives quite directly from the hypothesis that knowledge possessed by inventors but not embodied in patents can only be accessed through face-to-face interaction. This increases the probability that two agents close in terms of social space will invent within a close area of knowledge space, and then cite each other.

With respect to network structure, it turns out that, for the parameterisation of the baseline simulation, clustering is the most important property favouring

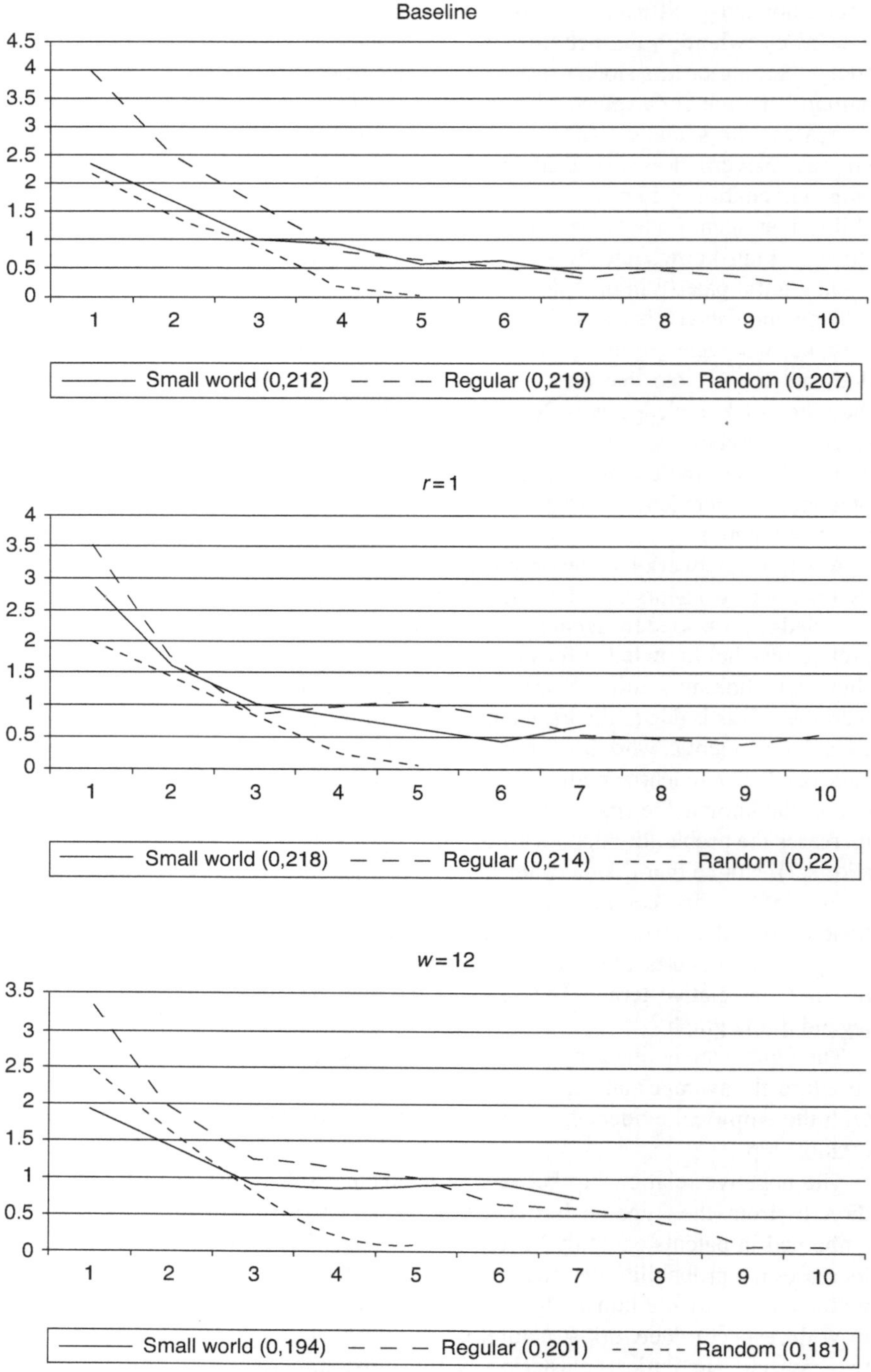

Figure 2.5 Social distance and knowledge diffusion (normalised to average citations)

knowledge diffusion. The regular network and the small world dominate, in terms of average number of citations, over the random network.

Somewhat paradoxically, the random network, although it reduces the average distance between inventors, is the worst structure for knowledge diffusion, because of its lack of multiple paths among inventors. Also, it is notable that it is the lower rate of knowledge diffusion that generates the more unequal distribution of patents: knowledge diffusions tend to mitigate the effect of the cumulativeness of learning.

In the parameterisation of the baseline simulation, we consider $r = 0.25$. Note that access to patents in individual learning can be considered as an imperfect substitute for 'shortcuts' in the network, because it allows access to knowledge created 'far' in the social space (this imperfection is due to the fact that $r < 1$ and access is limited to patented knowledge). One could expect that reducing r should increase the importance of social connections to distant areas in the network, and then favour small world and random networks over regular networks. The second plot of Figure 2.5 shows that this is in fact the case.

From the graph, we note that for a low value of r, the small world, which has both low average distance and high clustering, maximises the degree of knowledge diffusion. This is true for short social distances: indeed, for $r = 0.1$, the links to distant parts of the network are crucial to generate variety in knowledge within the clique, and then knowledge diffusion.

In the bottom of Figure 2.5, we display the results of when the level of competition is greater (i.e. $w = 12$). By assumption, in this case, inventors have smaller incentives to learn socially (this is also shown by the lower rates of average citation). As a consequence, social proximity is less important: being very close is not a determinant. Of course, having direct access to someone (distance equal to 1) is an advantage, but apparently it is less important than other scenarios: in order to obtain some gain it is enough to be connected indirectly, since the effect of social distance is still positive (i.e. not smaller than 1) until distance 5 for regular networks, and to a lesser degree for small world as well.

2.5 Conclusion

It is now common to look at knowledge as a key input in the growth process of firms and countries. This view is widely shared, not only by economists and other social scientists, but by policy-makers as well (think, for instance, of the Lisbon Agenda). Moreover, in the process of knowledge creation and diffusion, various forms of collaborations and social interactions, which we can refer to as 'networks' in a broad sense, are perceived to have a crucial role.

While both the theoretical and the empirical literatures on these themes have increased in recent years, the connections between the two have been limited. This chapter aims to fill this gap. Specifically, we developed a model of knowledge creation and diffusion within a population of inventors, where knowledge is represented in a way that makes it possible to consider theoretically the proxies of knowledge and knowledge flows, i.e. patents and patents citations, used in the empirical analysis.

Our analysis was focused on the network determinants of knowledge flows. The empirical analysis suggests a negative relationship between social distance and knowledge flows, as measured by the probability of citation. First of all, our model is able to reproduce such a negative relationship. Additionally, we produced other results which constitute testable (but, to the best of our knowledge, not yet tested) implications of our model. In particular, we showed that the relationship between social distance and knowledge flows is affected by the structure of the network in which the inventors are embedded. Both high clustering and low average distance positively affect knowledge diffusion, but access to the pool of existing patents can be a substitute for connection between otherwise distant parts of the network. Furthermore, we showed that when incentives to use the social network are lower (in our case, because of higher competition), then social distance is less relevant in explaining patent citations.

Implications of our model

While we believe that the specific point addressed in this chapter is important, our model could provide insights for other issues as well. We intend to explore these avenues for further research in the future. First of all, the model could be enriched by considering the network as endogenous. In this case, the past patents of an inventor could work as signal driving the formation of links, when knowledge is not otherwise observable. Second, we could explore how different interpretations of patenting requirements (e.g. the novelty criterion) can affect the validity of the proxies used in the empirical work.

Appendix

List of parameters

Parameter	*Description*	*Value*
N	Number of inventors	100
n	Average number of links per agent	6
q	Equation 2.1, 2.2	0.025
α	Equation 2.2	0.5
λ	Equation 2.1, 2.2	0.5
x	Novelty criterion	0.6
w	Equation 2.3	8
l	Minimum size of patents (side)	5
T	Number of periods	50,000

Notes

1 If one inventor has participated in both teams, the social distance is equal to 0.
2 Von Hippel himself makes claims for the applicability of his arguments 'to any situation in which individuals or organisations are involved in a competition where

possession of property know-how represents a form of competitive advantage' (von Hippel, 1987).

3 The complete set of parameters that are constant across simulations is reported in the Appendix. Furthermore, the knowledge space is given by a square grid of 100 cells. As for the initial conditions, for each agent the probability that a cell (k_1, k_2) belongs to the initial knowledge set is 0.01 if $30 \leq k_1(k_2) \leq 70$, and 0 otherwise.

4 Moreover, another parameter has been modified in order to check the robustness of the results, namely the degree of novelty each patent should be satisfy in order to be granted (i.e. x, the fraction of cells not overlapping with other patents). As a further check for robustness, we considered different seeds for the generation of random numbers. The results are qualitatively unchanged. Further information is available on request.

5 In both Cowan and Jonard (2004) and Cassi and Zirulia (2008), low average distance matters for diffusion even in the long run because of dispersion in each passage.

References

Agrawal, A., Cockburn, I. and McHale, J. (2006) 'Gone but not forgotten: knowledge flows, labour mobility and enduring social relationships', *Journal of Economic Geography*, 6(5): 571–91.

Balconi, M., Breschi S. and Lissoni, F. (2004) 'Networks of inventors and the role of academia: an exploration of Italian patent data', *Research Policy*, 33(1): 127–45.

Baum, J., Cowan, R. and Jonard, N. (2010) 'Network-independent partner selection and the evolution of innovation networks', *Management Science*, 56(11): 2094–110.

Breschi, S. and Lissoni, F. (2004) 'Knowledge networks from patent data: Methodological issues and research targets', in W. Glänzel, H. Moed and U. Schmoch (eds) *Handbook of Quantitative S&T Research*, Dordrecht: Kluwer Academic Publishers.

Breschi, S. and Lissoni, F. (2005) 'Cross-firm inventors and social networks: localised knowledge spillovers revisited', *Annales d'Economie e de Statistique*, Issue 79/80: 189–209.

Burt, R.S. (1992) *Structural Holes: The Social Structure of Competition*, Cambridge: Harvard University Press.

Cassi, L. (2004) *Conoscenza e Reti Sociali: tre saggi*, PhD Dissertation, Ancona University.

Cassi, L. and Zirulia, L. (2008) 'The opportunity cost of social relations: the effectiveness of small world', *Journal of Evolutionary Economics*, 18(1): 77–101.

Cohen, W.M. and Levinthal, D.A. (1989) 'Innovation and learning: the two faces of R&D', *Economic Journal*, 99(397): 569–96.

Coleman, J. (1988) 'Social capital in the creation of human capital', *American Journal of Sociology* 94 (Supplement): S95–S120.

Cowan, R. and Jonard, N. (2003a) 'The dynamics of collective invention', *Journal of Economic Behavior and Organization*, 52(4): 513–32.

Cowan, R. and Jonard, N. (2003b) 'The workings of scientific communities', in A. Geuna, A.J. Salter and W.E. Steinmueller (eds) *Science and Innovation. Rethinking the Rationales for Funding and Governance*, Cheltenham: Edward Elgar.

Cowan, R. and Jonard, N. (2004) 'Network structure and the diffusion of knowledge', *Journal of Economic Dynamics and Control*, 28(8): 1557–75.

Cowan, R., Jonard, N. and Ozman, M. (2004) 'Knowledge dynamics in a network industry', *Technological Forecasting and Social Change*, 71(5): 469–84.

Cowan, R., Jonard, N. and Zimmermann, J.B. (2002) 'The joint dynamics of networks and knowledge', in R. Cowan and N. Jonard (eds) *Heterogeneous Agents and Interactions and Economic Performance,* vol. 521, Lecture Notes in Economics and Mathematical Systems, Berlin: Springer.

Cowan, R., Jonard, N. and Zimmermann, J.B. (2003) 'On the creation of networks and knowledge', in M. Marsili and A.P. Kirman (eds) *Proceedings of the 8th WEHIA Conference on Heterogeneous Interacting Agents*, Berlin: Springer.

Fleming, L., King, C. and Jud, A.I. (2007) 'Small worlds and regional innovation', *Organization Science*, 18(6): 938–54.

Giuliani, E. and Bell, M. (2005) 'The micro-determinants of meso-level learning and innovation: evidence from a Chilean wine cluster', *Research Policy*, 34(1): 47–68.

Griliches, Z. (1990) 'Patent statistics as economic indicators: a survey', *Journal of Economic Literature*, 28(4): 1661–707.

Jaffe, A.B., Trajtenberg, M. and Henderson, R. (1993) 'Geographic localisation of knowledge spillover as evidenced by patent citations', *The Quarterly Journal of Economics*, 108(3): 577–98.

Machlup, F. (1980) *Knowledge: Its Creation, Distribution and Economic Significance, Volume I: Knowledge and Knowledge Production*. Princeton: Princeton University Press.

Morone, P. and Taylor, R. (2010) *Knowledge Diffusion and Innovation: Modelling Complex Entrepreneurial Behaviours*, Cheltenham: Edward Elgar Publishing.

Newman, M.E.J. (2000) 'Who is the best connected scientist? A study of scientific coauthorship networks', SFI working paper 00-12-64.

Powell, W.W. and Grodal, S. (2005) 'Networks of innovators', in J. Fagerberg, D. Mowery and R. Nelson (eds) *The Oxford Handbook of Innovation*, Oxford: Oxford University Press.

Simon, H.A. (1991) 'Bounded rationality and organizational learning', *Organisation Science*, 2(1): 125–34.

Singh, J. (2005) 'Collaborative networks as determinants of knowledge diffusion patterns', *Management Science*, 51(5): 756–70.

Verspagen, B. and Duysters, G. (2004) 'The small worlds of strategic technology alliances', *Technovation*, 24(7): 563–71.

von Hippel, E. (1987) 'Cooperation between rivals: informal know-how trading', *Research Policy*, 16(6): 291–302.

Watts, D. and Strogatz, S.H. (1998) 'Collective dynamics of 'small world' networks', *Nature*, 393(6684): 440–42.

Zirulia, L. (2009) 'The dynamics of networks and the evolution of industries: a survey of the empirical literature', in F. Malerba and N. Vonortas (eds) *Innovation Networks in Industries*, Cheltenham: Edward Elgar Publishing.

3 Knowledge and innovative entrepreneurship

Social capital and individual capacities

Uwe Cantner and Michael Stuetzer

3.1 Introduction

A central development within management literature has been the growth of nascent entrepreneur research analysing on-going venture start-up efforts and/or firms in gestation over time (Davidsson, 2006). New ventures have an important effect upon economic development. They are credited for the transfer of innovations into the market (Schumpeter, 1934; Acs and Plummer; 2005) and with creating regional employment (e.g. Fritsch and Mueller, 2004).

Central questions in nascent entrepreneurship research concern the characteristics of the venture creation process and the factors affecting the performance of these firms (for an overview, see Davidsson, 2006). Among other factors considered in the literature, the social embeddedness of the entrepreneur has been found to play a pivotal role (Davidsson and Honig, 2003). Social capital enables entrepreneurs to access resources (Florin *et al.*, 2003) or novel information (Uzzi, 1997) in order to create opportunities (Baker and Nelson, 2005). During the venture creation process, most firms suffer from substantial resource constraints (Shepherd *et al.*, 2000) and use their personal networks as a means to access resources and information far below market price (Elfring and Hulsink, 2003).

However, a sizeable gap exists in the burgeoning social capital literature on the subject of team start-ups. A most prominent finding is that team start-ups are more successful than solo start-ups (e.g. Lechler, 2001). One of the explanations offered for this is that entrepreneurs can combine their abilities and financial capital in a team, giving them an advantage over solo entrepreneurs (e.g. Gartner, 1985; Stam and Schutjens, 2006). Sometimes explicitly (e.g. Colombo and Grilli, 2005; Stam and Schutjens, 2006), but more often implicitly (e.g. Davidsson and Honig, 2003; van Gelderen *et al.*, 2005), the same argument is applied to the usage of social capital, i.e. that the social capital of individual team members is combined to provide an advantage for teams over solo entrepreneurs. As yet, to our knowledge, no study has explicitly analysed whether more social capital is found within teams, as compared to solo entrepreneurs, and whether this leads to their better performance.

This chapter addresses these two questions, and empirically explores the use of the social capital of solo entrepreneurs and entrepreneurial teams during the venture creation process. In doing so, we refine the empirical concept of social capital in that we do not look at its mere existence, but focus instead upon its use in terms of concrete support (e.g. advice on the business plan, marketing, or research and development – R&D) for entrepreneurs. We address two major research questions. The first concerns the differential use of social capital. Do solo entrepreneurs rely on social capital more often than new venture teams, or is it the other way around? And how does each type of start-up use social capital? More precisely, we investigate the relationship between social capital and other characteristics of a new venture and its founders (e.g. human capital). The second research question then turns to the effect of social capital upon subsequent new venture performance. Appropriate hypotheses for this study are tested using a data-set of 456 start-ups in innovative industries in the German state of Thuringia.

The remainder of this chapter is organised as follows. In Section 2, we review the theory and previous research on social capital in order to generate six testable hypotheses. In Section 3, we describe the data-set and the methods employed to measure the use of social capital. We then present the results of our analysis (Section 4). The chapter concludes with Section 5, where we interpret and discuss the results and draw some conclusions.

3.2 Theoretical background

3.2.1 New firm creation and social capital

Creating a new firm, in comparison to being an employee, involves high levels of risk and uncertainty (Lumpkin and Dess, 2001). Entrepreneurs may consider alleviating the effects of risk and uncertainty by approaching others for help and advice, which can be broadly captured by the concept of social capital. While there are various definitions of social capital in the literature (for an overview, see Adler and Kwon, 2002) we follow the integrative approach of Nahapiet and Goshal (1998). They define social capital at the individual level 'as the sum of the actual and potential resources embedded within, available through and derived from the network of relationships possessed by an individual or a social unit' (Nahapiet and Goshal, 1998, p. 243). Social capital is multidimensional, encompassing a structural, a relational and a cognitive dimension (Nahapiet and Goshal, 1998). While the structural dimension is concerned with the properties of the social network, such as the density and the connectivity among actors (Burt, 1992), the relational aspect of social capital refers to the quality and type of interpersonal relationships (Granovetter, 1992). The cognitive dimension of social capital captures the shared representations and systems of meaning that individuals have with one another. Another distinction in social capital literature is the distinction between bridging and bonding social capital (Putnam, 2000). Bridging social capital refers to links between individuals and organisations representing different expertise, views of the world, and cultural habits (e.g. Samuelsson and Davidsson, 2009). In contrast, bonding social capital refers to the positive (but sometimes also negative) effects of cohesion and trust between actors which enables collective actions (Putnam, 2000). In a closely related classification of social capital, theorists distinguish between weak and strong ties (Granovetter, 1973). Here, weak ties describe loose relationships to actors, providing non-redundant information (e.g. Davidsson and Honig, 2003), whereas strong ties refer to close relations with a limited set of actors, featuring trust and its positive by-products (e.g. Samuelsson and Davidsson, 2009).

Using the definition of Nahapiet and Goshal (1998) as a starting point, different implications arise for solo and team-started ventures. We will return to that point immediately after introducing the concept of new venture teams. We define a venture as a team start-up where more than one person is actively involved in the venture creation process and where these persons own, or had owned, a part of the venture (Kamm *et al.*, 1990). With regard to mastering the venture creation process, the superiority of team start-ups compared to solo start-ups is one feature readily acknowledged in entrepreneurship research (e.g. Cooper and Bruno, 1977; Lechler, 2001). In particular, for high technology firms (the sample group of interest here), there is a higher skills requirement, making team start-up a necessity. Gartner argues that 'individuals combine their abilities in teams in order to start an organisation successfully' (Gartner, 1985, p. 703). Hence, the advantage of a team lies in the bundling of human and financial capital (Stam and Schutjens, 2006).

Upon initial investigation, the argument for bundling human and financial capital can also be applied to a solo entrepreneur's use of social capital, considered as the ability of an actor to mobilize useful resources from his social network (Bourdieu, 1986; Burt, 1992; Coleman, 1988). Teams combine and integrate the social capital of their members, which possibly provides them with an advantage over solo entrepreneurs (Davidsson and Honig, 2003). As yet, to our knowledge, little is known about whether more social capital is found within teams, as compared to solo entrepreneurs, and whether this leads to their superior performance. Comparing venture teams and solo entrepreneurs with respect to the structural dimension of social capital, the former may have an advantage through broader access to critical resources due to the larger number of contacts within their social network. The decision to create a venture team or to add an additional team member has the potential to increase the social capital base of the start-up and, as a result, may improve the resource profile of the new venture, leading to increased new venture persistence and success., This argument is implicitly made in a considerable number of studies, as belonging to a start-up team is considered to be an indicator of social capital (e.g. Colombo and Grilli, 2005; Davidsson and Honig, 2003; van Gelderen *et al.*, 2005).

Looking at the relational dimension of social capital, a contrary argument can be put forward. While a positive correlation may exist between team size and the possible access to resources via entrepreneurs' contacts, the actual use of those contacts may not be correlated with team size. Compared to a solo entrepreneur, a new venture team can complete more venture creation activities in-house through combining the (often different) skills of its members (Gartner, 1985). Thus the actual use of social capital may in fact decrease.

In our empirical analysis, we explore whether the use of social capital differs between solo and team start-ups. With respect to team start-ups, two counteracting arguments are to be considered: First, the strengthening and broadening of the social network in a team increases (ceteris paribus) the likelihood of using social capital. Second, the ability of a team to perform more tasks on its own decreases the likelihood of using social capital. Both of these effects work in opposite directions. With due care, we therefore test whether the use of social capital differs at all between the two types of founded venture via the following hypothesis:

H1: Solo entrepreneurs and entrepreneurial teams differ regarding their respective use of social capital in the venture creation process.

3.2.2 The effects of social capital

A further focus of our analysis is the differential effect of social capital use upon venture performance between solo entrepreneurs and new venture teams. Given the nature of the data-set consisting of start-ups in innovative industries, we assess the literature concerning the social capital of tech-based as well as knowledge-based start-ups. The review of that literature reveals that social capital influences the venture creation process via three different channels: (1) it assists (nascent) entrepreneurs in accessing resources; (2) it provides trusted feedback to the entrepreneurs; and (3) it provides access to novel information.

Access to resources is of critical importance to small and young companies in innovative industries which traditionally suffer from a range of resource constraints, such as limited financial capital, skilled workforce, or equipment necessary for R&D and production (Aldrich and Martinez, 2001), all of which are critical for growth. Anderson *et al.* (2007) analysed ten technology companies in Aberdeen and found evidence that the use of entrepreneurs' contacts with former business partners provided support in recruiting their workforce. Much more work has been done on studying the relationship between social capital and the financing of start-ups. Shane and Cable (2002) argue that potential investors were able to screen and to evaluate entrepreneurs and their business ideas via network ties, which formed the basis of their investment decision. Florin *et al.* (2003) report, in relation to a sample of US firms seeking to float on the stock exchange, that the level of social capital is positively and significantly related to the level of attracted funds and return on sales. However, this result could only be partially confirmed by Honig *et al.* (2006), who found some evidence for a relationship between social capital and the amount of sales, but no links between social capital and financial capital.

Furthermore, social capital affects growth aspirations among nascent entrepreneurs (Liao and Welsch, 2003), which is considered to be a precursor of subsequent venture growth (Baum *et al.*, 2001). Using a sample of Swedish technascent entrepreneurs, Samuelsson and Davidsson (2009) found that projects which make extensive use of social capital make significant progress in the venture creation process. Taken together, we propose the hypothesis:

H2: Social capital in the venture creation process has a positive impact on later new venture performance.

Trusted feedback is the second transfer channel of social capital. Its theoretical foundations lie in the relational dimension of social capital (Nahapiet and Goshal, 1998), which deals with the quality or the kind of ties that an actor possesses (Granovetter, 1990). Within the relational aspect of social capital, tie strength has attracted great interest from the research community. Although it is a simplification of Granovetter's (1973) original argument that tie strength is a continuum, ties are typically categorised as being either weak or strong. Thereby, Granovetter characterises strong ties, in contrast to weak ties, by a combination of high emotional intensity and intimacy, much time spent with the network contact, and high reciprocity of services.

The strength of strong ties lies in the high level of trust between the network members. It is well known that networks with a high proportion of strong ties are 'dense' networks (Burt, 2000), which indicates that many network members are directly connected to each other. Scholars highlight the importance of trusted feedback and the transfer of tacit knowledge (Aldrich and Martinez, 2001) which stems from such networks as necessary components of entrepreneurial learning for entrepreneurs (Zahra *et al.*, 2006). We thereby understand learning as the process of accumulating the knowledge required to be effective in starting up and managing new ventures (Politis, 2005).

Learning takes place throughout the venture creation process. Bhave (1994) was one of the first researchers to recognise the complex nature of the venture creation process, which he described as nonlinear and iterative. Feedback loops between the different stages of venture creation are a key feature of his model, as they allow for changes in the business concept after receiving corresponding feedback and information from, for example, customers and financiers. Other scholars also emphasise the importance of learning and adapting in the venture creation process (Aldrich and Ruef, 2006; Ronstadt, 1988; Shane and Venkataraman, 2000) in order to allow the development of routines and capabilities (Zahra *et al.*, 2006) necessary to run a business successfully (Teece *et al.*, 1997).

A well-known example of the benefits of learning from strong ties is the study by Elfring and Hulsink (2003). They report that high-tech start-ups benefit from the trusted feedback of their strong ties in order to better recognise opportunities. Studying 23 cases in New York's apparel industry, Uzzi (1997) finds that companies profit from information transfer on strategies, prices and products from a dense network which enables them to take advantage of fast-changing market opportunities. However, Uzzi (1997) acknowledges that there are serious drawbacks in relying solely on strong ties and high-density networks. It is argued that information and ideas coming from too-densely connected networks lack newness. Entrepreneurs, who receive information only from inside such insulated networks may experience below-average performance. This disadvantage is of particular importance for high-tech start-up projects with innovative products (as shown by Presutti *et al.*, 2007), as they operate within global markets and require greater diversity of ideas, information and feedback concerning their business ideas, in line with the greater complexity and requirements of their numerous international markets. In evaluating these mixed findings on the effects of strong ties and dense networks on entrepreneurial performance, we postulate the following hypothesis:

H3: Strong ties in the venture creation process have a positive impact on later new venture performance.

Access to novel information – the third transfer channel – is beneficial for entrepreneurs because ventures in gestation often do not possess information about relevant facets of an industry, such as prices, production processes, inputs, and competition (Aldrich and Ruef, 2006), which are critical requirements of the entrepreneurial learning described above (Zahra *et al.*, 2006). This information is widely dispersed among individual actors within the market (customers and suppliers), as well as among people seemingly unrelated to the market (engineers, technicians, or financiers).

In general, in order to access this dispersed information, weak ties are considered important, because through them it is possible to reach distant subgroups of the network via a close network partner. In contrast to strong ties, which have a tendency toward closure (Coleman, 1988), weak ties can serve as bridges to indirect ties (Granovetter, 1973). Therefore, weak ties enlarge the network of an entrepreneur and provide the nascent entrepreneur with access to novel information, which may assist in the discovery of more profitable business opportunities (Elfring and Hulsink, 2003), the development of products (Lechner and Dowling, 2003), the reduction of production costs (Yli-Renko *et al.*, 2001), and the formation of potential investor contacts (Shane and Cable, 2002). Therefore, we hypothesise that:

H4: Weak ties in the venture creation process have a positive impact on later new venture performance.

To access social capital in general, and strong and weak ties in particular, requires that the entrepreneur or the new venture team show the appropriate ability to do so. This leads to the concept of human capital. A number of empirical studies report that human capital variables (e.g. entrepreneurial experience, leadership experience, or business experience) have positive significant effects on the progress of nascent entrepreneurs and subsequent venture success (e.g. Honig *et al.*, 2006; Samuelsson and Davidsson, 2009). Being more specific in our discussion on the effects of social capital on venture performance, we argue that an entrepreneur or a new venture team learns more successfully if human capital aligns with social capital. More precisely, entrepreneurs with pronounced human capital variety should have a higher level of 'absorptive capacity' to tap a broad array of relevant information (Cohen and Levinthal, 1990). With human capital variety, we refer mainly to an entrepreneur's or a venture team's breadth of experience over different functional activities. Following Hayton and Zahra (2005), we argue that, because of their broader experience, these entrepreneurs should be more able to rate new information on its usefulness, and incorporate this new information more easily into their existing knowledge stock.[1] Furthermore, we suggest that entrepreneurs with higher human capital variety should have a larger social network to draw upon, giving them broader choices and opportunities for selecting the most appropriate helpers within their networks. This latter argument is considered within the context of weak ties, because the strong tie network of an entrepreneur consists of only a limited number of persons (Lechner and Dowling, 2003).

To the best of our knowledge, thus far only the study by Batjargal (2007) on internet start-ups in China has examined the moderating effect of human capital on the linkage between social capital and venture success. Although the econometric findings are mixed, Batjargal (2007) concludes that the combined effect of human capital and social capital enhances the survival chances of newly founded businesses. We, therefore, propose the following hypotheses:

H5: The relationship between social capital in the venture creation process and subsequent venture performance is stronger for solo entrepreneurs and entrepreneurial teams with a higher level of human capital variety; and

H6: The relationship between weak ties in the venture creation process and subsequent venture performance is stronger for solo entrepreneurs and entrepreneurial teams with a higher level of human capital variety.

3.3 Data-set and methodology

3.3.1 Data-set and interview strategy

For our empirical analysis, we use data from the 'Thuringian Founder Study' – an interdisciplinary research project on the success and failure of innovative start-ups. One part of this study is a data-set of innovative young firms. This unique data-set comprises the entries of private and commercial companies in the commercial register (Handelsregister) between the years 1994 and 2006 in the German state of Thuringia. This design made it possible to interview not only the founders of active companies, but also the founders of ventures that had failed. The database is restricted to entries in innovative industries (Grupp *et al.*, 2000).[2] The first registered owner-managers for each new entry form the survey population. From this population, a random sample of 2,604 start-ups was generated. From January to October 2008, the research team conducted 639 face-to-face interviews with the solo entrepreneur or the leading entrepreneur of a start-up team (response rate: 25 per cent). As some companies were not genuinely new, but were in fact subsidiaries or the result of a diversification of an existing company into a new business field, we removed 76 companies from the sample.

Thirteen companies were removed from the sample due to interview quality concerns. For this chapter, we restricted the analysis to observations within the complete data-set and therefore dropped 78 observations with missing values in one or more used variables. Furthermore, to avoid censoring, we dropped 16 observations which started later than 2005.[3] Our empirical analysis evaluates the effect of social capital use in the venture creation process on subsequent venture performance in the third business year. The final sample consists of 456 companies, which can be further classified as 182 solo entrepreneurs and 274 new venture teams.

The structured interviews were conducted by members of the research project and student research assistants who were trained in various sessions in December 2007. The research team used a retrospective design to collect the data. To overcome the bias of hindsight as well as memory decay (Davidsson, 2006), the research team adapted the 'Life History Calendar' tool from psychology in order to obtain information on the venture creation process. The Life History Calendar is a useful tool for constructing individual processes and developments (Caspi *et al.*, 1996; Freedman *et al.*, 1988). It gains more reliable and valid retrospective information than traditional questionnaires (Belli *et al.*, 2001). The participants, together with the interviewer, filled in major life events and sequences in the Life History Calendar (family life, working sequences, historical events, and important dates of the business history) at the commencement of the interview. During the interview, the Life History Calendar was visible to the participants. Before each retrospective item was discussed (e.g. team composition, human and social capital questions), we asked the interviewee to look at the relevant time point in the Life History Calendar and verbally recall any special events that took place during that time in order to assist them in remembering. The interview strategy and the Life History Calendar are in line with the recommendation of Belli *et al.* (2004).

Restricting this study to the German state of Thuringia has the major advantage of reducing sample heterogeneity stemming from factors such as regional differences, for example. This is of particular importance in Germany, where there are still pronounced differences in the determinants of new venture success between Eastern and Western Germany (Fritsch, 2004).

3.3.2 Dependent variable

Our dependent variable attempts to measure the performance of start-up firms. We approximate this by the absolute number of employees in the third year of operation of the firm. The solo entrepreneur, members of the new venture team, and the board of directors were not counted as employees. As our sample consists only of new firms and does not include franchises or corporate ventures, the vast majority of firms have zero employees in the venture creation phase. For that reason, employment growth rates could not be computed (for a similar approach, see Baum *et al.*, 2000). If a new venture did not survive the third business year, the number of employees remained coded as zero. Traditional outcome variables such as firm value, profitability, and turnover are not applied in this study, for two reasons. First, the self-reported measure of sales turned out to be unreliable. While respondents could assess the amount of sales generated in the first three business years, monetary reforms in Germany replaced the Deutschmark with the Euro over several steps between 1999 and 2001, which made it difficult for the entrepreneurs to correctly attribute the sales to either currency. Second, secondary data from business information providers could not be used, because such databases tend to focus on larger, surviving firms, substantially reducing the overlap with the data-set.

Nevertheless, we checked the validity of the dependent variable. Two business information providers (Creditreform and Bureau van Dijk) provided data regarding employment growth in the first three business years for 66 start-ups in our data-set. We found that our measure of number of employees and the corresponding information provided by Creditreform and Bureau van Dijk (2009) correlated highly ($r = 0.78, p < 0.001$).

3.3.3 Independent variables

Our independent variables attempt to measure the actual use of social capital in the venture creation process, which comprises the time from the first concrete steps in the venture creation until the start of the first business year.[4]

Typically, researchers use the name generator or the position generator in social capital measurement. The name generator (McCallister and Fischer, 1978) maps the ego-centred social network of an entrepreneur, comprising persons who were most helpful in establishing and running an entrepreneur's new venture. However, the name generator has a tendency to focus on strong ties (van der Gaag and Snijders, 2004), and therefore we opted not to use this method.

The position generator (Lin and Dumin, 1986) uses the occupations of network members as an indicator of the access to valuable resources and information. The usefulness of this instrument hinges on the relative importance and relatedness of the individuals role to the type of start-up being created. For a biotech start-up, knowing bankers or a professor in biology may be more useful than knowing a poet, but the opposite may be true for an entrepreneur opening a bookstore. Hence, this approach has limited value for studies which are not focused on a single industry with a clear hierarchy of useful contacts.

We therefore attempted to improve the measurement of social capital in the field of entrepreneurship by applying a more recently used measurement procedure: the resource generator, as developed by van der Gaag and Snijders (2005). This approach focuses on *potential* helpful flows of resources and typically asks a battery of questions such as 'Do you know any people who can lend you 5,000€?' The main advantage of this measurement concept is that it measures social capital from a 'general' base (van der Gaag and Snijders, 2004), which refers to the potential to access different, concrete and restricted sub-domains of social capital. For our analysis, we adapted the methodology of the resource generator to *concrete* resource flows instead of potential resource flows because our approach is based on the 'use' of social capital rather than its mere existence.

In order to quantify social capital use, we asked the entrepreneurs if they received advice, support or help from a third party, free or for less than the usual charge, during the venture creation process, in relation to nine different fields. These fields are derived from the nascent entrepreneurship literature (see Davidsson, 2006 for an overview), where important activities in the venture creation process, such as R&D, market exchange, financing and management, are addressed. We chose the items to cover the activities that are important in enabling a business to get up and running, focusing primarily on the recognition and exploitation of the business opportunity (Shane and Venkataraman, 2000).

Table 3.1 displays our measures of social capital use. For solo entrepreneurs and the interviewee of a start-up team, we accessed information on whether the advice, support or help came from the circle of close friends and family (*strong ties*) or from acquaintances (*weak ties*). Following the suggestions of Marsden and Campbell (1984), closeness or emotional intensity serves as an indicator of the tie strength.[5] Note that in the case of a new venture team, the interviewee was briefed not to report the help which he received from the other members of the team. We count only help from outside the new venture team. Consequently, the interviewee was asked whether his team members received any outside advice or support, regardless of whether the helpers were family, friends or acquaintances.[6]

Table 3.1 Social capital variables for predicting venture success

Social capital variable	*Operationalisation*
Social capital use	
Strong ties (Interviewee)	We asked the solo entrepreneur or the interviewee of a start-up team if he received help, support or advice from a third party, free or for less than the usual charge, during the venture creation process. More precisely, we asked: How many people from your circle of close friends and family members . . . (1) . . . have helped to write the business plan? (2) . . . have supported the project with experience in the specific industry? (3) . . . have conveyed contacts to potential customers? (4) . . . brought knowledge and experience needed for the development of products and services? (5) . . . brought knowledge and experience needed for producing products/delivering services? (6) . . . have helped the project with contacts for potential investors and lenders? (7) . . . have helped in marketing and promotion? (8) . . . have helped the project with their contacts to the administration and policy or their reputation? (9) . . . have helped with the refinement of the business idea? However, we did not simply use the received advice. Instead, dummy variables for each field were created, indicating whether the entrepreneurs used social capital at all. The measure of help from strong ties is then the count of fields with received help, support or advice.

(Continued)

Table 3.1 (Continued)

Social capital variable	*Operationalisation*
Weak ties (Interviewee)	Count of fields with received help, support or advice (same procedure as with strong ties) from the circle of acquaintances. Acquaintances were defined as people the entrepreneur knew and would talk to when meeting on the street.
Overall social capital (Interviewee)	Count of fields with received help, support or advice (same procedure as with strong ties) from either their circle of acquaintances or their circle of close friends and family members.
Overall social capital (Team)	In the case of a start-up team, we asked the interviewee if the other team members received help, support or advice from a third party in any of the nine respective fields. To ensure answerability of the questions, these are only binary items of whether the other members used social capital. The measure of overall social capital is an aggregation of the help received by the interviewee and the other team members. For overall social capital we computed the count of fields with received help, support or advice across all members of the start-up team.
Any social capital (Interviewee)	Dummy: 1=Use of social capital in any of the nine different categories; otherwise=0; data at the interviewee level.
Any social capital (Team)	Dummy: 1=Use of social capital across all members of the start-up team in any of the nine different categories; otherwise=0.
Social capital traditional	
Knowing other managers and business owners	Dummy: 1=Knowing other managers and business owners from the first steps in the venture creation process until the start of the first business year; otherwise=0; data at the venture level.
Encouragement and social support	Dummy: 1=Received encouragement and social support in the venture creation process until the start of the first business year; otherwise=0, data at the interviewee level.
Public advice	Dummy: 1=Received advice from public institutions for different activities in the venture creation process until the start of the first business year; otherwise=0; data at the venture level.

To verify the information provided by the interviewee, for a random sample of 55 cases the research team conducted an additional face-to-face interview with another member of the start-up team and received 42 matchable and usable responses. We performed dependent t-tests for paired samples on the equality of means concerning our main social capital variables, the *overall social capital* use (indicated by the number of fields with received advice) for the complete team ($t = -0.48$, $p = 0.63$) and the propensity to use *any social capital* in at least one field ($t = -0.37$, $p = 0.71$). The tests reveal no statistical differences in either case, providing evidence for the reliability of our social capital variables.

As suggested in the literature (e.g. Delmar and Gunnarsson, 2000; Vivarelli, 2004), we also collected data on whether the entrepreneurs' networks contained *other managers and business owners* (whether they provided support or not), whether the entrepreneur received *public advice* from public consulting agencies, and whether people provided *encouragement or social support* in starting their business. These social capital variables serve as a standard of comparison to our measures of social capital use and are measured at the venture level (Table 3.1), with the exception of *encouragement and social support*. This variable is based on the interviewee only, because the respective question for the other team members can hardly be answered by the interviewee in a reliable way.

As an indicator for human capital variety, we used the *variety of functional background* of the entrepreneur(s), which is measured by the number of functional areas in which the founder (team) has prior work experience (Table 3.2). In the

Table 3.2 Independent variables for predicting venture success

Independent variable	*Operationalisation*
Human capital	
Number of team members	Count of all team members who were actively involved in the venture creation process until the start of the first business year + ownership of a part of the venture.
Variety of functional background	Count of categories with working experience prior to the first steps into the venture creation process across all team members (Six categories: 1=Management, 2=Marketing/Sales/Promotion, 3=Accounting/Controlling/Financing, 4=Engineering/R&D, 5=Production, 6=Personnel); data at the venture level.
Leadership experience	Count of years with experience in executive positions prior to the first steps into the venture creation process across all team members; data at the venture level.
Entrepreneurial experience	Count of companies (registered in the commercial register) prior to the first steps into the venture creation process across all team members; data at the venture level.
Others	
Service company	Dummy: 1=Company offers mainly services; otherwise=0.
Innovativeness	Dummy: 1=Conducting R&D in the venture creation phase and the first three years of business was a major activity for the start-up; otherwise=0.
Start-up capital	Financial capital (equity + debt) at the start of the first business year, Categorical variable: 1=1,000 euros or less, 2=1,000 euros to 9,999 euros, 3=10,000 euros to 49,999 euros, 4=50,000 euros to 99,999 euros, 5=100,000 euros to 249,999 euros, 6=250,000 euros to 499,999 euros, 7=more than 500,000 euros.
Time dummies	Start-up year, 4 dummy variables: (1) start-up prior to 1994, (2) start-up between 1994 and 1997, (3) start-up between 1998 and 2000, (4) start-up between 2000 and 2006.
Industry dummies	NACE, 1-digit: (1) Chemical industry, metalworking industry, engineering, (2) Electrical engineering, fine mechanics, optics, (3) Information and communication technology, R&D, services, (4) Miscellaneous.

case of a new venture team, we count as team members all persons who were actively involved in the venture creation process and owned, or had part-ownership, of the venture. Persons entering into and exiting from the team were also counted as team members. As additional indicators for human capital, we included, at the venture level the *number of team members*, years of *leadership experience*, and prior *entrepreneurial experience* since, in similar studies, they have been found to have a significant impact on the development and performance of new ventures (Colombo and Grilli, 2005; Cooper *et al.*, 1994; Eisenhardt and Schoonhoven, 1990).

To control for the effect of financial capital, we included the *start-up capital* in the first year of operation. Final controls refer to industry, the start-up year, the possible differences between industrial and *service companies*, and the *innovativeness* of the start-up. The descriptive statistics and correlation matrices are separately displayed for solo entrepreneurs and entrepreneurial teams in Tables 3.3 and 3.4 respectively.

3.4 Results

3.4.1 Do solo entrepreneurs and new venture teams differ in the use of social capital?

We start with a test of Hypothesis 1: Do solo entrepreneurs and new venture teams differ in the use of social capital and, if so, in which fields? To answer this question, we distinguish between two cases. In the first, we compare the interviewees from the different modes of firm founding (solo start-up versus team start-up), henceforth called the interviewee level. In the second case, we compare the solo start-up with the aggregate of all members of a team start-up, henceforth called the venture level. On the one hand, these comparisons are accomplished by using our measure for *overall social capital use*, representing the number (out of nine fields) in which social capital can be used, and by the propensity to use *any social capital*. On the other hand, we compare both start-up modes on the basis of the traditional social capital variables. We apply Wilcoxon-Man-Whitney and Chi-square tests in order to find differences in those counts and probabilities.

With respect to the interviewee level, we find (Table 3.5) that a solo entrepreneur uses, in general, more social capital than a member of a team start-up. More precisely, the solo entrepreneur uses, with a probability of 76 per cent, *any social capital* and at the mean *overall social capital* in 3.0 fields, compared to 68 per cent and 2.3 fields in the sub-sample of the interviewees of a team start-up. These differences are significant at least at the 5 per cent level. Looking at the traditional indicators of social capital, we find no statistically significant differences between the two modes of firm founding on the interviewee level.

With regard to the venture level (Table 3.5), we find no statistically significant difference between solo start-ups (76 per cent; 3.0 fields) and team start-ups using *any social capital* in 73 per cent of the cases representing *overall social capital* use in 2.7 fields. Testing also for two of the three traditional indicators for social

Table 3.3 Descriptive statistics and intercorrelation matrix for solo entrepreneurs

Variable		*(1)*	*(2)*	*(3)*	*(4)*	*(5)*	*(6)*	*(7)*	*(8)*	*(9)*	*(10)*	*(11)*	*(12)*	*(13)*	*(14)*
(1)	Number of employees, 3rd year	–	–	–	–	–	–	–	–	–	–	–	–	–	–
(2)	Overall social capital (Interviewee)	.00	–	–	–	–	–	–	–	–	–	–	–	–	–
(3)	Any social capital (Interviewee)	.05	**.67**	–	–	–	–	–	–	–	–	–	–	–	–
(4)	Weak ties (Interviewee)	.09	**.78**	**.51**	–	–	–	–	–	–	–	–	–	–	–
(5)	Strong ties (Interviewee)	–.09	**.59**	**.38**	.04	–	–	–	–	–	–	–	–	–	–
(6)	Knowing other managers and business owners	.02	**.26**	**.26**	**.19**	**.16**	–	–	–	–	–	–	–	–	–
(7)	Encouragement and social support	.05	**.41**	**.34**	**.26**	**.33**	**.24**	–	–	–	–	–	–	–	–
(8)	Public advice	.12	**.21**	.07	**.25**	**.03**	–.06	.13	–	–	–	–	–	–	–
(9)	Variety of functional background	.04	–.05	.02	.06	**–.15**	.07	–.07	–.00	–	–	–	–	–	–
(10)	Leadership experience	**.28**	–.11	**–.25**	.04	**–.15**	–.02	–.13	.03	**.28**	–	–	–	–	–
(11)	Entrepreneurial experience	–.07	.07	.04	.11	–.05	–.03	–.06	–.10	.11	**.15**	–	–	–	–
(12)	Service company	–.14	**.19**	**.25**	.12	.12	**.20**	–.04	.02	–.05	**–.21**	.04	–	–	–
(13)	Innovativeness	–.01	–.10	–.20	–.07	–.05	–.07	–.09	.03	.09	**.19**	.12	–.12	–	–
(14)	Start-up capital	.23	–.05	–.01	–.02	–.04	–.06	–.10	.12	.17	.13	.08	**–.22**	.13	–
Mean		6.77	2.98	0.76	2.02	1.36	0.58	0.52	0.42	3.02	6.73	0.18	**0.48**	0.29	3.18
SD		11.56	2.45	0.43	2.18	2.01	0.50	0.50	0.50	1.74	7.67	0.48	**0.50**	0.45	1.40

Note: Correlation coefficients displayed in bold are significant at the 5 per cent level.

Table 3.4 Descriptive statistics and intercorrelation matrix for entrepreneurial teams

Variable		*(1)*	*(2)*	*(3)*	*(4)*	*(5)*	*(6)*	*(7)*	*(8)*	*(9)*	*(10)*	*(11)*	*(12)*	*(13)*	*(14)*	*(15)*	*(16)*	*(17)*
(1)	Number of employees, 3rd year	–	–	–	–	–	–	–	–	–	–	–	–	–	–	–	–	–
(2)	Overall social capital (Team)	.04	–	–	–	–	–	–	–	–	–	–	–	–	–	–	–	–
(3)	Overall social capital (Interviewee)	.05	**.91**	–	–	–	–	–	–	–	–	–	–	–	–	–	–	–
(4)	Any social capital (Team)	–.10	**.66**	**.60**	–	–	–	–	–	–	–	–	–	–	–	–	–	–
(5)	Any social capital (Interviewee)	–.06	**.66**	**.69**	**.89**	–	–	–	–	–	–	–	–	–	–	–	–	–
(6)	Weak ties (Interviewee)	.04	**.78**	**.84**	**.50**	**.57**	–	–	–	–	–	–	–	–	–	–	–	–
(7)	Strong ties (Interviewee)	.03	**.50**	**.61**	**.34**	**.38**	.08	–	–	–	–	–	–	–	–	–	–	–
(8)	Knowing other managers and business owners	–.01	**.41**	**.41**	**.33**	**.36**	**.33**	**.26**	–	–	–	–	–	–	–	–	–	–
(9)	Encouragement and social support	.10	**.36**	**.35**	**.24**	**.24**	**.30**	**.20**	**.34**	–	–	–	–	–	–	–	–	–
(10)	Public advice	–.07	.10	**.13**	**.13**	**.14**	.12	.07	–.00	–.00	–	–	–	–	–	–	–	–
(11)	Number of team members	–.03	.05	.05	.05	.07	.07	–.02	.07	.05	.05	–	–	–	–	–	–	–
(12)	Variety of functional background	**.14**	.04	.04	–.07	–.04	.01	.07	.04	–.12	–.12	.11	–	–	–	–	–	–
(13)	Leadership experience	.07	.08	.12	.06	.08	**.14**	.00	.06	.07	.07	**.32**	**.32**	–	–	–	–	–
(14)	Entrepreneurial experience	.01	–.01	–.04	–.04	–.02	.01	–.08	–.03	–.14	**–.14**	**.28**	**.28**	**.39**	–	–	–	–
(15)	Service company	–.05	–.01	.02	–.03	–.02	.01	.02	.06	–.05	–.05	**–.15**	–.09	–.03	–.05	–	–	–
(16)	Innovativeness	.00	.09	.05	.04	.05	.02	.06	–.03	.09	.03	.16	.10	.03	.08	**–.15**	–	–
(17)	Start–up capital	**.32**	.04	.04	–.02	.01	.05	.00	.01	.07	.07	.06	**.23**	**.18**	**.17**	**–.13**	**.18**	–
Mean		9.12	2.71	2.30	0.73	0.68	1.74	0.80	0.61	0.52	0.44	2.77	4.33	16.81	1.13	0.49	0.34	3.31
SD		14.84	2.51	2.32	0.45	0.47	2.12	1.46	0.49	0.50	0.50	0.90	1.64	17.74	1.78	0.50	0.47	1.34

Note: Correlation coefficients displayed in bold are significant at the 5 per cent level.

Table 3.5 Use of social capital between solo entrepreneurs and entrepreneurial teams

		Solo entrepreneurs (mean values)	*Entrepreneurial teams (mean values)*	*Wilcoxon-Mann-Whitney test*[a] *Chi-square test*[b]
Social capital use				
Overall social capital	IL[c]	3.0	2.3	2.920 (0.004) ***
	VL	3.0	2.7	1.244 (0.210)
Any social capital	IL	0.76	0.68	4.167 (0.041) **
	VL	0.76	0.73	0.800 (0.371)
Social capital traditional				
Knowing other managers and business owners	IL	0.58	0.53	1.165 (0.281)
	VL	0.58	0.61	0.482 (0.488)
Encouragement and social support[d]	IL	0.52	0.52	0.001 (0.971)
	VL	0.52		
Public advice	IL	0.42	0.42	0.015 (0.903)
	VL	0.42	0.44	0.185 (0.667)
Number of observations		182	274	

Note: [a] Wilcoxon-Mann-Whitney test on overall social capital use with prob > |t| in parentheses; [b] Chi-square test on any social capital use and on social capital traditional with prob > |z| in parentheses; [c] data in first row on interviewee level (IL), data in second row on the venture level (VL), for solo entrepreneurs both levels are identical; [d] encouragement and social support is based on the interviewees response only; *** (**,*) denotes a significance level of 1 per cent (5 per cent, 10 per cent).

capital[7] does not deliver significant differences between the solo and team responses.

To summarise, we find no support for Hypothesis 1, according to which solo entrepreneurs and new venture teams differ in their use of social capital.[8]

3.4.2 The effects of social capital

In testing the effects of social capital on venture performance, we refer to hypotheses (H2) on overall social capital, (H3) on strong ties, and (H4) on weak ties. Each of these factors is supposed to have a positive impact on new venture performance, as expressed in the absolute number of employees in the third year of firm operation. We ran regressions for a sample containing all start-ups, including both solo and team start-ups. We again distinguished two ways of representing team start-ups, namely the venture level and the interviewee level. As the dependent variable is a count variable and there is evidence for the presence of over-dispersion, we use negative binomial regression models for the analyses which follow. The regression results are displayed in Models 1–3 in Table 3.6. Looking at the venture level in Model 1, *overall social capital* turns out to be insignificant. In Model 2 and Model 3, relying upon variables at the interviewee

Table 3.6 The effect of social capital use

	Dependent variable: Number of employees in the third year of operation										
	All start-up projects			*Solo entrepreneurs*				*Entrepreneurial teams*			
	Venture level	*Interviewee level*						*Venture level*		*Interviewee level*	
	(1)	*(2)*	*(3)*	*(4)*	*(5)*	*(6)*	*(7)*	*(8)*	*(9)*	*(10)*	*(11)*
Social capital use											
Social capital (Team)	0.01	—	—	—	—	—	—	0.03	–0.02	—	—
Social capital (Interviewee)	—	0.03	—	0.10	0.10	—	—	—	—	—	—
Weak ties (Interviewee)	—	—	0.03	—	—	0.18**	0.16**	—	—	0.01	–0.02
Strong ties (Interviewee)	—	—	–0.06	—	—	–0.12	–0.11	—	—	0.02	0.02
Social capital traditional											
Knowing other managers and business owners	0.02	0.01	0.02	–0.05	–0.05	–0.05	–0.05	0.04	0.04	0.04	0.05
Encouragement and social support	0.07	0.06	0.08	0.04	0.04	0.05	0.05	0.05	0.08	0.05	0.70
Public advice	0.01	0.01	0.01	0.09	0.09	0.05	0.05	–0.09	–0.10	–0.09	–0.10
Human capital and controls											
Number of team members	0.03	0.04	0.02	—	—	—	—	–0.01	–0.02	–0.01	–0.02
Variety of functional background	0.18***	0.18***	0.18***	0.04	0.05	0.01	0.02	0.18***	0.17**	0.17**	0.16**
Leadership experience	0.04	0.04	0.03	0.27***	0.27***	0.24***	0.24***	–0.09	–0.08	–0.08	–0.06

(Continued)

Table 3.6 (Continued)

	Dependent variable: Number of employees in the third year of operation										
	All start-up projects			*Solo entrepreneurs*				*Entrepreneurial teams*			
	Venture level	*Interviewee level*						*Venture level*		*Interviewee level*	
	(1)	*(2)*	*(3)*	*(4)*	*(5)*	*(6)*	*(7)*	*(8)*	*(9)*	*(10)*	*(11)*
Entrepreneurial experience	–0.07	–0.07	–0.07	–0.15**	–0.15**	–0.17	–0.17**	–0.07	–0.06	–0.07	–0.06
Service company	0.01	0.01	0.01	–0.02	–0.02	–0.01	0.01	0.05	0.05	0.05	0.07
Innovativeness	–0.04	–0.04	–0.04	–0.08	–0.07	–0.06	–0.05	–0.02	–0.03	–0.02	–0.03
Start–up capital	0.32***	0.32***	0.33 ***	0.24***	0.25***	0.25	0.26***	0.38***	0.39***	0.38***	0.38***
Time/Industry dummies	Yes/ Yes	Yes/ Yes	Yes/ Yes	No/ Yes	No/ Yes	No/ Yes	No/ Yes	Yes/ Yes	Yes/ Yes	Yes/ Yes	Yes/ Yes
Interaction terms											
Variety of functional background × Overall social capital	—	—	—	—	0.12	—	—	—	0.15 ***	—	—
Variety of functional background × Weak ties (Interviewee)	—	—	—	—	—	—	0.08	—	—	—	0.14 **
Constant	1.97***	1.97***	1.97***	1.72***	1.72***	1.70***	1.70***	2.05***	2.03***	2.05***	2.05***
Chi^2	108.8	109.0	110.9	67.60	69.63	74.27	75.18	74.56	81.12	74.57	80.13
Pseudo R^2	0.04	0.04	0.04	0.06	0.06	0.07	0.07	0.04	0.05	0.04	0.05
Number of observations	456	456	456	182	182	182	182	274	274	274	274

Note: Negative binomial regression; *** (**,*) denotes a significance level of 1 per cent (5 per cent, 10 per cent).

level, we do not get significant coefficients for either individual *overall social capital* or for *weak ties* and *strong ties*. Furthermore, in all three models, the traditional social capital variables *knowing other managers and business owners*, *encouragement and social support*, and *public advice* show no significant effects. With regard to human capital, we only find significant positive effects for *variety of functional background* at the 1 per cent level. With regard to the controls, we find significantly positive effects for *start-up capital* at the 1 per cent level, as well as significant time and industry dummies.

Based upon these results, we are forced to reject Hypotheses 2 to 4. This is an unexpected outcome and, combined with the unexpected result of no difference in using social capital between solo start-ups and entrepreneurial teams, leaves us with quite a puzzle. A solution to this puzzle may be found in analysing whether the two types of start-ups differ in their respective use of social capital. This may give some explanation for the results found so far.

3.4.3 The differential use of social capital

Looking at the ways in which the two types of start-ups use social capital, as a dependent variable we use various binary measures for the general use of *any social capital*. As independent variables, we include the controls, as well as one of the traditional social capital measures, *knowing other managers and business owners*. We start by analysing solo entrepreneurs.

Table 3.7 provides the results of the logistic regression. Model 1 refers to solo entrepreneurs. We find *knowing other managers and business owners* to have a positive significant effect on the use of social capital at a level of 1 per cent. A significantly negative effect at the 1 per cent level is found for *leadership experience*. In addition, service companies are significantly more likely to use social capital, whereas more innovative ventures use significantly (at the 10 per cent level) less social capital.

Performing the same analyses for entrepreneurial teams, we ran two models distinguishing between the venture level (Model 2 in Table 3.7) and the interviewee level (Model 3 in Table 3.7). For Models 2 and 3, as for solo entrepreneurs, *knowing other managers and business owners* shows as significantly positive at the 1 per cent level for entrepreneurial teams. At the venture level in Model 2, higher *innovativeness* and higher *leadership experience* contribute significantly to the usage of social capital in the complete team at the 5 per cent and 10 per cent levels, respectively. The effect of the *variety of functional background*, however, is significantly negative at the 5 per cent level. The level of the interviewee in Model 3 reveals significantly positive effects from the number of team members (5 per cent) and the leadership experience (1 per cent).

Comparing these two sets of results, we find major differences in the use of social capital between the two types of start-ups. *Leadership experience* reduces the use of social capital for solo entrepreneurs, but increases the use of social capital in start-up teams. For new venture teams only, a higher *variety of functional*

Table 3.7 The differential use of social capital

	Dependent variable: Any social capital use		
		Entrepreneurial teams	
	Solo entrepreneurs	*Venture level*	*Interviewee level*
	(1)	*(2)*	*(3)*
Social capital traditional			
Knowing other managers and business owners	0.601***	0.841***	0.861***
Human capital and controls			
Number of team members	—	–0.088	0.325**
Variety of functional backgrounds	0.221	–0.360**	–0.054
Leadership experience	0.592***	0.330*	0.322*
Entrepreneurial experience	0.211	–0.125	–0.334
Service company	0.496**	–0.135	0.244
Innovativeness	–0.379*	0.428**	0.172
Start-up capital	0.144	–0.067	0.115
Time/Industry dummies	No/No	No/No	No/No
Constant	1.453***	1.234***	–1.537***
Chi^2	35.701	47.673	41.487
Pseudo R^2	0.182	0.148	0.142
Number of observations	182	274	274

Note: Logistic regressions; *** (**,*) denotes a significance level of 1 per cent (5 per cent, 10 per cent).

background significantly reduces the use of social capital. In addition, the *number of team members* is positively correlated with the use of social capital in entrepreneurial teams.

This difference in the way that the use of social capital is determined between solo entrepreneurs and entrepreneurial teams is both remarkable and unexpected, given the existing literature on social capital. One may ask whether this can be explained by the significant differences between the two groups in terms of major features such as *innovativeness* or their assignment to certain *industries* and start-up years. However, Chi-square tests on equality and Wilcoxon-Mann-Whitney tests could not be rejected for *innovativeness* ($\chi^2 = 1.27$, $p = 0.26$), *industry assignment* ($\chi^2 = 0.66$, $p = 0.88$), *service company* ($\chi^2 = 0.94$, $p = 0.76$), or start-up year ($\chi^2 = 15.99$, $p = 0.45$). The only difference we found between both start-up modes with regard to the independent and control variables is the *variety of functional background* ($z = -2.05$, $p = 0.04$). Hence, we can conclude that the purpose of accessing social capital differs between solo entrepreneurs and entrepreneurial teams. For the former, it is more a matter of whether the entrepreneur is convinced that they are capable of mastering the task successfully, as expressed by *leadership experience*. In entrepreneurial teams, the focus is more on getting the portfolio of competences right, as expressed by the *variety of functional background.*

3.4.4 The differential effect of social capital

Based on these results, we now return to the first analysis of the effects of social capital on firm performance, as expressed in terms of total employment three years after foundation. We ran separate regressions for the two types of venture founding and integrated an interaction term accounting for the manner in which social capital is used in both groups.

SOLO ENTREPRENEURS

Table 3.6 (middle section) displays the results of the negative binomial regressions. We distinguish between the case of social capital in general (Models 4 and 5) and the case of disaggregated social capital in terms of weak and strong ties (Models 6 and 7). Using identical controls in all four models, we find *start-up capital* and *leadership experience* to be significant predictors (at the 1 per cent level) of venture performance. The regressions also show significantly negative effects for *entrepreneurial experience* on venture success. This result is most unusual and can only be understood in light of the transformation process of Eastern Germany from a planned to a market economy (Fritsch, 2004). During this process, from 1990 onwards a considerable number of West German entrepreneurs founded businesses in the eastern part of Germany. The data suggests that these western entrepreneurs failed more often than eastern entrepreneurs if they did not team up with people from East Germany. It could be argued that these entrepreneurs lacked relationships with suppliers and critical contacts to facilitate access to customers and were vulnerable in the face of fast-changing market conditions. Furthermore, western entrepreneurs often ran businesses in their home region, which they could easily return to if the new businesses venture in Eastern Germany failed.

Looking at our hypotheses which state that overall *social capital* (H2), *weak ties* (H3), and *strong ties* (H4) have a positive impact on new venture performance, we find only Hypothesis 3 (Model 6) to be supported at the 5 per cent level. Insignificant coefficients for overall *social capital* (Model 4) as well as *strong ties* (Model 6) force us to reject Hypotheses 2 and 4. In contrast, the traditional social capital indicator variables of *knowing other managers and business owners, encouragement and social support* and *public advice* show no significant effects in all models.

In order to test Hypotheses 5 and 6, which suggest the moderating effects of the *variety of functional background* (H5) on the relationship between overall *social capital* and performance as well as (H6) on the relationship between *weak ties* and performance, we include respective interaction terms in Models 5 and 7 in Table 3.6. However, both hypotheses have to be rejected due to insignificant coefficients for the respective interaction terms.

ENTREPRENEURIAL TEAMS

Turning to entrepreneurial teams, Table 3.6 (right section) delivers the results of the negative binomial regressions. We again distinguish between the venture level in Models 8 and 9 and the level of the interviewee in Models 10 and 11. With regard to the human capital variables, the results differ from those of the solo entrepreneurs: *leadership experience* and *entrepreneurial experience* are not essential for the success of entrepreneurial teams. Instead, team *variety of functional background* is highly significant at the 5 per cent level. With respect to the traditional social capital indicators, all results for solo entrepreneurs are confirmed: *knowing other managers and business owners*, *encouragement and social support*, and *public advice* all failed to show significant effects.

Again, examining the hypotheses which state that overall *social capital* (H2), *strong ties* (H3), and *weak ties* (H4) will have a positive impact on new venture performance, we find all hypotheses rejected (Models 8 and 10) due to insignificant coefficients. In contrast, the interaction term of *variety of functional background* × *social capital* in Model 9 is positive and highly significant at the 1 per cent level. Hence, *variety of functional background* moderates the effect of team *social capital* on firm performance. This result is also found at the interviewee level (Model 11). Here again, the interaction term *variety of functional background* × *weak ties* is significantly positive at the 5 per cent level. Hence, quite distinct from the evidence on solo entrepreneurs, we find here a moderating effect for the *variety of functional background*. Running an OLS regression instead of a negative binomial regression confirms these results, albeit at a lower level of significance of 10 per cent.

We examine the impact of the *variety of the functional background* in more detail in Figure 3.1.[9] As illustrated in the left part of the figure, entrepreneurial teams which had greater variety in their functional background enjoyed a higher employment level when employing *social capital* more often, supporting

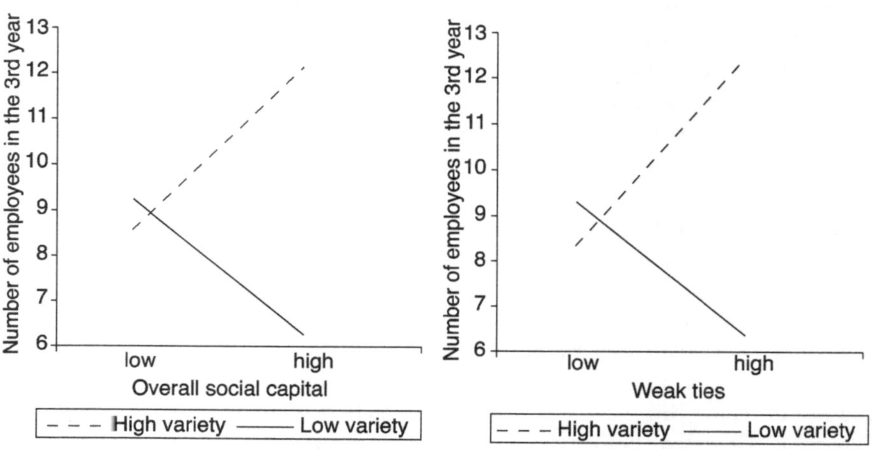

Figure 3.1 The moderating effect of teams' variety of functional background

Hypothesis 5. This result holds if we focus on social capital in terms of *weak ties* (right part of Figure 3.1), supporting Hypothesis 6.[10]

3.5 Discussion and conclusion

3.5.1 Interpretation and discussion of the results

This study empirically examined the use of social capital among solo entrepreneurs and entrepreneurial teams in the venture creation process. Based on a sample of 456 start-ups in innovative industries, we tried to answer two research questions: First, do entrepreneurial teams use social capital more often than solo entrepreneurs? Second, what are the effects of social capital use in the venture creation process upon subsequent venture performance? Table 3.8 summarises the results.

In answer to the first question, we found that venture teams do not use more social capital than solo entrepreneurs in the venture creation process. This unexpected result is due to the fact that the two links explained below have reverse but quantitatively co-equal impacts upon social capital use.

The standard proposition concerning the social capital issue is that a team start-up (as compared to a solo entrepreneur), or a larger team (as compared to a smaller team), has more social capital. This proposition is sometimes made explicitly (e.g. Colombo and Grilli, 2005), but is more often implicitly applied (e.g. Davidsson and Honig, 2003; van Gelderen *et al.*, 2005). Its validity depends upon how we define social capital. If we define social capital as the *potential*

Table 3.8 Summary of results

	Results		
Hypotheses	*All start-up projects*	*Solo entrepreneurs*	*Entrepreneurial team*
H1: Solo entrepreneurs and entrepreneurial teams differ in social capital use		Supported	
H2: Overall social capital positive for performance	Not supported	Not supported	Not supported
H3: Strong ties positive for performance	Not supported	Not supported	Not supported
H4: Weak ties positive for performance	Not supported	Supported	Not supported
H5: Human capital variety moderating the effect of overall social capital on performance	Not tested	Not supported	Supported
H6: Human capital variety moderating the effect of weak ties on performance	Not tested	Not supported	Supported

access to resources and information, the standard proposition holds true, because the number of team members will be positively correlated with the overall number of contacts and hence with possible access to resources or information. However, when we focus on the actual *use* of the network contacts, the proposition is questionable at least, if not unfounded. In a start-up team, members combine their (often) different skills, abilities, information and resources, enabling in-house handling of more activities in the start-up process. Hence, the actual use of social capital will be negatively correlated with the corresponding heterogeneity of the start-up team.

Looking at the empirical results, we find evidence for both links affecting the use of social capital of new venture teams. First, team size is positively correlated with social capital use, suggesting that a new venture team compared to a solo entrepreneur, as well as a larger team compared to smaller one, has more contacts to use. Second, the variety of functional backgrounds in a team is negatively correlated with social capital use. This result suggests that the use of those contacts is dependent upon other characteristics of the entrepreneurs. Previous empirical literature has paid limited attention to that second link. The study by Renzulli and Aldrich (2005) is an exception, and it complements our results. They focus on the determinants of tie activation for business start-ups and find that heterogeneity within the social network of an entrepreneur significantly increases the probability of using those contacts for business purposes. In contrast to the present study, they evaluate the characteristics of network ties and the resulting impact upon social capital use, while we are concerned with the characteristics of the team or solo entrepreneur and their impact upon social capital use. In both cases, heterogeneity among actors is positively correlated with the use of social capital.

Despite the evidence that new venture teams and solo entrepreneurs do not differ in their use of social capital, there are pronounced differences in the way in which each start-up mode uses social capital in the venture creation process. We find that the human capital characteristics influencing social capital use are different for each group. For solo entrepreneurs, there are clear indications of a substitutive relationship between human capital in terms of the leadership experience of the founder and social capital use. For start-up teams, in contrast, no such clear relationship was found. Leadership experience correlates positively with social capital use. Team size and the variety of a team's knowledge base have reverse effects upon social capital, as described above.

With regard to the second question, we find that social capital use affects new venture performance differently for each start-up mode. The results of Section 4.4 lead to the conclusion that, for entrepreneurial teams, there are rather indirect effects of social capital use on firm performance, moderated by *human capital variety*. The more that teams are specialised in their functional background, the more the team members work with and learn from each other and the less reliant they are upon accessing social capital. A more diversified team complements the human capital available by increasingly relying upon social capital. In contrast, for solo entrepreneurs, there appears to be a direct relationship of social capital upon performance. Solo entrepreneurs profit from information provided by their weak ties; however, their human capital variety (*variety of functional background*) does not contribute significantly to any employment effects.

The results of the analysis lead to the conclusion that solo start-ups and team start-ups differ beyond the pure number of entrepreneurs. Although the difference in the significance level of the interaction term between human capital and social capital variables is only indirect evidence, we argue that one of the key characteristics which differentiates solo entrepreneurs from entrepreneurial teams is the learning process. We therefore understand the need for the development of necessary knowledge in starting up and managing new ventures (Politis, 2005). This process is more complex for teams because as they work together in the start-up project, they also learn together. Consider the case of a solo entrepreneur. He can directly evaluate any information stemming from his personal contacts and integrate this into his knowledge base. By way of contrast, a member of a new venture team may not use such contacts directly; he will probably first ask his team members if he should approach his personal contact for help or information. Thereafter, the team members probably consult this outside help collectively and then together evaluate the usefulness of the information and their further actions.

This supposed model fits well with the data and the descriptions the entrepreneurs gave during the interviews. We suppose that, for a team which has a broad knowledge base, it is more likely that they will opt against external help. However, if such a team does use social capital, it profits considerably from the information transfer as a result of two mechanisms. First, their learning and evaluation procedures enable them to detect more valuable information. Second, because of the breadth of their knowledge base, they can integrate and exploit the new knowledge more efficiently. This view of organisational learning and the importance of a diverse knowledge base are in line with recent studies on venture teams (e.g. Hayton and Zahra, 2005).

3.5.2 Implications for practice

This study has several implications for practice. For those who have chosen to start up alone, access to novel information about markets, prices and competitors is of critical importance. This information is best accessed via weak ties, which include (former) colleagues, friends and former employers, as well as people at conferences and trade fairs. We find that help, advice and support from such weak ties have positive effects upon venture performance. In contrast, help from close friends and family members does not appear to have measurable effects upon performance. Entrepreneurs may value trusted feedback from such sources highly, but the information lacks breadth and scope.

For those who have chosen to team up with other people in order to start a venture, our implications are somewhat counter-intuitive. We observe a high level of human capital in the new venture teams. On average, in four out of six predefined categories, the team as a whole benefits from the work experience of its members. Teams with a high variety of skills tend not to use their contacts, instead relying heavily upon the knowledge base within the start-up team. However, these teams would gain the most from making full use of their network contacts. It seems that such teams have several advantages, as compared to less-equipped teams. First, they can better evaluate outside information with regard to its usefulness. Second, they probably have a choice of different helpers, leading to a higher quality of help.

3.5.3 Implications for theory

The results make one particular contribution to entrepreneurship theory, concerning discussion of the nature of an entrepreneurial team. What is an entrepreneurial team? Is it a case of the leading entrepreneur dominating the other team members? Is it the sum of its parts? Is it more than, or something rather different to, the sum of its parts? These questions are of crucial importance to the understanding of entrepreneurship, since a substantial share of new venture projects are started by teams. The answers given to those questions by the research community have changed over recent decades.

The trait approach treated the entrepreneur as a lonely hero and mainly paid attention to the psychological characteristics of the single actor (for an overview, see Gartner, 1988). The entrepreneurial team was not part of the research agenda. Over the past few years, the majority of research has used the venture as the level of analysis (Davidsson and Wiklund, 2001). Team-related variables are often treated by summing the individual responses of the entrepreneurs. In our view, this is progress, because it at least accepts the existence of the new venture team. However, studies focusing on team issues in entrepreneurship are scarce, albeit with some notable exceptions (e.g. Chandler *et al.*, 2005; Chowdhurry, 2005). These studies find evidence that internal team processes such as communication, co-labouring and common decision-making are important predictors for team success. This contradicts the view that teams are purely the sum of their parts, but does not answer the question of whether the team is more than, or different to, the sum of its parts.

We found interaction effects between human and social capital variables for team start-ups but not for solo start-ups, suggesting that the team start-ups are something different than the sum of their parts. We argue that this interaction effect stems from collective work and information-sharing between the team members in the venture creation process, fostering learning at the individual and the collective levels. Our view is supported by research on teams operating in a range of contexts, such as primary care teams (Bunniss and Kelly, 2008), new product development teams (Bourgeon, 2007), and multidisciplinary working teams in the oil and gas industry (van der Vegt and Bunderson, 2005). All these studies emphasise the roles of collective work and information-sharing in the learning process of a team.

In the field of entrepreneurship, some work has already been done concerning collective cognition (West, 2007; Shepherd and Krueger, 2002; Ensley and Pearce, 2001). For example, West (2007, p. 83) argues that in team start-ups 'decisions are not left up to the individual'. Instead, the team often makes the decision. For West, it is important to understand how the individual perspectives of the entrepreneurs translate into a collective understanding, triggering collective decisions and actions. His model of collective cognition contains the individual cognitions of the team members, as well as team internal processes and the environment external to the team.

As Weick and Roberts (1993) suggest, we want to emphasise that we use the word 'collective' instead of 'group', because we do not think that the team members merge into one group, and we deny the existence and importance of neither the individuals nor the collective. Both levels – the individual as well as the collective – are present in an entrepreneurial team. Thus, research combining the individual and the collective level should yield valuable results for entrepreneurship. Future research may address in more detail how individual skills and individual social network contacts translate into the knowledge base of the emerging venture and which factors, such as communication and trust, influence this process. Process research techniques could shed light upon these transfer mechanisms.

3.5.4 Limitations

Naturally, our analysis has its limitations. First and most important, the study is retrospective in nature. Although we use the above-described techniques to gain reliable information about the venture creation process from the entrepreneur, we cannot completely rule out memory decay and hindsight bias. In one extreme case, there was a time span of 20 years between the first steps in the venture creation process and the interview. Second, we use self-reported measures of the number of employees as a dependent variable. Thus, the results suffer from a self-report bias. However, we checked the reliability of the data using secondary information from a business information provider. The market value or turnover of the start-up would be more appropriate dependent variables, but unfortunately this information was inaccessible for the present data-set, which consisted of young and small enterprises. Third, concerning the independent variables, we relied upon information from only one member of a start-up team. We then checked the reliability of the respondent information by interviewing an additional member of the entrepreneurial team. Regardless of whether these efforts confirm the overall reliability of our social capital use variables, we still lack disaggregated information on the use of weak and strong ties for the other team members.

Notes

1 Principally, one could think of different human capital variables affecting the learning process. However, the approximation of human capital by the heterogeneity of the functional background of top management teams in high-tech ventures is suggested by Hayton and Zahra (2005), who argue that the absorptive capacity of a new venture team is better measured by the breadth of the knowledge base rather than its depth (e.g. heterogeneity of functional background versus the average number of years of leadership experience of the entrepreneurial team).

2 Grupp *et al.* (2000) define innovativeness at the level of the industry. On average, companies in innovative industries spend more than 3.5 per cent of their turnover on research and development.

3 Firms founded in 2006 cannot answer any question on their third year of business activity.

4 The first business year is defined as the time when accounting started either because of obligations from the commercial register or because of first revenues. This does not necessarily correspond to the date of registration in the commercial register.

5 In their seminal work, Marsden and Campbell (1984) identify educational differences, kinship and the fact that two persons work together as important predictors of tie strength. They suggest closeness or emotional intensity as the best available indicators for evaluating the strength of a tie. The majority of the empirical studies apply this concept (for an overview, see Kim and Aldrich, 2005), either intentionally or unintentionally, for practical reasons, since this measurement procedure is easy to administer and straightforward.

6 In the case of team founders, the distinction between weak and strong ties cannot be made, as the interviewee was usually not able to classify his co-founders' contacts as weak or strong. Therefore, we only have information about tie strength with regard to the interviewee of the new venture team.

7 Since we are operating at the venture level, we cannot perform a comparison with respect to the variables of encouragement and social support, because we only possess this data for the interviewee member of the start-up team.

8 Interestingly, this result holds for the traditional indicators of social capital. What empirically distinguishes these traditional indicators from the nine fields of used social capital is that they occur with a much higher probability. Furthermore, the traditional indicators do not show the observed pattern with higher occurrence for a solo entrepreneur compared to the interviewee of a new venture team. This confirms our argument that traditional indicators cannot disentangle social capital from team issues.

9 These figures are computed using the regression coefficient of a respective OLS-regression.

10 The results do not hold true if we run a regression on the moderated effect of strong ties. In this case, the respective interaction term is insignificant. These regressions are not shown here, but are available from the author upon request.

References

Acs, Z. and Plummer, L. (2005) 'Penetrating the Knowledge Filter in Regional Economies', *The Annals of Regional Science*, 39(3): 439–56.

Adler, P. and Kwon, S. (2002) 'Social Capital: Prospects for a New Concept', *Academy of Management Review*, 27(1): 17–40.

Aldrich, H.E. and Martinez, M.A. (2001) 'Many are Called, but Few are Chosen: An Evolutionary Perspective for the Study of Entrepreneurship', *Entrepreneurship Theory and Practice*, 25(4): 41–56.

Aldrich, H.E. and Ruef, M. (2006) *Organizations Evolving*, 2nd edn, London, Thousand Oaks: Sage Publications.

Anderson, A.R., Park, J. and Jack, S. (2007) 'Entrepreneurial Social Capital: Conceptualizing Social Capital in New High-tech Firms', *International Small Business Journal*, 25(3): 245–72.

Baker, T. and Nelson, R.E. (2005) 'Creating Something from Nothing: Resource Construction through Entrepreneurial Bricolage', *Administrative Science Quarterly*, 50(3): 329–66.

Batjargal, B. (2007) 'Internet Entrepreneurship: Social Capital, Human Capital and Performance of Internet Ventures in China', *Research Policy*, 36(5): 605–18.

Baum, J.A.C., Calabrese, T. and Silverman, B.S. (2000) 'Don't go it Alone: Alliance Network Composition and Startups' Performance in Canadian Biotechnology', *Strategic Management Journal*, 21(3): 267–94.

Baum, J.R., Locke, E.A. and Smith, K.G. (2001) 'A Multidimensional Model of Venture Growth', *Academy of Management Journal*, 44(2): 292–303.

Belli, R.F., Shay, W.L. and Stafford, F.P. (2001) 'Event History Calendars and Question List Surveys', *Public Opinion Quarterly*, 65(1): 45–74.

Belli, R.F., Lee, E.H., Stafford, F.P. and Chou, C. (2004) 'Calendar and Question-List Survey Methods: Association between Interviewer Behaviors and Data Quality', *Journal of Official Statistics*, 20(2): 185–218.

Bhave, M.P. (1994) 'A Process Model of Entrepreneurial Venture Creation', *Journal of Business Venturing*, 9(3): 223–42.

Bourdieu, P. (1986) 'The Forms of Capital', in J.G. Richardson (ed.) *Handbook of Theory and Research for Sociology of Education*, New York: Greenwood Press.

Bourgeon, L. (2007) 'Staffing Approach and Conditions for Collective Learning in Project Teams: The Case of New Product Development Projects', *International Journal of Project Management*, 25(4): 413–22.

Bunniss, S. and Kelly, D.R. (2008) "The Unknown becomes the Known': Collective Learning and Change in Primary Care Teams', *Medical Education*, 42(12): 1185–94.

Burt, R.S. (1992) *Structural Holes: The Social Structure of Competitions*, Cambridge, Massachusetts: Harvard University Press.

Burt, R.S. (2000) 'The Network Structure of Social Capital', in R.I. Sutton and B.M. Staw, (eds) *Research in Organizational Behavior*, Vol. 22, Greenwich: JAI Press.

Caspi, A., Moffitt, T.E., Thornton, A., Freedman, D., Amell, J.W., Harrington, H., Smeijers, J. and Silva, P.A. (1996) 'The Life History Calendar: A Research and Clinical Assessment Method for Collecting Retrospective Event-History Data', *International Journal of Methods in Psychiatric Research*, 6(2): 101–14.

Chandler, G.N., Honig, B. and Wiklund, J. (2005) 'Antecedents, Moderators and Performance Consequences of Membership Change in New Venture Teams', *Journal of Business Venturing*, 20(5): 705–25.

Chowdhurry, S. (2005) 'Demographic Diversity for Building an Effective Entrepreneurial Team: Is it Important?', *Journal of Business Venturing*, 20(6): 727–46.

Cohen, W.M. and Levinthal, D.A. (1990) 'Absorptive Capacity: A New Perspective on Learning and Innovation', *Administrative Science Quarterly*, 35(1): 128–52.

Coleman, J.S. (1988) 'Social Capital in the Creation of Human Capital', *American Journal of Sociology*, 94(S): 95–120.

Colombo, M.G. and Grilli, L. (2005) 'Founders' Human Capital and the Growth of New Technology-based Firms: A Competence-based view', *Research Policy*, 34(6): 795–816.

Cooper, A.C. and Bruno, A.V. (1977) 'Success among High-Technology Firms', *Business Horizons*, 20(2): 16–22.

Cooper, A.C., Gimeno-Gascon, F.J. and Woo, C.Y. (1994) 'Initial Human and Financial Capital as Predictors of New Venture Performance', *Journal of Business Venturing*, 9(5): 371–95.

Creditreform and Bureau van Dijk (2009) MARKUS database, online available: http://www.creditreform.de/Deutsch/Creditreform/Unser_Leistungsspektrum/Marketing_Services/Business_Marketing/MARKUS_Firmenprofile.jsp

Davidsson, P. (2006) 'Nascent Entrepreneurship: Empirical Studies and Developments', *Foundations and Trends in Entrepreneurship*, 2(1): 1–79.

Davidsson, P. and Honig, B. (2003) 'The Role of Social and Human Capital Among Nascent Entrepreneurs', *Journal of Business Venturing*, 18(3): 301–31.

Davidsson, P. and Wiklund, J. (2001) 'Levels of Analysis in Entrepreneurship Research: Current Research Practice and Suggestions for the Future', *Entrepreneurship Theory and Practice*, 25(4): 81–100.

Delmar, F. and Gunnarsson, J. (2000) 'How Do Self-Employed Parents of Nascent Entrepreneurs Contribute?', *Frontiers of Entrepreneurship Research*, Wellesley: Babson.

Eisenhardt, K.M. and Schoonhoven, C.B. (1990) 'Organizational Growth: Linking Founding Team, Strategy, Environment and Growth Among U.S. Semiconductor Ventures, 1978-1988', *Administrative Science Quarterly*, 35(3): 504–29.

Elfring, T. and Hulsink, W. (2003) 'Networks in Entrepreneurship: The Case of High-Technology Firms', *Small Business Economics*, 21(4): 409–22.

Ensley, M.D. and Pearce, C.L. (2001) 'Shared Cognition in Top Management Teams: Implications for New Venture Performance', *Journal of Organizational Behavior*, 22(2): 145–60.

Florin, J., Lubatkin, M. and Schulze, W. (2003) 'A Social Capital Model of High-Growth Ventures', *Academy of Management Journal*, 46(3): 374–84.

Freedman, D., Thornton, A., Camburn, D., Alwin, D. and Young-DeMarco, L. (1988) 'The Life History Calender: A Technique for Collecting Retrospective Data', *Sociological Methodology*, 18(2): 37–68.

Fritsch, M. (2004) 'Entrepreneurship, Entry, and Performance of New Businesses in Two Growth Regimes: East and West Germany', *Journal of Evolutionary Economics*, 14(5): 525–42.

Fritsch, M. and Mueller, P. (2004) 'Effects of New Business Formation on Regional Development over Time', *Regional Studies*, 38(8): 961–75.

Gartner, W.B. (1985) 'A Conceptual Framework for Describing the Phenomenon of New Venture Creation', *Academy of Management Review*, 10(4): 696–706.

Gartner, W.B. (1988) ''Who Is an Entrepreneur?' is the Wrong Question', *Entrepreneurship Theory and Practice*, 12(4): 11–32.

Granovetter, M.S. (1973) 'The Strength of Weak Ties', *American Journal of Sociology*, 78(6): 1360–80.

Granovetter, M.S. (1990) 'The Old and The New Economic Sociology: A History and an Agenda', in R. Friedland and A. Robertson (eds) *Beyond the Market Place: Rethinking Economy and Society*, New York: Walter de Gruyter.

Granovetter, M.S. (1992) 'Problems of Explanation in Economic Sociology', in N. Nohria and R. Eccles (eds) *Networks and Organizations: Structure, Form and Action*, Boston, MA: Harvard Business School Press.

Grupp, H., Jungmittag, A., Schmoch, U. and Legler, H. (2000) *Hochtechnologie 2000: Neudefinition der Hochtechnologie für die Berichterstattung der technologischen Leistungsfähigkeit Deutschlands*, Karlsruhe/Hannover: Fraunhofer-Institut für Systemtechnik und Innovationsforschung (ISI) und Niedersächsisches Institut für Wirtschaftsforschung (NIW).

Hayton, J.C. and Zahra, S.A. (2005) 'Venture Team Human Capital and Absorptive Capacity in High Technology New Ventures', *International Journal of Technology Management*, 31(3–4): 256–74.

Honig, B., Lerner, M. and Raban, J. (2006) 'Social Capital and the Linkages of High-Tech Companies to the Military Defense System: Is there a Signaling Mechanism', *Small Business Economics*, 27(4–5): 419–37.

Kamm, J.B., Schuman, J.C., Seeger, J.A. and Nurick, A.J. (1990) 'Entrepreneurial Teams in New Venture Creation: A Research Agenda', *Entrepreneurship Theory and Practice*, 14(4): 7–17.

Kim, P.H. and Aldrich, H.E. (2005) 'Social Capital and Entrepreneurship', *Foundations and Trends in Entrepreneurship*, 1(2): 55–104.

Lechler, T. (2001) 'Social Interaction: A Determinant of Entrepreneurial Team Venture Success', *Small Business Economics*, 16(4): 263–78.

Lechner, C. and Dowling, M. (2003) 'Firm Networks: External Relationships as Sources for the Growth and Competitiveness of Entrepreneurial Firms', *Entrepreneurship & Regional Development*, 15(1): 1–26.

Liao, J. and Welsch, H. (2003) 'Social Capital and Entrepreneurial Growth Aspirations: A Comparison of Technology- and Non-technology-based Nascent Entrepreneurs', *Journal of High Technology Management Research*, 14(1): 149–70.

Lin, N. and Dumin, M. (1986) 'Access to Occupations Through Social Ties', *Social Networks*, 8(4): 365–85.

Lumpkin, G.T. and Dess, G.G. (2001) 'Linking Two Dimensions of Entrepreneurial Orientation to Firm Performance: The Moderating Role of Environment and Industry Life Cycle', *Journal of Business Venturing*, 16(5): 429–51.

McCallister, L. and Fischer, C.S. (1978) 'A Procedure for Surveying Personal Networks', *Sociological Methods and Research*, 7(2): 415–44.

Marsden, P.V. and Campbell, K.E. (1984) 'Measuring Tie Strength', *Social Forces*, 64(2): 482–501.

Nahapiet, J. and Goshal, S. (1998) 'Social Capital, Intellectual Capital, and the Organizational Advantage', *The Academy of Management Review*, 23(2): 242–66.

Politis, D. (2005) 'The Process of Entrepreneurial Learning: A Conceptual Framework', *Entrepreneurship Theory and Practice*, 29(4): 399–424.

Presutti, M., Boari, C. and Fratocchi, L. (2007) 'Knowledge Acquisition and the Foreign Development of High-tech Start-ups: A Social Capital Approach', *International Business Review*, 16(1): 23–46.

Putnam, R. (2000) *Bowling Alone: The Collapse and Revival of American Community*, New York, NY: Simon & Schuster.

Renzulli, L.A. and Aldrich, H.E. (2005) 'Who Can You Turn to? Tie Activation within Core Business Discussion Networks', *Social Forces*, 84(1): 323–41.

Ronstadt, R. (1988) 'The Corridor Principle', *Journal of Business Venturing*, 3(1): 31–40.

Samuelsson, M. and Davidsson, P. (2009) 'Does Venture Opportunity Variation Matter? Investigating Systematic Differences between Innovative and Imitative New Ventures', *Small Business Economics*, 33(2): 229–55.

Schumpeter, J.A. (1934) *The Theory of Economic Development*, Cambridge (MA): Cambridge University Press; first published in German: *Die Theorie der Wirtschaftlichen Entwicklung*, Berlin 1911: Duncker & Humblot.

Shane, S.A. and Cable, D. (2002) 'Network Ties, Reputation, and the Financing of New Ventures', *Management Science*, 48(3): 364–81.

Shane, S.A. and Venkataraman, S. (2000) 'The Promise of Entrepreneurship as a Field of Research', *Academy of Management Review*, 25(1): 217–26.

Shepherd, D.A. and Krueger, N.F. (2002) 'An Intentions-based Model of Entrepreneurial Teams' Social Cognition', *Entrepreneurship Theory and Practice*, 27(2): 167–85.

Shepherd, D.A., Douglas, E.J. and Shanley, M. (2000) 'New Venture Survival: Ignorance, External Shocks, and Risk Reduction Strategies', *Journal of Business Venturing*, 15 (5–6): 393–410.

Stam, E. and Schutjens, V. (2006) 'The Fragile Success of Team Startups', in A. Groen, R. Oakey, P. van der Sijde and S. Kauser (eds) *New Technology-based Firms in the New Millennium*, 5th edn, Amsterdam: Elsevier.

Teece, D.J., Pisano, G. and Shuen, A. (1997) 'Dynamic Capabilities and Strategic Management', *Strategic Management Journal*, 18(7): 509–33.

Uzzi, B. (1997) 'Social Structure and Competition in Interfirm Networks: The Paradox of Embeddedness', *Administrative Science Quarterly*, 42(1): 35–67.

Van der Gaag, M.P.J. and Snijders, T.A.B. (2004) 'Proposals for the Measurement of Individual Social Capital', in H. Flap and B. Völker (eds) *Creation and Returns of Social Capital*, London: Routledge.

Van der Gaag, M.P.J. and Snijders, T.A.B. (2005) 'The Resource Generator: Social Capital Quantification with Concrete Items', *Social Networks*, 27(1): 1–29.

Van der Vegt, G.S. and Bunderson, J.S. (2005) 'Learning and Performance in Multidisciplinary Teams: The Importance of Collective Team Identification', *Academy of Management Journal*, 48(3): 532–47.

Van Gelderen, M., Thurik, R. and Bosma, N. (2005) 'Success and Risk Factors in the Pre-Startup Phase', *Small Business Economics*, 24(4): 365–80.

Vivarelli, M. (2004) 'Are All the Potential Entrepreneurs So Good?', *Small Business Economics*, 23(1): 41–9.

Weick, K.E. and Roberts, K.H. (1993) 'Collective Mind in Organizations: Heedful Interrelating on Flight Decks', *Administrative Science Quarterly*, 38(3): 357–81.

West, G.P. III (2007) 'Collective Cognition: When Entrepreneurial Teams, Not Individuals, Make Decisions', *Entrepreneurship Theory and Practice*, 31(1): 77–102.

Yli-Renko, H., Erkko, A. and Sapienza, H.J. (2001) 'Social Capital, Knowledge Acquisition and Knowledge Exploitation in Young Technology-Based Firms', *Strategic Management Journal*, 22(6/7): 587–613.

Zahra, S.A., Sapienza, H.J. and Davidsson, P. (2006) 'Entrepreneurship and Dynamic Capabilities: A Review, Model and Research Agenda', *Journal of Management Studies*, 43(4): 917–55.

4 Agglomeration economies, forward linkages and regional specialisation in business services in the EU27

Valentina Meliciani and Maria Savona

4.1 Introduction

The growth of service sectors represents one of the most important changes in the structure of advanced economies over recent decades and has sparked a large literature (such as Peneder, 2003; Peneder *et al.*, 2003; Parrinello, 2004; Savona and Lorentz, 2005; Schettkat and Yocarini, 2006; Montresor and Marzetti, 2011, among the most recent studies). A substantial part of this literature acknowledges that processes of tertiarisation have largely been driven by a particularly dynamic branch of services – business services (BS, in what follows).[1] In fact, business services have exhibited higher rates of employment, value added, and productivity growth with respect to other branches of services and to the rest of the economy, contributing to cross-country differences of growth patterns (Francois, 1990; Rowthorn and Ramaswamy, 1999; Guerrieri *et al.*, 2005; Kox and Rubalcaba, 2007a and 2007b; Muller and Doloreux, 2009). Moreover, these services have been found to benefit industries that use them as intermediate inputs through efficiency gains and knowledge diffusion (Kox and Rubalcaba, 2007a and 2007b).

Cross-country growth divergences in Europe are also due to regional polarisation patterns of employment and productivity growth (Guerrieri *et al.*, 2005; Fagerberg *et al.*, 1997; Meliciani, 2006; Sterlacchini, 2008, among others). The regional literature has often focused on 'clubs' of knowledge-intensive regions (Moreno *et al.*, 2005; Crescenzi *et al.*, 2007; Verspagen, 2007; Sterlacchini, 2008), overlooking the close link between regional sectoral specialisation and knowledge intensity. In this context, therefore, disentangling the factors that drive BS specialisation at the regional level is of great importance.

Agglomeration economies, since Marshall (1920), have traditionally been best placed to explain the sectoral concentration of economic activities at the local level.[2]

The classical sources of *agglomeration economies* – localisation, urbanisation and Jacob's externalities – are strongly associated with the spatial distribution of industries, as shown by the traditional models of economic geography (Fujita *et al.*, 1999; Midelfart-Knarvik *et al.*, 2000). In line with the literature on urbanisation economies (Glaeser *et al.*, 1992 and 1995; Henderson *et al.,* 1995), urban areas are more likely to attract and host highly qualified human capital and feed the knowledge endowment of the region. We expect that highly urbanised, densely populated and highly skilled regions are ceteris paribus more prone to specialise in BS (Kox and Rubalcaba, 2007a; Cainelli *et al.*, 2006; Evangelista, 2006; Doloreux and Shearmur, 2011).

A further aspect of Marshallian agglomeration economies is linked to the presence of local markets of consumers and suppliers. This is often acknowledged in the regional literature as the presence of *Hirschmann forward and backward linkages*, which shape regional specialisation and are likely to drive growth polarisation (Hirschmann, 1958; McCann and van Oort, 2009). Fast growing industries – and business services in particular, which are highly dependent on the intermediate demand of manufacturing industries and other services – tend to be spatially concentrated, depending upon the structure of backward and forward linkages. We expect, therefore, that regions characterised by high shares of service-intensive user industries (dense in forward linkages) are ceteris paribus more likely to specialise in BS.

Agglomeration economies also occur at the cross-regional level: in line with some of the most recent regional literature (van Oort 2007, Raspe and van Oort, 2007) we claim that the above factors have a strong spatial dependence. We expect that BS specialised regions tend to cluster; that being surrounded by highly urbanised and densely populated regions, as well as by regions specialised in service-intensive user industries with high endowment of human capital, also positively affect the BS specialisation of the typical region.

The contribution of this work is to reprise and empirically extend the traditional Marshallian agglomeration economies explanation of BS sectoral specialisation in two directions: first, by explicitly accounting for the presence of Hirschmann forward linkages as an important determinant of BS specialisation, and second, by accounting for the spatial dependence of these determinants in a spatial econometric framework. Our findings also add to the literature on services by providing a spatial picture of the knowledge and intermediate demand determinants of BS specialisation, which thus far have mainly been considered at the country level (Guerrieri and Meliciani, 2005; Savona and Lorentz, 2005; Kox and Rubalcaba, 2007a).

This chapter is organised as follows: Section 2 reviews the literature relevant to choosing the variables related to specialisation in business services and supporting our empirical model; Section 3 gives a broad picture of the geography of specialisation in business services and its spatial correlation; Section 4 presents the empirical model and the econometric strategy; Section 5 discusses the estimation results; and, finally, Section 6 summarises the findings and draws the main conclusions of the chapter.

4.2 The 'holy trinity' of Marshallian agglomeration economies revisited and regional specialisation in business services

4.2.1 Urbanisation externalities and knowledge density

The classical theories of agglomeration economies date back to the contribution of Marshall in the late nineteenth century (Marshall, 1920), and have since sparked a substantial amount of theoretical and empirical work (for a historical review, see McCann and van Oort, 2009; see also van Oort, 2004 and 2007; Burger *et al.*, 2008).

Marshall's 'holy trinity' of sources of agglomeration economies include (i) the 'industrial atmosphere' which generates knowledge spillovers; (ii) a labour market conducive to a significant pool of specialised skills and knowledge endowment; and, finally, (iii) a substantial presence of local consumer and suppliers markets – i.e. density of backward and forward linkages (Hirschmann, 1958). Marshall's sources of agglomeration economies have been variously reprised over time, though they have mainly been analysed with respect to their impact on regional growth and development. They have been associated with the presence of:

- *localisation externalities* stemming from sectoral density, which favours internal and external economies of scale, though these depend on the specific sector (see, for instance, Combes, 2000; van Oort, 2007);
- *urbanisation externalities* which, while independent from the sectoral structure, are due to urban and population density, which facilitate knowledge spillovers (Glaeser *et al.*, 1992 and 1995; Henderson *et al.*, 1995);
- *Jacobs' externalities*, deriving from the variety of activities within urban contexts (Jacobs, 1969; Duranton and Puga, 2000). This type of externality tends to be higher in regions with a relatively higher *related*, rather than *unrelated*, variety of urban activities (Frenken *et al.*, 2007; McCann and van Oort, 2009).

Agglomeration economies have rarely been accounted for as determinants of sectoral specialisation – even when the sectoral dimension has been explicitly taken into account (Combes, 2000; van Oort, 2007). Raspe and van Oort (2007) argue that *geographic*, *dynamic*, and *sectoral* context-dependency in the analysis of agglomeration effects has been overlooked and is deserving of major attention. More specifically, on services, Combes (2000), van Oort (2007) and Burger *et al.* (2008) find that localisation externalities are positively related to services employment growth more than to employment growth in other sectors, arguing that service sectors benefit more from concentration than other economic activities.

In line with this evidence, an interesting study was carried out for US counties over the period 1972–2000 (Desmet and Fafchamps, 2005), in order to test the spatial distribution of service versus non-service jobs. The results show that non-service jobs tend to be spread out, whereas service employment is seen to cluster in areas of high aggregate employment, thus supporting the conjecture that agglomeration economies – and particularly localisation and urbanisation externalities – work strongly for service employment.

The literature highlights a specific role for large urban areas as attractors of business services (Jacobs, 1969, Duranton and Puga, 2000). Duranton and Puga (2005) present a model of functional specialisation, where multinational firms locate their 'headquarter functions' in large urban regions. Such locations make it possible for the headquarters to buy local inputs from specialised business service firms in areas such as R&D, marketing, financing, law, exporting, logistics, etc. Moreover, the location of the headquarters favours the co-location of specialised intermediate business service firms from which the headquarters can buy locally when they choose to outsource various services.

A further inducement for knowledge-intensive services to be located in regions with large urban areas is that they need to employ skilled labour and human capital, which tend to be concentrated in cities (Glaeser, 1999; Karlsson *et al.*, 2009; see Faggian and McCann, 2009 for a review). More specifically, urbanisation externalities are also linked to more intensive access to ICT in urban areas and the availability of specialised technology skills (Acs *et al.*, 2002; Raspe and van Oort, 2007; Doloreux and Shearmur, 2011). Raspe and van Oort explicitly take into account a series of proxies for the knowledge intensity of urban contexts, including educational levels, ICT, and R&D. They find these factors to be strongly related to the growth of knowledge-intensive activities in services in Dutch urban areas.

Kox and Rubalcaba (2007b) find, in fact, that business services tend to employ highly qualified people more often than most other industrial or service sectors, relatively speaking. The European Labour Force Surveys, as reported in Kox and Rubalcaba (2007b), indicate that in European countries the educational profile of employees in aggregate manufacturing and services is dominated by the intermediate educational level. In the manufacturing sector there is also a high proportion of workers with low educational levels, while in the service sector high levels of education prevail over low levels. When we look at the three categories of services considered in this study, both computer services and R&D services show very high shares of highly educated employees.

All in all, the arguments for agglomeration economies put forward by the literature suggest that localisation and urbanisation externalities favour specialisation in BS, which tend to cluster in regions with dense urban areas and a strong functional specialisation in knowledge-intensive and highly skilled activities. Knowledge flows more fluidly where both *spatial* and *sectoral* contiguity are relatively high. While the importance of spatial contiguity has largely been acknowledged in the regional literature, less attention has been devoted to empirically addressing the dimension of agglomeration economies linked to Hirschmann linkages, and to the particular structure of the intermediate demands of BS in general. We turn to this in the next section.

4.2.2 Backward and forward linkages

Reprising Marshall's explicit acknowledgement of consumer and supplier markets as an important dimension of agglomeration economies, Hirschmann (1958) – in a seminal contribution – identifies different types of externalities, depending upon whether activities are related to one another by backward or forward linkages, i.e. whether certain sectors concentrate where their clients are located or, rather, they migrate where new or growing sectors that provide them with inputs are located.[3] Hirschmann's linkages have been reprised in some contributions on local development theories (Jones, 1976; Loviscek, 1982) which focus on the relative strength of the 'inducement mechanisms' (Jones, 1976, p. 323) of particular sectors – the high-linkages ones – in relation to local development.

These aspects of agglomeration economies are particularly important in the context of this work, in which we account for the determinants of regional specialisation in BS from the perspective of agglomeration economies and, in particular, of forward linkages à la Hirschmann.

In the case of business services, the location of customer industries is particularly relevant, as these services are typically supplied to firms through strong supplier–user interactions (Muller and Zenker, 2001; Miles, 2005), which are crucially reliant upon geographical proximity. Consistently with this, Antonietti and Cainelli (2008) find that spatial agglomeration – where the probability of finding specialised external providers, face-to-face contact with clients, and close spatial interaction is high – is an important factor in the location of business services.

The importance of Hirschmann's forward linkages for BS specialisation at the local level is corroborated by a different stream of literature focusing on the role of intermediate versus final demand in explaining the growth of services. Several authors have argued that the rise of services, particularly of business services, in the last thirty years is mostly due to changes in production processes in many sectors and to the ensuing increase in the demand for services as intermediate goods (Francois, 1990; Rowthorn and Ramaswamy, 1999; Guerrieri and Meliciani, 2005; Savona and Lorentz, 2005; Francois and Woerz, 2007). The growing complexity in the organisation of manufacturing production and distribution resulting from new technologies, and the significant increase in coordination problems, has raised the service content of many manufactured goods, which goes well beyond the simple 'outsourcing' or 'contracting out' of services (Ten Raa and Wolff, 2001; Miozzo and Soete, 2001).

Some recent studies investigate the pattern of inter-sectoral linkages between business services and manufacturing. Guerrieri and Meliciani (2005), using Input-Output data, show regularities across countries in the intensity of use of Financial, Communication and Business services (FCB). In particular, they find that knowledge-intensive manufacturing industries make considerable use of FCB services, while labour and scale-intensive industries are, on average, low or medium users of FCB services. Similar results are found by Francois and Woerz (2007), who show how business services are in high demanded, especially by knowledge-intensive industries, including both manufacturing and service activities. Empirical evidence in support of the key role of intermediate demand – rather than final consumption or trade – in business services growth is also provided by Savona and Lorentz (2005) (see also Kox and Rubalcaba, 2007a and 2007b, and Montresor and Marzetti, 2011).

Overall this evidence suggests that the sectoral composition of regional economies and the presence of a high density of forward linkages à la Hirschmann strengthen the effects of urbanisation externalities on regional BS specialisation. We therefore expect regions with a high share of knowledge-intensive manufacturing and service industries to experience a higher demand for business services and, therefore, to be more likely to specialise in these activities.

4.3 The geography of regional specialisation in BS

In order to have a clear view of the spatial distribution of regional specialisation in business services, an index of comparative advantage in business services (employment in business services in region *i* over total employment of region *i* divided by employment in business services for all regions over total employment for all regions) was used to map the EU regions in 2003 (Figure 4.1). The map suggests the presence of an agglomeration pattern in the regional distribution of BS specialisation, with the main exceptions of the capital cities.

Many of the regions which are highly specialised in business services are regions where capital cities are located, in line with the urbanisation literature (Glaeser *et al.*, 1992; Glaeser, 1999). This is the case not only in high income countries, but also in Spain, Portugal, Greece and some new entrant Eastern European countries (Kozép-Magyarorszàg: the region of Budapest; Praha).

When we exclude regions with capital cities, there appears to be some 'country effects' in the spatial map of specialisation in business services. In fact, all of the Dutch regions and many regions of the UK (with some exceptions, especially in the western part of the country) and Germany appear to be highly specialised in these branches. On the other hand, none of the regions from new entrant countries, Portugal, Greece, and Finland (with the exception of regions with capital cities, as mentioned above) show a comparative advantage in business services. Regions in Spain, France and Italy show a more mixed pattern. In particular, Italy shows a north-south divide, while French and Spanish regions, being on average de-specialised, show relatively higher values for specialisation at their borders.

A clear clustering effect emerges in the comparative advantage indicator, as mapped in Figure 4.1, showing that there are factors which explain the sectoral composition of regional employment in BS which seems to spread to neighbouring regions.

We further assess spatial correlation by means of the Moran's I statistic (a measure of global spatial correlation) and the Moran scatterplot (Anselin, 1996). Statistical inference is based on the permutation approach with 10,000 permutations (Anselin, 1995). As for the spatial weight matrix – following, among others Dall'Erba and Le Gallo (2007) – we rely upon a distance matrix using the great circle distance between regional centroids.[4]

Figure 4.2 shows Moran's scatter and reports the associated global Moran's coefficient based upon the distance matrix defined above for the share of employment in business services.

The Moran coefficient indicates a positive and significant spatial autocorrelation in the regional specialisation of BS, although not too high (0.34). A negative contribution to the Moran's I is given by a certain number of capital cities which are highly specialised in BS but are sometimes surrounded by de-specialised regions. Based on the presence of spatial correlation, we adopt a spatial model as described in the next section.

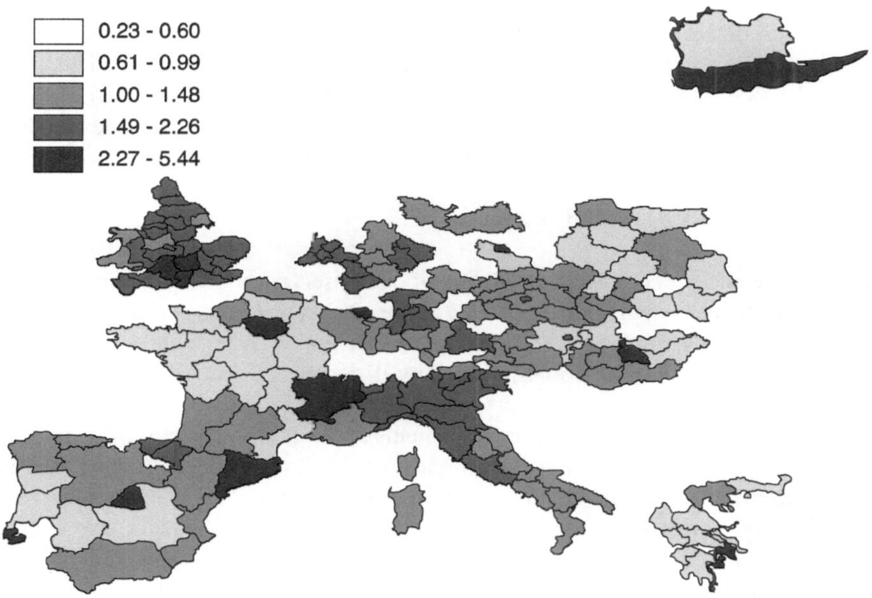

Figure 4.1 Specialisation in business services in EU regions – comparative advantage in 2003

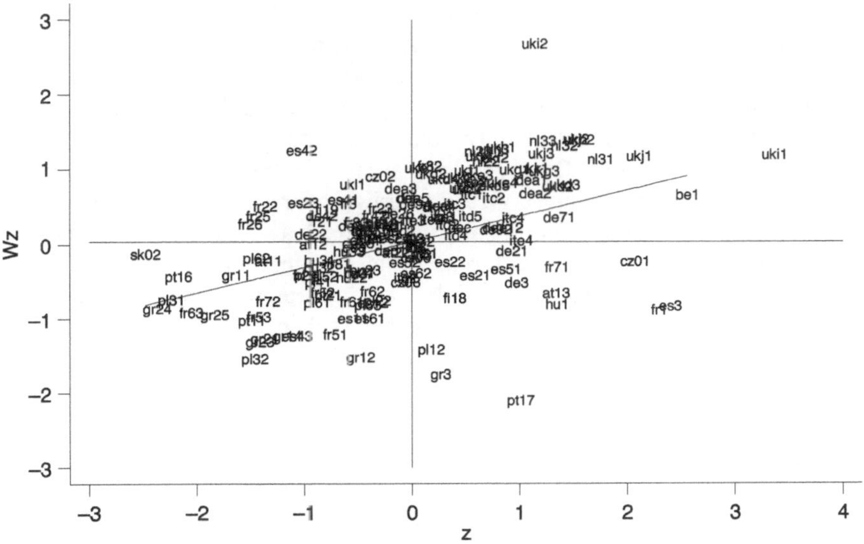

Figure 4.2 Moran scatterplot of specialisation in business services

Source: EUROSTAT Region database

Notes: Distance band between 0.0 and 2.5; z = vector of each the variable in deviation from the regional mean; Wz = vector of spatial lags

4.4 The empirical model

In order to detect the determinants of specialisation in business services allowing for spatial correlation, we estimate a Spatial Durbin Model (SDM). This is a general model that includes amongst the regressors not only the spatial lagged dependent variable, but also the spatial lagged set of independent variables. In the context of panel data, it can be represented as follows:[5]

$$Y_t = \rho W Y_t + X_t \beta_1 + W X_t \beta_2 + \lambda_t e_N + v_t \tag{4.1}$$

where Y_t denotes a $Nx1$ vector consisting of one observation for every spatial unit of the dependent variable in the t-th time period, X_t is a NxK matrix of independent variables, W is an NxN non-negative spatial weights matrix with zeros on the diagonal. A vector or matrix premultiplied by W denotes its spatially lagged value, ρ, β_1 and β_2 are response parameters, and λ_t denotes a time specific effect, which is multiplied by a $Nx1$ vector of units elements and v_t is a $Nx1$ vector of residuals for every spatial unit with zero mean and variance σ^2. This model nests most models used in the regional literature: with $\beta_2 = 0$ we have the spatial autoregressive model (SAR); with $\beta_2 = -\rho\beta_1$ the spatial error model (SEM); with $\rho = 0$ the spatially lagged X regression model (SLX); and finally with $\rho = 0$ and $\beta_2 = 0$ we have a non-spatial regression model. We chose the appropriate model on the basis of hypotheses testing.[6]

Based on the literature discussed in Section 2, our dependent variable (the regional share of employment in business services over total regional employment[7]) is determined by urbanisation economies, Hirschmann forward linkages/intermediate demand, and human capital. As proxies of urbanisation economies we take: i) the share of population over the regional area (population density); and ii) dummies for regions where capital cities are located. Hirschmann forward linkages/intermediate demand coming from manufacturing industries (FLM) and from services industries (FLS) respectively are proxied by the weighted share of employment in manufacturing/service industries that are strong users of business services over total employment. In particular, we take a vector measuring the use of services on output for manufacturing/services sectors that are high BS users and, for each region and year, we multiply it by total employment in each respective manufacturing/service sector; this number is then divided by the region's i total employment in year t:

$$FL_{it} = \frac{\sum_{j=1}^{m} W_j E_{ijt}}{\sum_{j=1}^{n} E_{ijt}}$$

where:
i=region, j=sector, t=time, m=number of high BS users manufacturing/service sectors, n=total number of sectors, E=employment, W=weight given by the

average (across European countries) share of business services in total industry output as computed from Eurostat symmetric Input-Output tables in 2000. The indicator is higher where regional employment in manufacturing/services sectors that are strong users of BS with respect to total regional employment for each year is higher.[8]

Finally, human capital is measured as the share of human resources specialised in S&T, as measured in terms of persons with a university degree (HRSTE).

All variables are in logarithms and the model is estimated for a panel of 164 NUTS2 EU27 regions drawn from the Regio database pooled over the period 1999–2003.[9]

In our spatial regression that includes a spatial lag of the dependent and independent variables, a change in a single explanatory variable in region *i* has a *direct impact* on region *i*, as well as an *indirect impact* on other regions (see LeSage and Fischer, 2008 for a discussion). This result arises from the spatial connectivity relationships that are incorporated in spatial regression models and increases the difficulty of interpreting the resulting estimates. LeSage and Pace (2009) provide computationally feasible means of calculating scalar summary measures of these two types of impacts that arise from changes in the explanatory variables. These routines have been extended by Elhorst (2010) to the panel data model. In this chapter we use the Elhorst (2010) Matlab routines that allow us to compute direct and indirect effects.

4.5 Econometric results

Since all the restrictions were rejected (see tests at the end of the table) we report results based upon the more general model (spatial Durbin). Coefficients, direct, indirect, and total effects of each variable with their asymptotic t-values are reported in Table 4.1.

Looking at the direct effects, all the coefficients have the expected signs and are significant at conventional levels. Urbanisation economies, manufacturing forward linkages and services' forward linkages are all relevant factors in explaining specialisation in business services, as suggested by the literature.

Looking at the variables capturing urbanisation externalities, i.e. the dummy for regions with capital cities and population density, they are both positively related to BS specialisation, showing that urban and densely populated areas create a favourable environment for the localisation of business services. High population density, as well as the presence of urban centres, can also be interpreted as a (final) demand determinant of BS specialisation.

As far as forward linkages are concerned, they emerge as strategic factors favouring a high share of BS at the regional level. This is the case both for linkages between BS and manufacturing sectors and between BS and other service sectors. These results confirm that for business services the location of customer industries is particularly relevant, as these services are typically supplied to firms

Table 4.1 The determinants of specialisation in business services 1999–2003 – Spatial Durbin Model estimates

Variables	*Coefficient*	*Direct effect*	*Indirect effect*	*Total effect*
Regions with capital cities	0.087** (1.930)	0.080* (1.763)	–0.092 (–0.876)	–0.012 (–0.100)
Population density	0.094*** 7.234)	0.094*** (7.394)	–0.001 (–0.045)	0.093*** (2.828)
Manufacturing forward linkages	0.129*** (6.580)	0.135*** (7.151)	0.069* (1.739)	0.204 (4.997)
Services forward linkages	0.941*** (18.431)	0.961*** (18.760)	0.263** (2.002)	1.224*** (8.895)
Share of human resources with science & technology degree	0.351*** (9.684)	0.343*** (9.888)	–0.130** (–2.148)	0.213*** (3.828)
BS specialisation in neighbour's regions	0.467*** (12.043)			
Lagged regions with capital cities	–0.094 (–1.398)			
Lagged population density	–0.045** (–2.338)			
Lagged manufacturing forward linkages	–0.021 (–0.755)			
Lagged services forward linkages	–0.291*** (–3.247)			
Lagged share of human resources with science & technology university degree	–0.239*** (–5.143)			
LM spatial lag	99.25***			
Robust LM spatial lag	7.30***			
LM spatial error	195.48***			
Robust LM spatial error	103.53***			
Wald spatial lag	77.75***			
Wald spatial error	12.23***			
R-squared	0.808			
Log-likelihood	–48.314			
Observations	820			

Notes: *, **, *** indicate significant at the 10 per cent, 5 per cent and 1 per cent levels respectively. Time dummies are included.

through strong supplier–user interactions (Muller and Zenker, 2001; Miles, 2005), which are crucially reliant upon geographical proximity.

Finally, the strategic role played by highly educated human capital for BS specialisation (as suggested by Kox and Rubalcaba, 2007b, among others) is also confirmed by the positive and highly significant coefficient of the variable measuring human resources in science and technology.[10]

Overall, agglomeration factors affecting specialisation in BS within the region are well captured by the direct effects.

Turning to spatial dependence, the highly significant coefficient of the lagged dependent variable (and some lagged independent variables) suggest the presence of clustering effects behind the determinants of BS specialisation. The positive coefficient of the spatial lag confirms the descriptive picture provided by the Moran scatter plot discussed above, and establishes the spatial dependence of BS specialisation.

However, in order to disentangle the contribution made to spatial dependence of each explanatory variable we have to look at indirect effects.

The signs and significance of indirect effects in the spatial Durbin specification provide interesting insights into the different role of spatially lagged independent variables. There are two possible (equivalent) interpretations of these effects. One interpretation (the one that we adopt in our discussion) reflects how changing each explanatory variable of all neighbouring regions by some constant amount would affect the dependent variable of a typical region. LeSage and Pace (2009) label this as the average total impact on an observation. The second interpretation measures the cumulative impact of a change in each explanatory variable in region *i* over all neighbouring regions, which LeSage and Pace (2009) label the average total impact from an observation (see also LeSage and Fischer, 2008).

Interesting results emerge from the indirect effects of our variables. In fact, while being surrounded by highly populated regions and/or capital cities does not affect specialisation in a given region, we find a positive significant role of forward linkages coming from surrounding regions. It appears that, also at the regional level of aggregation, having potential intermediate customers of BS (both manufacturing and services firms) located in areas that are geographically contiguous is a factor that favours one region's specialisation in BS. On the other hand, the negative indirect effect of human resources in science and technology (measured as the share of people with a university degree) indicates the existence of a 'displacing' effect, with service-based activities moving away from the region towards surrounding areas where human resources are concentrated.

Finally, it is interesting to underline that, while the estimated coefficients of the SDM do not differ substantially from the direct effects, the coefficients of spatially lagged variables are misleading (they point to a negative impact of both services' forward linkages and population density) because they do not take into account the whole set of connectivity relationships that are incorporated into the spatial regression model. These can only be assessed by looking at the size and significance of indirect effects.

4.6 Summary of the findings and conclusions

Our study investigates the structural and spatial determinants of specialisation in business services at the regional level. We reprise the traditional Marshallian agglomeration economies explanation of spatial distribution of industries, apply it to the case of BS sectoral specialisation, and extend it in two directions: first, by explicitly accounting for the presence of Hirschmann forward linkages as an important determinant of BS specialisation, and, second, by accounting for the spatial dependence of sources of agglomeration economies in a spatial econometric framework. Most of the conjectures derived from the literature and illustrated in Section 2 are corroborated by the empirical analysis.

In line with the urbanisation literature, we find that business services tend to concentrate in capitals and urbanised areas that ensure high levels of final demand. This has important implications for the evolution of income disparities at the regional level and calls for regional cohesion policies. In fact, while the concentration of valued-added and knowledge-intensive activities in cities may foster regional growth, it could also cause negative externalities in surrounding areas. Also, the hypothesis of BS concentration in regions with a high endowment of skilled human resources is supported by our findings. The particular econometric specification chosen here allows us to disentangle whether the presence of skilled human capital in neighbouring regions also exerts a positive impact upon the typical region specialisation. Interestingly, we find a 'displacing' effect, with services activities moving away from the region towards surrounding areas where human resources are concentrated. This might lead to a vicious cumulative mechanism, whereby BS is driven to concentrate in regions where human capital is relatively easier to access, and at the same time acts as an attractor factor for further human capital to move to regions where the labour market is more likely to absorb skilled resources, reinforcing the polarisation effect mentioned above.[11]

The hypothesis of the particular importance of Hirschmann (forward) linkages as a determinant of BS specialisation is largely supported by our findings. A substantial presence of BS users sectors – both manufacturing and services themselves – affects regional specialisation in business services. This result has important implications since it suggests that a region's ability to develop an efficient and dynamic service economy is linked to the prior structure of knowledge-intensive industries as the main users of business services. As a consequence, regions specialised in these industries are in a favourable position for developing a comparative advantage in business services. Interestingly, we also find that the presence of dense forward linkages in neighbouring regions also proves to be a driving factor of BS specialisation, once again supporting the Marshallian argument of industrial agglomeration economies via forward linkages.

Overall, the 'holy trinity' of Marshallian economies, and the way they are revisited here in order to explain BS regional specialisation within a spatial econometric framework, are corroborated by the evidence. This leads us to infer that specialisation advantages are likely to show that they are counterfactual of de-specialisation disadvantages. Our findings provide supporting arguments for the determinants of regional polarisation of growth – driven by favourable (or unfavourable) sectoral specialisation. This is all the more so as we have highlighted that a local concentration of knowledge tends to be an important determinant of specialisation in BS and is likely to lead to a cumulative – either virtuous or vicious – mechanism, leading to inter-regional as well as intra-regional (cities versus surrounding areas) disparities.

This polarisation effect and the virtuous/vicious mechanisms linked to the location of business services – which, in this work, was found at the national and regional levels – are likely to be further reinforced when we take internationalisation factors into account. As Nefussi and Schwellnus (2010) show, FDI in services tends to 'follow' FDI in manufacturing, due to the Input-Output linkages and complementarity illustrated in our work. Similarly, location choices of MNCs in services tend to co-locate with their high-user manufacturing MNCs, though they are found to be more sensitive to market access and skills rather than cost factors, as in their manufacturing counterparts (Py and Hatem, 2010).

All in all, this chapter aims to add to both the regional economic literature and the literature on structural change and tertiarisation, which has so far lacked a regional perspective. Providing a spatial picture of the determinants of BS specialisation contributes to an understanding of the underlying reasons behind employment productivity and growth divergences at the regional level, in line with other recent contributions addressing these issues (Fagerberg *et al.*, 1997; Guerrieri *et al.*, 2005; Meliciani, 2006; Verspagen, 2007; Sterlacchini, 2008, among others), which are likely to become more important after the enlargement to Eastern EU countries.

Although a detailed discussion of this issue is outside the scope of this chapter, the analysis presented here is also aimed at informing European industrial and innovation policies, which are increasingly designed at the regional level – and rightly so (Verspagen, 2007).

Notes

1 There is little consensus on the exact boundaries of Business Services, though they have been defined as 'a set of service activities that – through their use as intermediary inputs – affect the quality and efficiency of the production activities, by complementing or substituting in-house service functions' (Kox and Rubalcaba, 2007a, p. 4). In this study BS include computer and related services (ISIC 72), R&D (ISIC 73) and other business services (ISIC 74).

2 For a recent reassessment of agglomeration theories in a historical perspective, see McCann and van Oort (2009).

3 According to Hirschmann (1958, p. 100), quoted in Jones (1976):

> The input-provision, derived demand, or *backward linkage effects*, i.e. every non primary economic activity, will induce attempts to supply through domestic production the inputs needed in that activity. The output-utilisation or *forward*

linkage effects, i.e., every activity that does not by its nature cater exclusively to final demands, will induce attempts to utilise its outputs as inputs in some new activities.

4 The functional form and the cut-off distance are chosen relying on *a priori* considerations on the scope of spatial spillovers in our sample and on comparisons of the overall explanatory power of the model (as measured by the R-squared and Log-likelihood) estimated with different spatial matrices as suggested by Lee (2009). Since our regions are already large (NUTS 2) we choose the minimum bandwidth which allowed each region to have at least one neighbour and we take the inverse of the distance (this is the matrix that maximises the R-squared and Log-likelihood in regression analysis). We also test for robustness using larger distance bands and the inverse of the squared distance ($k = 2$).

5 Elhorst (2005) presents a more general panel model, which also includes fixed effects and a dynamic specification. Due to the short time series available (1999–2003), we treat the data as a repeated cross-section (pooled estimation).

6 Lagrange Multiplier tests and their robust versions are used to test the OLS versus the SAR and SEM, Wald tests are used for testing the SAR and SEM versus the SDM, while the test of the SLX versus the SDM is a t-test on the coefficient of the spatial lag of the dependent variable.

7 From now on we omit the explanation of the suffix i and t which refer respectively to the region and to the time period. All the regressors included in the equation specification refer to region i at time t.

8 Coefficients are obtained by regressing the share of business services in total output on industry dummies for all European countries included in the analysis for the year 2000. Manufacturing sectors that make considerable use of business services are all (with the exception of tobacco products) knowledge-intensive industries (printed matter and recorded media; chemicals and chemical products; office machinery and computers; radio, television and communication equipment and apparatus; medical, precision and optical instruments; watches and clocks), while labour and scale-intensive industries appear, on average, to be low or medium users of business services. Among service industries, above average users are the BS sectors themselves, while average users are mainly finance, insurance, post and communications, wholesale and retail trade. All these industries have a share of BS that is comparable to that of above average manufacturing users. Therefore, we include in the indicator of potential demand coming from service industries all service industries that are average BS users, while we do not consider above average users since these are the BS sectors themselves. Regression results are available on request.

9 The regions belong to the following countries: Austria, Belgium, Czech Republic, Germany, Spain, Finland, France, Greece, Hungary, Italy, Netherlands, Poland, Portugal, Slovakia and United Kingdom. Only regions for which there was enough data to construct a balanced sample by interpolating missing values were included.

10 We also introduced in the regressions the share of students enrolled in tertiary education over total population, but this variable was not significant, showing that what really matters for BS is education in science and technology and not general human capital.

11 With the data at our disposal, we are not able to test migration flows across regions and therefore add to the literature on brain drain and gain at the regional level. We leave this for our research agenda.

References

Acs, Z.J., Anselin, L. and Varga, A. (2002) 'Patents and innovation counts as measures of regional production of new knowledge', *Research Policy*, 31(7): 1069–85.

Anselin, L. (1995) 'Local indicators of spatial association–LISA', *Geographical Analysis*, 27(2): 93–115.

Anselin, L. (1996) 'The Moran scatterplot as an ESDA tool to assess local instability in spatial association', in M. Fisher, H.J. Scholten and D. Unwin (eds) *Spatial analytical perspectives on GIS*, Taylor & Francis, London.

Antonietti, R. and Cainelli, G. (2008) 'Spatial agglomeration, technology and outsourcing of knowledge intensive business services. Empirical insights from Italy', *International Journal of Service Technology and Management*, 10(2–4): 273–98.

Burger, M.J., van Oort, F.G. and van der Knaap, B. (2008) 'A treatise on the geographical scale of agglomeration externalities and the modifiable areal unit problem', *Erasmus Research Institute on Management*, ERS-2008–076.

Cainelli, G., Evangelista, R. and Savona, M. (2006) 'Innovation and economic performance in services. A firm level analysis', *Cambridge Journal of Economics* 30(3): 435–58.

Combes, P.P. (2000) 'Economic Structure and Local Growth: France 1984–1993', *Journal of Urban Economics*, 47(3): 329–53.

Crescenzi, R., Rodriguez-Pose, A. and Storper, M. (2007) 'The territorial dynamics of innovation: a Europe–United States comparative analysis', *Journal of Economic Geography*, 7(6): 673–709.

Dall'Erba, S. and Le Gallo, J. (2007) 'Regional convergence and the impact of European structural funds over 1989–1999: a spatial econometric analysis', Department of Geography and Regional Development, University of Arizona Working Paper.

Desmet, K. and Fafchamps, M. (2005) 'Changes in the spatial concentration of employment across US counties: a sectoral analysis 1972–2000', *Journal of Economic Geography*, 5(3): 261–84.

Doloreux, D. and Shearmur, R. (2011) 'Collaboration, information and the geography of innovation in knowledge intensive business services', *Journal of Economic Geography*, Advance access; published April 8, 2011: 1–27.

Duranton, G. and Puga, D. (2000) 'Diversity and specialisation in cities: Why, where and when does it matter?', *Urban Studies*, 37(3): 533–55.

Duranton, G. and Puga, D. (2005) 'From sectoral to functional urban specialisation', *Journal of Urban Economics*, 57(2): 343–70.

Elhorst, J.P. (2005) 'Models for dynamic panels in space and time: An application to regional unemployment in the EU', Paper presented at the European Regional Science Association, Amsterdam.

Elhorst, J.P. (2010) 'Matlab software for spatial panels', Paper presented at the IVth World Conference of the Spatial Econometrics Association (SEA), Chicago, June 9–12.

Evangelista, R. (2006) 'Innovation in the European service industries', *Science and Public Policy*, 33(9): 653–68.

Fagerberg, J., Verspagen, B. and Caniëls, M. (1997) 'Technology, growth and unemployment across European regions', *Regional Studies*, 31(5): 457–66.

Faggian, A. and McCann, P. (2009) 'Human capital and regional development', in R. Capello and P. Nijkamp (eds) *Handbook of Regional Growth and Development Theories*, London, Edward Elgar.

Francois, J.F. (1990) 'Producer services, scale, and the division of labor', *Oxford Economic Papers*, 42(4): 715–29.

Francois, J.F. and Woerz, J. (2007) 'Producer services, manufacturing linkages, and trade', *Tinbergen Institute Discussion Paper*, TI 2007 – 045/2.

Frenken, K., Van Oort, F.G. and Verburg, T. (2007) 'Related variety, unrelated variety and regional economic growth', *Regional Studies* 41(5): 685–97.

Fujita, M., Krugman, P. and Venables, A.J. (1999) *The Spatial Economy. Cities, Regions, and International Trade*, Cambridge, MA, MIT Press.

Glaeser, E.L. (1999) 'Learning in cities', *Journal of Urban Economics*, 46(2): 254–77.

Glaeser, E.L., Kallal, H.D., Scheinkman, A. and Shleifer, A. (1992) 'Growth in cities', *Journal of Political Economy*, 100(6): 1126–52.

Glaeser, E.L., Scheinkman, J.A. and Schleifer, A. (1995) 'Economic growth in a cross-section of cities', *Journal of Monetary Economics*, 36(1): 117–43.

Guerrieri, P. and Meliciani, V. (2005) 'Technology and international competitiveness: The interdependence between manufacturing and producer services', *Structural Change and Economic Dynamics*, 16(4): 489–502.

Guerrieri, P., Maggi, B., Meliciani, V., Padoan, P.C. (2005) 'Technology diffusion, services and endogenous growth in Europe. Is the Lisbon strategy useful?', *Rivista di Politica Economica*, 95: 271–317 and International Monetary Fund Working Paper (WP/05/103).

Henderson, J.V., Kuncoro, A., and Turner, M. (1995) 'Industrial development in cities', *Journal of Political Economy*, 103(5): 1067–90.

Hirschmann, A.O. (1958) *The strategy of economic development*, New Haven, Connecticut and London: Yale University Press.

Jacobs, J. (1969) *The Economy of Cities*, London, Jonathan Cape.

Jones, L.P. (1976) 'The Measurement of Hirschmanian Linkages', *Quarterly Journal of Economics*, 90(2): 323–33

Karlsson, C., Johansson, B. and Stough, R.R. (2009) 'Human capital, talent and regional growth', *CESIS Electronic working paper series 191*, The Royal Institute of Technology, Sweden.

Kox, H.L.M. and Rubalcaba, L. (2007a) 'Analysing the contribution of business services to European economic growth', *MPRA Paper* n. 2003.

Kox, H.L.M. and Rubalcaba, L. (2007b) 'Business services and the changing structure of European economic growth', *MPRA Paper* n. 3750.

Lee, S.Y. (2009) 'Bias from misspecified spatial weight matrices in SAR models: theory and simulation studies', Department of Economics, San Francisco State University.

LeSage, J.P. and Fischer, M.M. (2008) 'Spatial growth regressions. model specification, estimation and interpretation', *Spatial Economic Analysis*, 3(3): 1742–80.

LeSage, J.P. and Pace, R.K. (2009) *Introduction to spatial econometrics*, Taylor & Francis CRC Press, Boca Raton.

Loviscek, A.L. (1982) 'Industrial cluster analysis-backward or forward linkages?', *The Annals of Regional Science*, 16(3): 36–47.

McCann, P. and van Oort, F.G. (2009) 'Theories of agglomeration and regional economic growth: a historical review', in R. Capello and P. Nijkamp (eds) *Handbook of Regional Growth and Development Theories*, London, Edward Elgar.

Marshall, A. (1920) *Principles of Economics*, Library of Economics and Liberty.

Meliciani, V. (2006) 'Income and employment disparities across European regions. The role of national and spatial factors', *Regional Studies*, 40(1): 75–91.

Midelfart-Knarvik, K.H., Overman, H.G., Redding, S.J. and Venables, A.J. (2000) 'The location of European industry', Report prepared for the Directorate General for Economic and Financial Affairs, European Commission.

Miles, I. (2005) 'Innovation in services', in J. Fagerberg, D. Mowery and R. Nelson (eds) *The Oxford Handbook of Innovation*, Oxford University Press, Oxford.

Miozzo, M. and Soete, L. (2001) 'Internationalization of services: a technological perspective', *Technological Forecasting and Social Change*, 67(2): 159–85.

Montresor, S. and Marzetti, G.V. (2011) 'The deindustrialization/tertiarisation hypothesis reconsidered. A sub-system application to the OECD7', *Cambridge Journal of Economics*, 35(2): 401–21.

Moreno, R., Paci, R. and Usai, S. (2005) 'Innovation clusters in the European regions', CRENoS, University of Cagliari Working Paper 2003/10.

Muller, E. and Doloreux, D. (2009) 'What we should know about knowledge-intensive business services', *Technology in Society*, 31(1): 64–72.

Muller, E. and Zenker, A. (2001) 'Business services as actors of knowledge transformation: the role of KIBS in regional and national innovation systems', *Research Policy*, 30(9): 1501–16.

Nefussi, B. and Schwellnus, C. (2010) 'Does FDI in manufacturing cause FDI in business services? Evidence from French firm-level data', *Canadian Journal of Economics*, 43(1): 180–203.

Parrinello, S. (2004) 'The service economy revisited', *Structural Change and Economic Dynamics*, 15(4): 381–400.

Peneder, M. (2003) 'Industrial structure and aggregate growth', *Structural change and economic dynamics*, 14(4): 427–48.

Peneder, M., Kaniovsky, S. and Dachs, B. (2003) 'What follows tertiarisation? Structural change and the role of knowledge-based services', *The Service Industry Journal*, 23(2): 47–66.

Py, L. and Hatem, F. (2010) 'Internationalisation et localisation des services: une analyse sectorielle et fonctionnelle appliquee aux rmes multinationales en Europe', *Economie et Statistique*, 426: 67–95.

Raspe, O. and van Oort, F.G. (2007) 'The knowledge economy and urban economic growth', *Papers in Evolutionary Economic Geography*, 0607.

Rowthorn, R. and Ramaswamy, R. (1999) 'Growth, trade and deindustrialisation', *IMF Staff Papers*, 46(1): 18–41.

Savona, M. and Lorentz, A. (2005) 'Demand and technological contribution to structural change and tertiarisation. An Input-Output structural decomposition analysis', *LEM Working Paper Series*, 2005/25.

Schettkat, R. and Yocarini, L. (2006) 'The shift to services employment: A review of the literature', *Structural Change and Economic Dynamics*, 17(2): 127–47.

Sterlacchini, A. (2008) 'R&D, higher education and regional growth. Uneven linkages among European regions', *Research Policy*, 37(67): 1096–1107.

Ten Raa, T. and Wolff, E. (2001) 'Outsourcing of Services and the productivity recovery in U.S. manufacturing in the 1980s and 1990s', *Journal of Productivity Analysis*, 16(2): 149–65.

Van Oort, F.G. (2004) *Urban Growth and Innovation*, Aldershot Ashgate.

Van Oort, F.G. (2007) 'Spatial and sectoral composition effects of agglomeration economies in the Netherlands', *Papers in Regional Science*, 86(1): 5–30.

Verspagen, B. (2007) 'The spatial hierarchy of technological change and economic development in Europe', *UNU-MERIT Working Paper Series*, 012.

Part II
Innovation

5 Technological competences and regional innovation networks

Measurement and visualisation with patent data

Andreas Pyka, Bernd Ebersberger and Dominik Hartmann

5.1 Introduction

Knowledge, agglomeration effects and innovation are decisive drivers of regional development. Many qualitative and descriptive analyses on innovative milieus, clusters and regional innovation systems have been undertaken over the last decades (e.g. Camagni, 1991; Cooke, 1998; Saxenian, 1994; Breschi and Lissoni, 2003). However, few contributions elaborate upon the quantitative methodologies for measuring the structures and dynamics of regional systems (e.g. Cantner *et al.*, 2009), which would be necessary for a desirable comparative approach. The existent taxonomies on regional measurement of innovation (e.g. Hollanders *et al.*, 2009a) consider some widely available innovation outputs and inputs, but do not properly include the core features of regional systems, such as network structures and the sectoral technological competences and the processes through which they evolve. To improve upon this, recent work focuses upon the evolutionary foundations of the regional systems of innovation approach (Frenken and Boschma, 2007; Boschma and Martin, 2007, 2010; Uyarra, 2010).

In order to contribute to the elaboration of quantitative methodologies, in this chapter we discuss if and how patent data can be employed to measure technological competences and innovation networks for medium developed regions, which are not global innovation hubs such as Silicon Valley, Bangalore or Baden Württemberg, but which do show some important technological skills and rank in the middle of industrialised countries with regard to their innovative performance (e.g. measured by the European Regional Innovation Scoreboard; Hollanders *et al.*, 2009b). The pragmatic reason for using patent data is their availability for most regions of the OECD. Further advantages of patent data are evident in the multi-faceted information they provide: they not only include the postal codes of the inventors and applicants in order to identify the regions, but also the technological field the patent belongs to (OECD, 2009). Patents therefore allow for a sorting of inventors with regard to regions as well as technology classes. However, it is well known that patent data are also characterised by a series of shortcomings.

While they seem to be a good means of measuring the technological specialisations and advantages of regions, the capacity to draw a comprehensive picture of the regional innovation networks is limited, at least for medium to low developed regions, because of varying propensities to patent in different sectors (see Griliches (1990), Archibugi (1992), and Smith (2005) for a general discussion of the limits of patents as indicators of technological change). Nevertheless, some useful insights on the networks of inventors, applicants and technology fields can be derived.

We discuss the strength and weaknesses of patent data by analysing the case of Bremen, a 'Bundesland' and two-city state in the north of Germany. Bremen is an interesting case because, on the one hand, it suffered from typical structural problems, such as corporate downsizing and job losses in old industries (Audretsch and Thurik, 2000). On the other hand, however, the region profited from creative regional policy designs with a focus on cooperation and technological concentration (Stenke and Willms, 2006) to prepare Bremen for the transformation towards a knowledge-based economy.

In section 2, we discuss the importance of knowledge, innovation networks and technological competences for regional development. Section 3 blueprints a methodology to reveal the technological profiles of regions via patent data, and section 4 applies the methodology to the case of Bremen. Section 5 concludes.

5.2 Innovation and regional development

5.2.1 The regional dimension in the knowledge-based economy

What is a 'region'? So far, there is no specific and generally accepted definition of a region (e.g. Cooke and Memedovic, 2003). However, a certain common understanding concerning the geographical nature of the term, as well as the need to identify criteria which allow the drawing of boundaries between regions, can be found in the literature. Regions are characterised by a certain internal cohesion, as well as by differences in their environment which allow a differentiation from their bordering regions. Accordingly, the term region indicates a spatial area which has at least one homogenous criterion. Criteria might be a common administration, a geographic setup, a specific sectoral structure, or regional traditions, rules and norms. The size and boundaries of a region depends upon the selected criteria of homogeneity and heterogeneity. Due to endogenously created novelties, exogenous shocks, or structural changes, the boundaries and roles of regions changes over time, new regions emerge and old ones perish (e.g. Cooke and Memedovic, 2003).

Is the relevance of regions diminishing due to the emergence of ICT, multinational enterprises, global market places, open innovation and the knowledge-based economy? Indeed, internationalisation and new forms of transport and communication have led to profound changes in the global production of knowledge, of goods, and of services, as well as the coordination and interaction between different economic activities. Economies of scale have lessened in importance, while the role of entrepreneurial activities, innovation networks, and knowledge has gained momentum (Castells, 1996; World Bank, 1999; Audretsch and Thurik, 2000). The higher transferability and the importance of the accumulation of knowledge lead to both centripetal and centrifugal forces. On the one hand, multinational enterprises are outsourcing standardised and labour intensive production processes from high to low wage countries and regions. On the other hand, agglomeration effects and the cumulative and tacit character of knowledge favour a higher regional concentration of R&D and marketing activities, competitiveness and growth.

Hence, despite globalisation, ICT, and open innovation, regions are the places where innovation takes place, where competences are located in R&D departments of firms and research institutes, and where collective learning within closely connected network occurs. Regions with a profound knowledge infrastructure and a prolific innovation system based upon interactive learning are more likely to benefit from knowledge-intensive and high value added production processes and receive sophisticated R&D investments, thus allowing for the creation of a greater number of higher paid jobs.

5.2.2 Technological competences, sectoral development and internationalisation

The quantity and quality of investment attracted into a region, the quality of jobs being created, and the competitive position of the enterprises in a region in the global value chains are determined by the technological competences of that region. As Cantwell points out (2005, p. 561) 'competitiveness derives from the creation of the locally differentiated capabilities needed to sustain growth in an internationally competitive selection environment'.

Therefore, a strategic orientation concerning the creation and attraction of technological competences in future-oriented sectors is a necessary precondition to enable the virtuous circles of knowledge, innovation, high quality jobs, investment and competitiveness. It is of decisive importance for a sectoral adjustment aiming at the development of long-run growth potentials to (i) focus on endogenous capabilities and opportunities; (ii) not stay for too long on exhausted trajectories; and (iii) not reject promising new trajectories too early (Eliasson, 2005). Regional innovation policy is employed to find out what are the future technological strengths to be furthered and to identify relevant techno-economic opportunities. Also, the regional considerations matter for firm policies: due to the accelerated global innovation competition and rising complexity of value chains, an increasing activity of multinational enterprises (MNEs) engaged in asset-augmenting activities is observable. Today, MNEs not only concentrate on exploiting their competitive advantages, entering new markets and finding cheaper production places (asset-exploiting activities), but are increasingly engaged in R&D investments in foreign locations (asset-augmenting activities), in order to access complementary location-specific advantages, create completely new technological assets, enlarge their competitiveness and receive new impulses for their Corporate Innovation Systems (Dunning and Narula, 1995; Cantwell *et al.*, 2004;

Narula and Zanfei, 2005). The dimension of high-quality investment (both in terms of inflows as well as in expected returns on investment) has become less national and increasingly international, and at the same time regional. Without doubt, the technological competences of regions are decisive factors for attracting qualitative investment from national as well as international enterprises.

5.2.3 Innovation networks

Innovation is an interactive and collective process, which involves a varied range of actors from the public and private sectors. In the organisation of these complex knowledge creation and diffusion processes, innovation networks have proved to be advantageous (Pyka, 1999, 2002). Modern technical solutions are characterised by an increased interrelatedness between heterogeneous actors and knowledge fields. No single firm can keep pace with the development of all relevant technologies, and therefore firms have to seek access to external knowledge sources. In this respect, regional and international innovation networks of public and private actors have gained significant importance as a means of co-ordination of research and development (R&D) processes and competitiveness upgrading of regions and enterprises.

Because of the specific features of knowledge which are exchanged in the innovation networks (e.g. its technological local character, the tacitness as well as the cumulativeness), besides technological proximity (i.e. absorptive capacities) geographical proximity also plays an important role for an effective knowledge transfer and mutual learning in networks. Furthermore, the role of informal innovation networks for competence development and learning is stressed in the literature (Pyka 1997, 2000). Informal networks are also characterised by an important regional dimension, as they depend strongly on close personal (often face-to-face) contacts. In this light, informal networks provide channels for the exchange of regional knowledge.

5.3 Measurement and visualisation of technological competences and innovation networks of regions with patent data

Globalisation, structural change and the rising importance of the knowledge factor has led to substantial opportunities and threats for regions. All over the world, innovative regions forge ahead successfully, while less innovative ones lag behind for various reasons. However, the comparison of different regions and their dynamics, on a quantitative empirical basis, is rather difficult: first, qualitative differences in the design of the regional innovation systems often allows for descriptive approaches only. Second, data availability with respect to innovation activities already on the level of the nation state is difficult, and often the respective information cannot be found for the regional level. However, patent information is almost globally available and the information stored in a patent allows them to be broken down to the regional level.

5.3.1 Revealed Technological Advantage

The patterns of technological specialisation and competences of regions and countries can be measured by the so-called Revealed Technological Advantage Index, developed and used first by Soete (1987), Cantwell (1989), and Patel and Pavitt (1991). The indicator was constructed by reformulating Balassa's Revealed Comparative Advantage Index (Balassa, 1965) using patent data instead of export data. The regional Revealed Technological Advantage (RTA) can be defined as a region's share of all patents in a particular industry, relative to its share of all national patents (Cantwell *et al.*, 2004). An above (under) average amount of patents in a region in a certain industry indicates a high (low) technological competence of the region relative to the national profile. Denoting P_{ij} as the number of patents of the host region *j* in the industry *i*, the RTA for each region in that industry is defined as follows:

$$RTA = \frac{P_{ij} / \sum_j P_{ij}}{\sum_i P_{ij} / \sum_{ij} P_{ij}}$$

We normalise the revealed technological advantage RTA^n between −1 and 1: $RTA^n = (RTA - 1)/(RTA + 1)$.

A positive value suggests a comparative technological advantage in an industry; a negative value indicates a comparative technological disadvantage. Of course, the question arises whether patents are an adequate indicator for the technological competence profile of a region (e.g. Griliches, 1990; Archibugi and Pianta, 1992; Smith 2005). Obviously, not all the technological activities which shape the competence profile of a region can be found in patent statistics. Nevertheless, Hagedoorn and Cloodt (2003) show that patent statistics are a good indicator of innovation activities. In addition to being easily available, patents indicate the capability to appropriate returns from new knowledge and give a clue to the technological domains in which the firms in a region are active. For mapping the technological competences of a region we assume that competences can be approximated by the patents belonging to firms who have the ability to create, modify, imitate and implement technologies. For our regional analysis, patents show a further advantage: as each patent contains information on the persons involved in generating the invention as well as their residential addresses, one can identify the geographical location of the knowledge and expertise. This information can be used to map knowledge onto countries and regions based on international country identifiers and postal codes.

5.3.2 Innovation networks

Networks in general consist of nodes and edges which link the different nodes. In our regional innovation networks nodes are actors and their knowledge; the links between them are established by co-patenting, co-inventing and co-classification. Besides the indicator-based quantitative and statistical social network analysis, qualitative tools to visualise networks are also developed (e.g. Pyka and

Scharnhorst, 2008). For the purpose of this chapter we draw on a visualisation of the regional patent-based innovation network of Bremen. The visualisation allows us to disentangle the complex architecture of the networks stemming from the various and multifaceted character of the links.

Based on the information available in patent data, it is possible to construct various indicators for innovation networks: (i) in the case that more than one inventor is listed in a patent, these agents very likely collaborated in generating and developing the new knowledge covered in the patent application. Hence, this collaborative relationship has successfully contributed to the generation of new knowledge because, a third actor – the applicant – invested resources to attain the intellectual property rights; (ii) The inventor's relationship with the applicant is also an important feature of an innovation network as the location of competences is an essential information for managing successfully innovation projects: the inventor-applicant relationship can be described in a graphical representation of the network with two different types of nodes: the inventors and the applicants. Obviously, both play different roles in the generation, application and commercialisation of the new knowledge; (iii) Where more than one applicant is listed in the patent, more than one firm is engaged in achieving legal property rights for the new knowledge. This suggests a technology-related formal collaboration between the two applicants; (iv) The technological competences of actors can be identified with the technological classification in a patent document (IPC codes). Where more than one technological field is given to classify a patent document, a relationship can be established between these two fields (Saviotti, 2009). This relationship documents that the patent generated new knowledge which created an interface between two different knowledge fields.

To summarise, in a graphical representation of a regional patent-based innovation network the following linkages between actors and technology fields can be visualised:

- technological fields can be linked to each other (co-classification),
- technological fields can be linked to inventors,
- inventors can be connected to each other (co-invention),
- inventors can be connected to applicants,
- applicants can be connected to each other (co-application).

A combination of three different nodes (technologies, inventors, applicants) leads to a so-called tripartite network.

5.3.3 Degree of internationalisation

The patent documents contain information on the geographic location of both the applicants and the inventors. This allows for an analysis of the internationalisation of the innovation system of Bremen. Patents which are internationally applied for and are attributed to a regional inventor can be interpreted as regional inventions with international impact. Local and international applicants of a patent indicate national and international exploitation of knowledge. Patents which show two or more inventors from different countries indicate international collaboration and exchange of knowledge. On the contrary, regional knowledge production and application indicates a smaller degree of embeddedness within the international innovation system.

5.4 The case of Bremen

The socioeconomic setup and specific characteristics of the regions under study have to be taken into account in order to obtain reliable results and derive appropriate policy conclusions. Therefore we briefly present the main historical and structural characteristics of Bremen, before analysing the technological competences and network structures of Bremen.

5.4.1 Structural change and innovation policy in Bremen

The twin sister cities Bremen and Bremerhaven form a small 'Bundesland' in the north of Germany. Historically, the North Sea, fishing, and harbour activities have shaped the economy and life of the people in Bremen. Although Bremen lies some 60 kilometres away from its smaller twin sister Bremerhaven and the estuary mouth of the North Sea, a maritime atmosphere and spatial proximity to the sea is omnipresent. A series of historical sites indicate storage, shipping and trading activities. Bremerhaven has one of the biggest container terminals in the world and the Bremenports together are the third most important harbour in Europe after Rotterdam and Hamburg. Over time, enormous amounts of raw materials entered Europe through Bremen. Based upon its shipping and trading activities, it became a decisive logistics hub and diversified into related sectors such as the food industry. For example some important coffee brands (e.g. Haag and Jacobs), Milka chocolate, Philadelphia cheese, Becks beer and many other products and brand names have been and are processed or designed in Bremen.

Due to rising global and national competition, Bremen entered the 1980s and 1990s with deep *structural problems*, such as bankruptcies, rising debts, and skyrocketing unemployment rates. Negative culmination points have been the bankruptcies of the shipyards AG Bremen in 1983 and Bremer Vulkan AG in 1996. The local administration of Bremen had virtually no other choice than to promote the creation of jobs in new and knowledge-intensive activities. Despite scarce financial resources – partly received from national government subventions for less developed regions in Germany, and later from European Structural Funds for regions with structural problems – policy actors in Bremen did quite a successful job in concentrating resources, building up competence clusters, and fostering cooperation between private and public actors in new activities such as aerospace, material sciences, ICT and wind energy. Today, Bremen also domiciled the production of parts for the international space station and Airbus. Furthermore, it has one of the most successful technology parks in Germany, with currently over 6,000 employees in a concentrated spatial area close to the University of Bremen.

Of course, problems persist – such as a very huge unemployment and public debt rate – but some important steps have been undertaken to make Bremen ready for the knowledge-based economy and create new opportunities for growth and jobs.

Several policy programmes (such as the Innovation Programme for Bremen BIP 1984/1987, BIP 1988/1995 and the Special Investment Programme ISP 1994–2004; for an overview see Stenke and Willms, 2006), placed specific emphasis on the technological infrastructure, the attraction of leading research institutions (such as the Fraunhofer Institute for Manufacturing Technology and Applied Materials Research in Bremen), the concentration of efforts and clustering, as well as technology transfer between different actors from the public and private domains.

Recently, the programme Innovision 2010 aimed to foster – in line with the European Lisbon strategy – the knowledge-based competitiveness of Bremen through clustering and innovation networks. Logistics, Aerospace, New Materials, Environment and Energy, ICT, Biotechnology and Health have been identified as innovation fields with considerable technological capabilities and potential in Bremen. A Fraunhofer study about the regional development potentials in the Bremer Innovation Fields (Koschatzky *et al.*, 2004), as well as recent policy papers of the Delegate for Technology of the Bremer Senate (Timm, 2007), placed strong emphasis on knowledge spillovers and innovation opportunities at the interfaces of the innovation fields. The following priority areas for action at the interfaces between the innovation fields have been identified: mobile solutions, eLogistics, innovative materials, ecological intelligence, and the future market for (e-)health.

For several reasons, Bremen is a valuable case study. First, we can observe the typical structural problems and new opportunities of many industrialised regions throughout the 1980s and 1990s, with corporate downsizing, production outsourcing, bankruptcies and job losses in traditional manufacturing activities on the one hand, and new opportunities and jobs in services, high-tech start-ups and knowledge-intensive activities on the other hand (e.g. Audretsch and Thurik, 2000). Second, Bremen actively engaged in common policy approaches in order to foster regional competitiveness, such as cluster creation, upgrading of the technological infrastructure and the promotion of knowledge transfer and innovation networks. Third, Bremen – like many other regions in the OECD – is not a cutting edge technology and entrepreneurship region such as Silicon Valley, Cambridge, Route 128, but it definitely shows some significant technological strengths and possibilities, which can be identified as promising targets for policy intervention.

However, in order to evaluate the results of our case study of Bremen and the appropriateness of using patent data to analyse medium developed regions, we also need to consider the specific characteristics which distinguish Bremen from most other cases. First, Bremen is an important harbour, logistics and trading hub. But despite its high degree of internationalisation, the spatial boundaries and commuting patterns of the two-city state of Bremen can be rather easy defined.

Reasons for that are its small size and specific spatial setup. The cities Bremen and Bremerhaven total some 661,000 people and are surrounded by an area with rather low population density. Just a few people of Bremen work in the minor cities and villages of the surrounding areas or commute to other larger cities such as Oldenburg (50 km distance), Hamburg (110 km) or Hannover (120 km). Furthermore, the fact that Bremen and Bremerhaven form their own German Federal State *Bundesland* Bremen affords the local policy-makers higher autonomy and greater influence than in many other regions.

5.4.2 Data, regional boundaries and commuting

In this section we outline the technological competences of Bremen based on the patent applications which were filed at the European Patent Office (EPO) in the year 2002.[1] As patents bear the residential location of the inventor, the Bremen inventors are identified based on the German postal codes for the city of Bremen. Supposing that a proportion of the inventors and the human capital which can be attributed to Bremen is not resident within the narrow city boundaries, but also in the surrounding area, we include these by fractional accounting. We assume that the city of Bremen is the major commuting target for people living in the surrounding areas which are part of Lower Saxony. This assumption is consistent with the predictions of gravity models of commuting (see e.g. Thorsen and Gitlesen, 1998). The probability of commuting decreases with the distance of the target location. We further assume that this probability is Poisson distributed with mean of 22 kilometres, where the mean distance is retrieved from OECD (2007) estimations of German commuting patterns in rural areas. The probability of commuting is used for fractional weighting of identified patents from source locations other than the city of Bremen. Figure 5.1 illustrates the contribution of locations to the overall patent count for Bremen.

In total we identify roughly 240 patents, upon which the following analysis is based. With this restricted number of available patents, the analysis serves illustrative purposes only. Nevertheless, it shows what can be done with the analysis of patent data and illustrates the value of these types of analysis to policy-makers as it establishes linkages to current policy challenges.

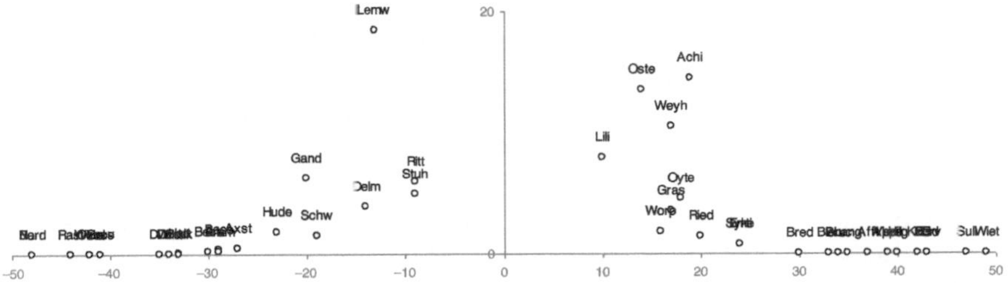

Figure 5.1 Regional municipalities and their contribution to the overall patent score

5.4.3 Technological competences of Bremen

In the next step we use the patent data to calculate the technological advantages and disadvantages, relative to national as well as international settings; in our case relative to Germany and to the OECD. This is useful in order to identify and consider 'local' strengths as well as international (dis-)advantages of regions when elaborating a taxonomy to compare and identify the innovative performance of regions (e.g. Hollanders, 2002, 2003, 2007; Hollanders *et al.*, 2009a).

5.4.3.1 Technological advantages relative to Germany

In order to investigate the technological competences of Bremen we compute the revealed technological advantage on the basis of International Patent Classification (IPC) classes. For our first overview we determine the technological advantage relative to Germany. This suggests that the findings can be interpreted as technological advantage/disadvantage on the national level. In a sense it captures the technological competitiveness of the analysed region – Bremen, in our case – in the German market for technologies.

First we provide an overview of the technological areas where Bremen enjoys advantages. In the second step we investigate the technology fields where Bremen reveals a technological disadvantage. In Figure 5.2 we display the technology fields with a positive RTA in descending order. Hence, the uppermost technology field – D02 – is the one with the most distinct technological advantage.

We find that the areas of innovation policy in Bremen match quite well with the empirically identified Revealed Technological Advantages (RTAs). The RTAs of Bremen can be directly connected to the innovation areas defined by the diverse policy programmes installed in Bremen in recent years. Some of the fields (most importantly logistics, aerospace and food industry) can be found with an RTA above the German average which indicates a pronounced specialisation in Bremen. In particular, the technological competences and overarching importance of logistics and related sectors for Bremen becomes evident. RTAs in transporting (B63: Ships, B64: Aircraft, aviation, cosmonautics, B65: Conveying, Packing, Storing), lightning and heating (F21, F24, F25: refrigeration or cooling; combined heating and refrigeration systems etc.), foodstuffs (A23), and agriculture (A01 fishing) indicate the strengths of Bremen in the logistics, aerospace and food industries. Furthermore, strengths in hydraulic engineering (E02: Hydraulic engineering; F15: Fluid pressure actuator, hydraulics and pneumatics) as well as in textile-related IPCs (D02: Yarns etc., D04: Braiding, Lace Making etc., D06: Treatment of Textiles etc.) may be closely related with the logistics sector.[2] Competences in material sciences (C09, C23) and measuring and checking devices (G07, GO1) contribute to the technological strengths of Bremen in the aerospace and logistics sectors. Despite the overarching importance of the logistics sector, other important sectors and declared innovation fields such as the health sector (A61, A62), ICT (G06) and biotechnology (C08, C01) also appear in the data. The only missing innovation field with respect to the RTAs is environmental technologies.

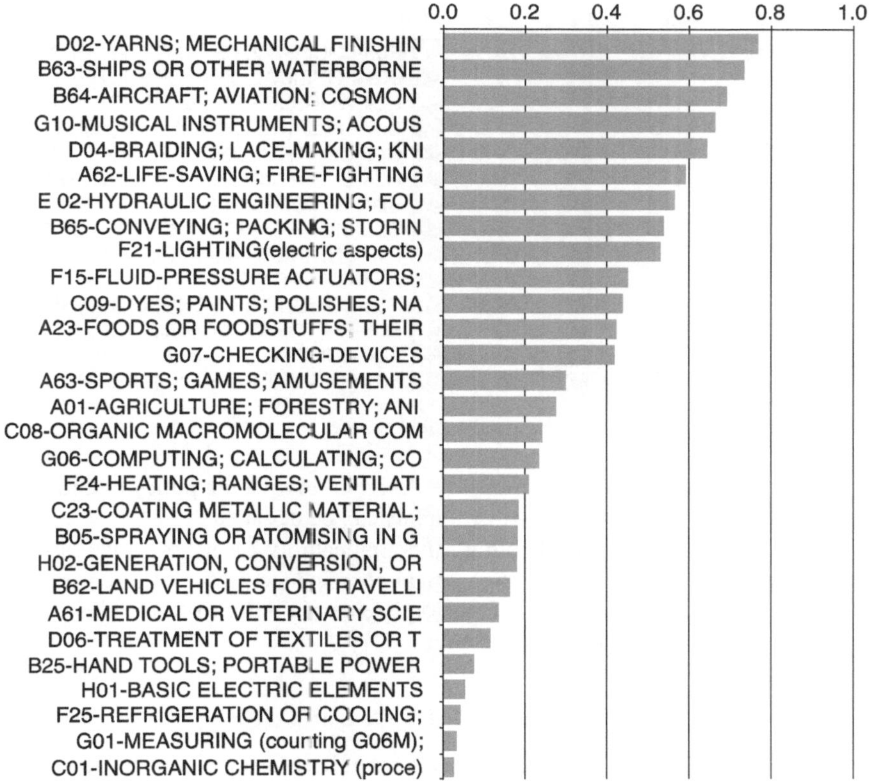

Figure 5.2 Revealed technological advantages in Bremen

In a number of other technology fields, Bremen exerts technological disadvantages relative to Germany as a whole (see Figure 5.3).

The revealed technological disadvantages of Bremen in machines and engines (F01, F02), electronic products and devices (H05, H04) and chemistry (C12, C07, B29, B01) – all of which are of major importance for Germany's industrial structure and technological competences – illustrates the different sectoral composition and technological competence profile of Bremen. On the one hand, these disadvantages have to be considered as technological weaknesses for Bremen, but, on the other hand, considering the evolutionary path of a harbour and trading city as well as the limited population of Bremen, the concentration upon endogenous capabilities and opportunities may be more common than trying to compete with the domains of other regions following other evolutionary traits of learning and competence accumulation.

Technological competences in other areas, emphasised by innovation policy such environmental technologies and blue biotechnologies, cannot be identified in our analysis. Blue biotechnology might still be close to basic science, in environmental technologies; however, a significant patent activity can be expected.

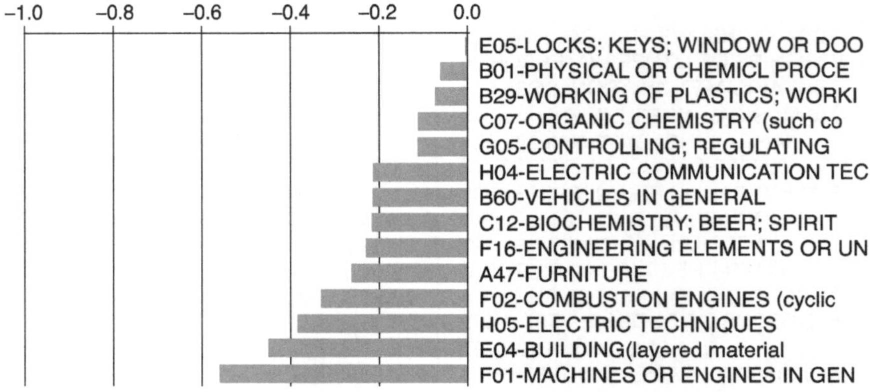

Figure 5.3 Revealed technological disadvantages in Bremen

From this result, some evidence for the re-design of innovation policy and for the implementation of specific policy instruments can be deviated.

Generally, the results underline both the applicability of our methodology in using patent data to reveal the technological competence profile of a region, as well as the adequacy of the policy focus on Bremen in order to create clusters in the so-called innovation fields of Bremen.

5.4.3.2 Technological advantages relative to the OECD

Having analysed the technological competences of Bremen in relation to Germany, we can also consider the competences relative to the most important international economies, as approximated by the OECD countries (Figure 5.4).

Figure 5.4 illustrates the RTA relative to Germany as discussed in the section above and the RTA relative to all OECD countries. The figure shows a high correlation between the national and the international competitive advantage, which is a typical feature of regions in large economies.[3] This high correlation indicates that technology fields which enjoy a technological advantage on the national level also internationally shape the profile of the region. Fields with a high technological disadvantage tend to be disadvantaged nationally as well as internationally.

The only remarkable exception in this consistent picture is shipbuilding technologies – B63 – where Bremen reveals a high technological national advantage. Yet, internationally the competences for Bremen are below average in this technological field. In spite of being a flagship sector for Bremen for centuries, in the last two to three decades Bremen has lost relative strength in the shipbuilding sector in a globalised world where new dynamic competitors are arising, especially in the East. A relative technological advantage in the shipbuilding sector – i.e. pronounced knowledge-orientation and innovativeness in this sector – will very likely be of outstanding importance with respect to the overall developments in the world economy. Harbours and related technologies are very

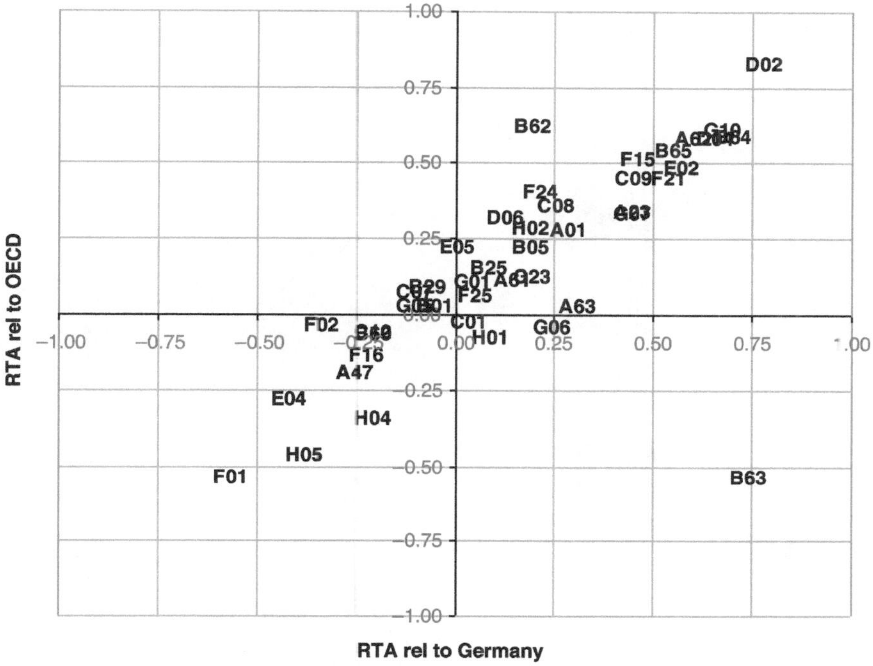

Figure 5.4 International comparison of Bremen's RTAs

likely to be key assets in the near future, providing tremendous support for the economic development of regions at the coastline which are able to master these technologies.

5.4.4 Networks of inventors, applicants and technology classes

Due to the relatively small number of patents analysed, our results on innovation networks are preliminary in nature, and provide a basic illustration of the possibilities of network analysis. As outlined in section 3.2, we can use patent data to analyse the linkages between inventors, applicants and technology fields. In Figure 5.5 we illustrate the tripartite innovation network of Bremen. The black nodes represent applicants, the dark grey nodes represent inventors, and the light grey nodes are technology classes on the IPC 2-digit level.

The network graph contains one large component in which firms (i.e. applicants), as well as inventors, also form linkages among each other, as well as shared knowledge fields. On the one hand, this illustrates a sound balance within Bremen's innovation networks, with the knowledge fields being available in order to generate prolific conditions for collective innovation processes. On the other hand, the links between the applicants are far from dense, and also quite a few

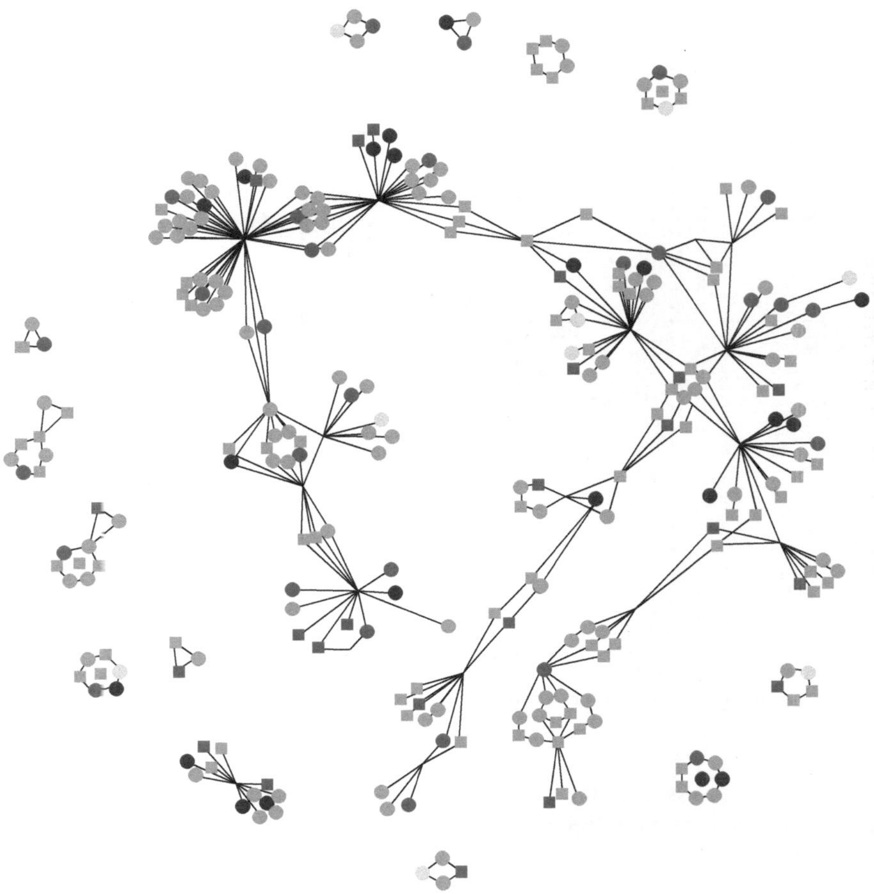

Figure 5.5 Patent-based innovation network of Bremen

firms (as well as technological fields) are not connected to the large component. Both observations indicate a broad scope for innovation policy to search for fruitful connections between different technological fields in order to exploit cross-fertilisations and to spur collective innovation by supporting regional collaborations. A large percentage of the patent applicants are multinational enterprises. This observation indicates that: (i) small local actors do not seem to have the financial and organisational resources to deal with the patenting process; (ii) the region, however, is able to attract MNEs by its technological competences profile and market and innovation opportunities; and (iii) there is both a need and scope for promoting the appropriation of innovation rents by local actors.

However, an assessment of the local innovation networks based on patent data should be done carefully. Only certain conclusions can be drawn. Due to strategic patenting, secrecy, the skewed distribution of patent values, and the typically limited number of overall patents in low to medium developed regions, patents do not provide a comprehensive picture of the regional knowledge flows and innovation networks. We do not get a proper picture of the whole set of knowledge transfer and exchange channels, through market mechanisms and strategic alliances, or through informal networks and knowledge spillovers. Subsequent work shows that the patent-based innovation network differs sharply from the series of publicly promoted networks in Bremen (Gundrum, 2007). While certainly many (if not most) publicly promoted networks fail, some of the programmes are successful and create social linkages and potential knowledge flows between agents. For example, in the case of Bremen, the promotion of clustering and networking within and between the wind energy sector, material sciences, and aerospace has been quite successful. In our case, the available patent data is not able to trace these potential knowledge channels and sectoral dynamics driven by policy interventions; consequently, innovation policy focusing on the condensation as well as the expansion of innovation networks has to adapt existent network structures with real knowledge flows in Bremen. Also, the linkages between the regional networks and the innovation network structures are to be considered in the design of regional innovation policies.

5.4.5 Degree of internationalisation

Figure 5.6 shows the degree of internationalisation in the relevant technologies in Bremen, where internationality is captured by the fraction of patents which are applied for by a Bremen-based and an international applicant (international co-invention).

We see a high degree of internationalisation in the biotech, aircraft and land-vehicles technologies. For the range of technologies displayed above, we observe a higher degree of internationalisation in the area of Bremen as compared to Germany as a whole. The only exceptions here are technologies related to medical and veterinary sciences. This result clearly underlines the comparatively high degree of internationalisation of research and development in Bremen, which offers an important strategic orientation of innovation policy.

5.5 Conclusions

Many qualitative and descriptive analyses on clusters and regional innovation systems have been conducted in recent decades. However, few contributions elaborate upon the quantitative methodologies employed to measure the structures and dynamics of regional systems, which would allow for a comparison of different regions. The extant taxonomies consider some widely available innovation outputs and inputs, but do not properly include the core features of these regional systems, such as network structures and sectoral technological competences.

This study explored the possibilities of mapping technological competences and innovation networks of regions by using patent data. The pragmatic reason

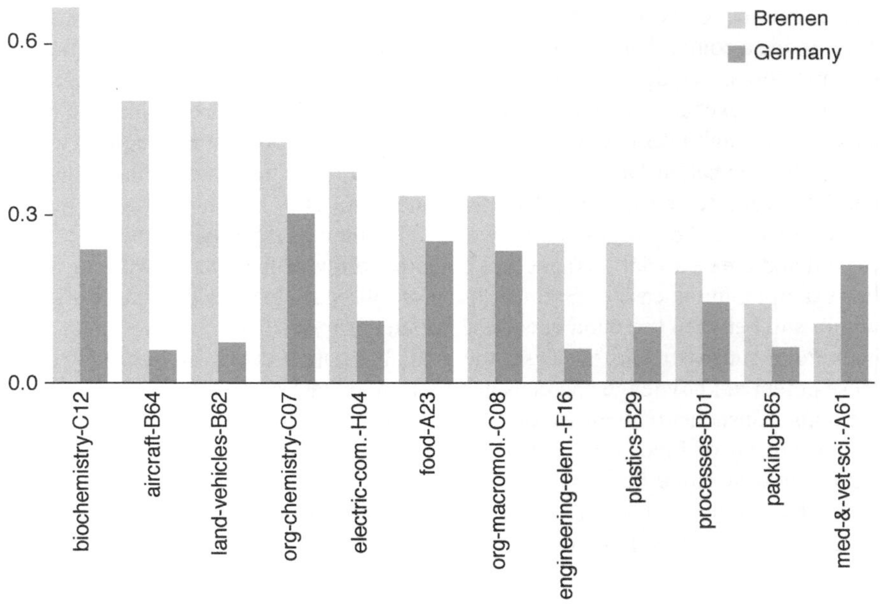

Figure 5.6 Internationalisation of the innovation profile in Bremen

for choosing this approach is the availability of patent data for most regions of the OECD. Further advantages of patent data lie in the information that they contain: they not only include the geographical postal codes of the inventors and applicants, but also the technological field or sector the patent belongs to. This allows for a matching of inventors within the different regions and technology classes: in this way, we can measure the technological advantages and disadvantages of regions by calculating the number of patents a region has in different IPC classes in comparison to the average contribution of other regions.

Our results show that patent data can be employed as theoretically reasonable and practically available proxy indicators for measuring the technological competence profile, network structures and degree of internationalisation of medium developed regions in industrial activities. For our case study on the northern German city state of Bremen the results are quite robust. The revealed (dis-)advantages are indeed well-matched with the intuitive and historically grown comparative technological strengths (in logistics, aerospace, food industry) and weaknesses (in machinery, electronics and chemistry) in comparison to Germany (e.g. advantage in shipbuilding) and the OECD (disadvantage in shipbuilding).

Additionally, the data applied allows us to measure intra- and inter-regional knowledge flows and cooperation by linking the inventors with the applicants and the technology classes (or any of the three combinations). Our study allows us to: (i) identify the agents (and key players) who are able to appropriate regional knowledge; (ii) visualise and map the networks between local and international inventors and applicants; and (iii) reveal promising intersections and linkages between different technology classes. All of this information is relevant for regional policy-makers. Hence, patent data certainly has limitations, but also important strengths.

Notes

1 The patents and related information were retrieved from the EPO bulletin database.
2 E.g. Deutsche Post, DHL holds patents on the treatment and transportation of suits.
3 This does not hold for the technological competitiveness of regions in small and open economies such as Austria (e.g. Ebersberger, 2006).

References

Archibugi, D. (1992) 'Patenting as an indicator of technological innovation: A review', *Science and Public Policy*, 16(6): 357–68.

Archibugi, D. and Pianta, M. (1992) *The technological specialization of advanced countries. A report to the EEC on international science and technology activities*, Kluwer, Boston.

Audretsch, D.B. and Thurik, A.R. (2000) 'Capitalism and democracy in the 21st century: From the managed to the entrepreneurial economy', *Journal of Evolutionary Economics*, 10(1): 17–34.

Balassa, B. (1965) 'Trade liberalization and 'revealed' comparative advantage', *The Manchester School of Economics and Social Studies*, 33(2): 99–123.

Boschma, R. and Martin, R. (2007) 'Special issue on evolutionary economic geography', *Journal of Economic Geography*, 7(5): 537–48.

Boschma, R. and Martin, R. (2010) *The Handbook of Evolutionary Economic Geography*, Cheltenham: Edward Elgar.

Breschi, S. and Lissoni, F. (2003) 'Mobility and social networks: localised knowledge spillovers revisited', *KITeS Working Papers* 142, KITeS, Centre for Knowledge, Internationalization and Technology Studies, Universita' Bocconi, Milano, Italy.

Camagni, R. (1991) *Innovation networks – spatial perspectives*, Belhaven Press: London, New York.

Cantner, U., Graf, H. and Meder, A. (2009) 'Urbane innovationssysteme: das innovationsnetzwerk in Jena', in B. Blättel-Mink and A. Ebner (eds) *Innovationssysteme: Technologie, Institutionen und die Dynamik der Wettbewerbsfähigkeit*, Wiesbaden: Verlag für Sozialwissenschaften.

Cantwell, J.A. (1989) *Technological Innovation and Multinational Corporations*, Blackwell: Oxford.

Cantwell, J.A. (2005) 'Innovation and competitiveness', in J. Fagerberg, D.C. Mowery and R.R. Nelson (eds) *The Oxford Handbook of Innovation*, Oxford University Press: Oxford and New York.

Cantwell, J.A., Dunning, J. and Janne, O. (2004) 'Towards a technology-seeking explanation of U.S. direct investment in the United Kingdom', *Journal of International Management*, 10(1): 5–20.

Castells, M. (1996) *The Information Age: Economy, Society and Culture. The Rise of the Network Society*, vol. I, Blackwell Publishers: Oxford and Malden, MA.

Cooke, P. (1998) 'Introduction: origins of the concept', in H.J. Braczyk, P. Cooke and M. Heidenreich (eds) *Regional Innovation Systems – The role of governance in a globalized world*, UCL Press: London.

Cooke, P. and Memedovic, O. (2003) 'Strategies for regional innovation systems: learning transfer and application', UNIDO policy papers, Vienna.

Dunning, J. and Narula, R. (1995) 'The R&D activities of foreign firms in the United States', *International Studies of Management & Organization*, 25(1–2): 39–73.

Ebersberger, B. (2006) Von Gipfeln und Tälern – Das technologische Kompetenzprofil Tirols', Vortrag 14. Sept. 2006, MCI, Innsbruck, http://www.mci.edu/fhwirtschaft/pdf/Praesentation_Bernd_Ebersberger.pdf

Eliasson, G. (2005) 'Competence blocs in the experimentally organized economy – On regional industrial transformation in the global economy', in G. Eliasson (ed.) *The Birth, the Life and the Death of Firms*, Ratio Institute Stockholm.

Frenken, K. and Boschma, R.A. (2007) 'A theoretical framework for evolutionary economic geography: industrial dynamics and urban growth as a branching process', *Journal of Economic Geography*, 7(5): 635–49.

Griliches, Z. (1990) 'Patent statistics as economic indicators: A survey', *Journal of Economic Literature*, 28(4): 1661–707.

Gundrum, U. (2007) Wissens – und Technologietransfer in Bremen. Grundlagen, Bestandaufnahme und Handlungsempfehlungen, Im Auftrag des Technologiebeauftragten des Senats der Freien Hansestadt Bremen.

Hagedoorn, J. and Cloodt, M. (2003) 'Measuring innovative performance: is there an advantage in using multiple indicators?', *Research Policy*, 32(8): 1365–79.

Hollanders, H. (2002) '2002 European Innovation Scoreboard', Technical Paper No.3 EU Regions – European Trendchart on Innovation.

Hollanders, H. (2003) '2003 European Innovation Scoreboard', Technical Paper No.3 Regional innovation performances – European Trendchart on Innovation.

Hollanders, H. (2007) '2006 European Regional Innovation Scoreboard (2006 RIS)', Maastricht Economic and Social Research and Training Center on Innovation and Technology.

Hollanders, H., Tarantola, S. and Loschky, A. (2009a) 'Regional Innovation Scoreboard (RIS) 2009 – Methodology Report', Pro Inno Europe, InnoMetrics

Hollanders, H., Tarantola, S. and Loschky, A. (2009b) 'Regional Innovation Scoreboard 2009', Methodology Report, Pro Inno Europe, InnoMetrics.

Koschatzky, K., Jappe, A., Hemer, J. and Wolf, B. (2004) *Technologietrends und regionale Entwicklungspotenziale in den bremischen Innovationsfeldern*, Handlungsorientierter Endbericht, Fraunhofer-Institut für Systemtechnik und Innovationsforschung, Karlsruhe

Narula, R. and Zanfei, A. (2005) 'Globalization of innovation: the role of multinational enterprises', in J. Fagerberg, D.C. Mowery, and R.R. Nelson (eds) *The Oxford Handbook of Innovation*, Oxford University Press, Oxford New York.

OECD (2007) 'Prüfbericht zur Politik für ländliche Räume – Deutschland', OECD.

OECD (2009) 'OECD Patent Statistics Manual', OECD publications, Paris.

Patel, P. and Pavitt, K.L.R. (1991) 'Europe's technological performance', in C. Freeman, M. Sharp and W. Walker (eds) *Technology and the Future of Europe: Global Competition and the Environment in the 1990s*, Thomson Learning, London.

Pyka, A. (1997) 'Informal networking', *Technovation*, 17(4): 207–20.

Pyka, A. (1999) *Der kollektive Innovationsprozeß – Eine theoretische Analyse informeller Netzwerke und absorptiver Fähigkeiten*, Duncker&Humblodt, Berlin.

Pyka, A. (2000) 'Informal networking and industrial life cycles', *Technovation*, 20(1): 25–35.

Pyka, A. (2002) 'Innovation networks in economics: From incentive-based to the knowledge-based approaches', *European Journal of Innovation Management*, 5(3): 152–63.

Pyka, A. and Scharnhorst, A. (2008) *Innovation Networks*, Springer: Berlin, Heidelberg, New York.

Saviotti, P.P. (2009) 'Knowledge networks: Structure and dynamics', in A. Pyka and A. Scharnhorst (eds) *Innovation Networks – New Approaches in Modelling and Analyzing*, Springer, Berlin, Heidelberg.

Saxenian, A.L. (1994) *Regional advantage: Culture and competition in Silicon Valley and Route 128*, Cambridge, MA: Harvard University Press.

Smith, K. (2005) 'Measuring innovation', in J. Fagerberg, D. Mowery, and R.R. Nelson (eds) *The Oxford Handbook of Innovation*, Oxford: Oxford University Press.

Soete, L.L.G. (1987) 'The impact of technological innovation on international trade patterns: the evidence reconsidered', *Research Policy*, 16(2–4): 101–28.

Stenke, G. and Willms, W. (2006) 'Innovationsbericht 2006', BAW Institut für regionale Wirtschaftsforschung GmbH, Bremen.

Thorsen, I. and Gitlesen, J.P. (1998) 'Empirical evaluation of alternative model specifications to predict commuting flows', *Journal of Regional Science*, 38(2): 273–92.

Timm, J. (2007) Schlussbericht des Technologie beauftragten des Senats der Freien Hansestadt Bremen: Die High-tech Region Bremen hat an Profil gewonnen, Bremen.

Uyarra, E. (2010) 'What is evolutionary about 'regional systems of innovation'? Implications for regional policy', *Journal of Evolutionary Economics*, 20(1): 115–37.

World Bank (1999) *World Development Report 1998–1999: Knowledge for Development*, Oxford University Press: New York.

6 Implementation cycles and the new economy in retrospect

Pasquale Scaramozzino, Jonathan Temple and Nir Vulkan

6.1 Introduction

In retrospect, the macroeconomic record of the USA in the 1990s was remarkable in a number of ways. The media attention given to the share prices of internet companies tended to obscure the achievements of the wider economy, including faster productivity growth, a rise in equipment investment, a reduction in output volatility, and an expansion that was sustained for exactly 10 years – one of the longest on record. There were some unwise predictions of the 'death' or 'taming' of the business cycle, but nevertheless, the volatility of US output was certainly lower in the late 1980s and 1990s than previously (McConnell and Perez-Quiros, 2000). This period of unusual stability, the subject of much formal and informal commentary, was dubbed the 'Great Moderation' (Stock and Watson, 2002).

The New Economy period directly inspired a number of papers which link business cycle fluctuations to the interaction between expectations and the implementation of new technologies. A particular focus has been the role of anticipated changes in productivity growth, and more general forms of 'news', in driving both output volatility and stock market fluctuations. The relevant papers include Beaudry and Portier (2006), Francois and Lloyd-Ellis (2008), Jaimovich and Rebelo (2009), Pástor and Veronesi (2009), and Beaudry, Collard and Portier (2011). The paper by Pástor and Veronesi (2009), in particular, is motivated by a particular feature of the New Economy period: despite the stability of output growth, there was uncertainty about whether this growth could be sustained, linked as it was to uncertainty about the productivity of new technologies. As the duration of the boom exceeded all expectations, forecasters repeatedly revised their predictions, as we document further below. This implies that the 1990s were an atypical period: unusually stable output growth was combined with a high degree of uncertainty about the trend growth rate.

In this paper, we use these observations to revisit a classic model of 'intrinsic' business cycles due to Shleifer (1986). The starting point for Shleifer's analysis is that firms must decide whether to implement innovations immediately, or wait for a period of higher aggregate demand, when the profitability of implementation may be greater. As well as an equilibrium within which firms implement immediately, there can also exist multiple short-cycle equilibria, and sometimes also longer cycles. The cycle is entirely driven by expectations about the timing of a boom.

Recent work on expectations and multiple equilibria in macroeconomics has tended to emphasize the fragility of similar multiplicity results when agents are uncertain about the beliefs of other agents. With this in mind, we extend Shleifer's model to incorporate the possibility of uncertainty about the underlying growth rate, motivated by the New Economy period. We will show that this uncertainty can eliminate cyclical equilibria, leaving immediate implementation as the only possible outcome. In Shleifer's model, such an outcome would tend to be associated with a period of unusually stable, non-cyclical productivity growth, and a reduction in the volatility of investments associated with implementation. These were arguably features of the American boom of the 1990s. For example, Leduc and Sill (2007) attribute the reduction in output volatility in the US to a decline in the size of shocks to total factor productivity. Fernández-Villaverde *et al.* (2010a, 2010b) characterize the period 1987–2006 in the US as one of unusually small, and positive, structural shocks, including productivity shocks.

The remainder of the paper is structured as follows. Section 2 provides a more detailed review of stylised facts about the New Economy, helping to motivate our extension of Shleifer's model. In section 3, we provide an overview of implementation cycles, emphasizing the role of expectations. Section 4 sets out the basic framework, before section 5 shows that uncertainty about the underlying growth rate leads to immediate implementation. Section 6 concludes.

6.2 Some stylised facts

In this section of the paper, we discuss evidence that is consistent with the model of business cycles presented by Shleifer (1986), and which will inform and motivate our later theoretical analysis. We are especially interested in evidence that supports a central result of Shleifer's model. In his model, even when inventions arrive evenly over time, they are implemented in waves. These waves arise because firms have an incentive to defer implementation, if other firms are similarly deferring, until aggregate demand is relatively high.

We first ask whether there is evidence to support the view that new ideas are implemented with delays, and in waves. We review previous research, and also provide some new indirect evidence by examining the behaviour of initial public offerings (IPOs) and productivity growth over the business cycle. We will argue that the cyclical patterns of these variables support the idea that innovations take place in waves.

More direct evidence on this point is hard to obtain. Survey-based accounts of the successful commercialisation of inventions sometimes reveal a pattern of distinct peaks and troughs, as pointed out by Van Reenen (1996) using the data set for the UK described in Robson, Townsend and Pavitt (1988). This does not establish, however, that innovation clustering is the outcome of strategic delays.

In this respect, some interesting evidence is provided by the behaviour of stock markets in the wake of technological changes. Hobijn and Jovanovic (2001) explain major changes in US stock market valuations in terms of a delay between the creation of new technologies (such as information and communications technologies) and their implementation by new entrants. They argue that the potential of new technologies may be widely known several years before the technologies are implemented. This helps to explain the substantial decline in US stock valuations in the 1970s, given declines in expected profitability for incumbents and the market's rational anticipation of entrants exploiting new technologies. This evidence is at least consistent with the view that the implementation of new ideas involves delays, perhaps because entrepreneurs await favourable economic conditions, although other interpretations of the delays are also possible.

In exploring this idea in more detail, we focus mainly on US time series for movements in multifactor productivity (MFP) and initial public offerings (IPOs). We use both of these as proxies for the extent of innovative activity in the economy. We will be able to show that, especially after 1980, these two alternative measures tend to fluctuate in similar ways. Their co-movements support the idea that new technologies are implemented in waves. Furthermore, the extent of volatility in each series was lower in the 1990s than previously, consistent with our claim that the clustering of innovations has diminished.

First of all, Figure 6.1 plots MFP growth for the private non-farm business sector in the USA between 1960 and 2001.[1] This shows the well-known tendency for marked year-to-year variations in MFP growth. This variation may simply reflect the random nature of technical progress. There could be sufficient randomness in the creation of new ideas that MFP growth varies substantially from year to year, even if implementation of a new idea is always immediate.[2] An alternative view attributes the variation in MFP growth to the measurement error of various kinds. Business cycles may be associated with systematic changes in measured MFP, notably through variations in factor utilization. There is evidence of this cyclical pattern in Figure 6.1, at least prior to the mid-1980s. Measured MFP growth will then vary at short horizons even when underlying technical progress follows a smooth path and new ideas are implemented without delay.

Given these limitations of the data on MFP growth, we combine this information with a more direct indicator of implementation, namely the number of initial public offerings (IPOs). Although IPOs vary in nature, a substantial number are clearly motivated by the desire to raise capital in the course of implementing a new business idea. Pástor and Veronesi (2005) note that around two-thirds of the leaders of IPOs cite the raising of capital as the main reason for an offering. Moreover, capital growth in the two years around the IPO is substantially higher than for comparable firms.

As with MFP growth, there is significant variation from year to year in the number of IPOs. Note that from the perspective of a finance textbook, the timing of an IPO should not matter, because any fairly-priced offering would not have a positive net present value. As Jenkinson and Ljungqvist (2001) note, this might suggest that the distribution of IPOs over time should be random. In fact, there is a well-known tendency for IPOs to cluster together in distinct waves. Both the year to year variation and the tendency for significant positive autocorrelation are

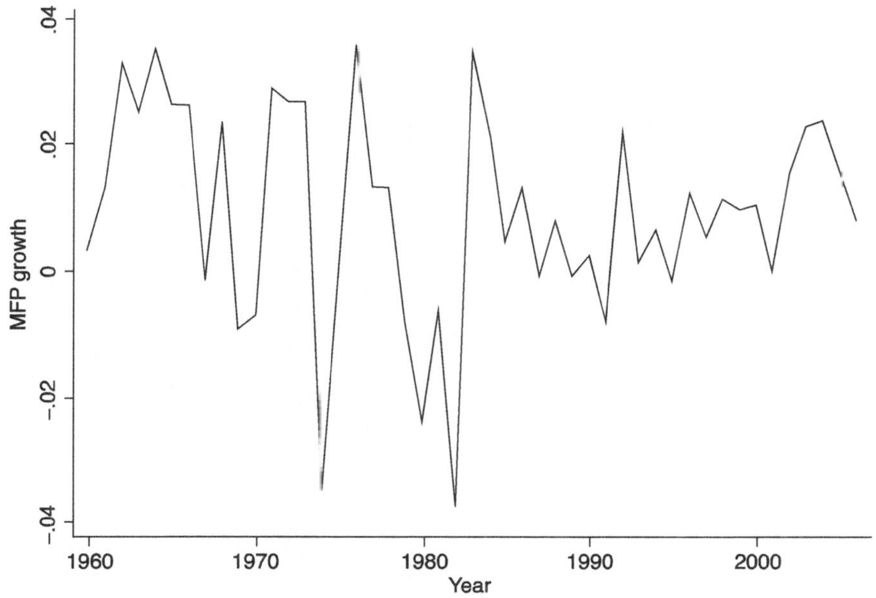

Figure 6.1 Annual MFP growth rate, non-farm private business

Notes: This figure shows annual data on MFP growth for the non-farm private business sector, excluding government enterprises, calculated from historical BLS data. See Appendix 1 for more information on the data.

apparent in Figure 6.2, which plots annual data on the number of IPOs in the USA since 1960 (see Appendix 1 for the source of these data). At first glance, this supports the narrative in which entrepreneurs are willing to defer bringing an idea to the market.

Again, there are several possible explanations for the observed waves in IPOs. These include the possibility that entrepreneurs wish to take advantage of mispricing in equity markets. As Pástor and Veronesi (2005) argue, it is not clear why the mispricing is clear to entrepreneurs but less readily observable to other market participants. Their preferred explanation is that the decision to go public can be seen as exercising a real option. Entrepreneurs might wish to delay an IPO, exercising the option only when there is a favourable change in market conditions. They present evidence that movements in expected aggregate profitability, including revisions to analysts' earnings forecasts, are one determinant of the positive net present value. As Jenkinson and Ljungqvist (2001) note, this might suggest that the distribution of IPOs over time should be random. In fact, there is a well-known tendency for IPOs to cluster together in distinct waves. Both the year to year variation and the tendency for significant positive autocorrelation are apparent in Figure 6.2, which plots annual data on the number of IPOs in the USA since 1960 (see Appendix 1 for the source of these data). At first glance, this

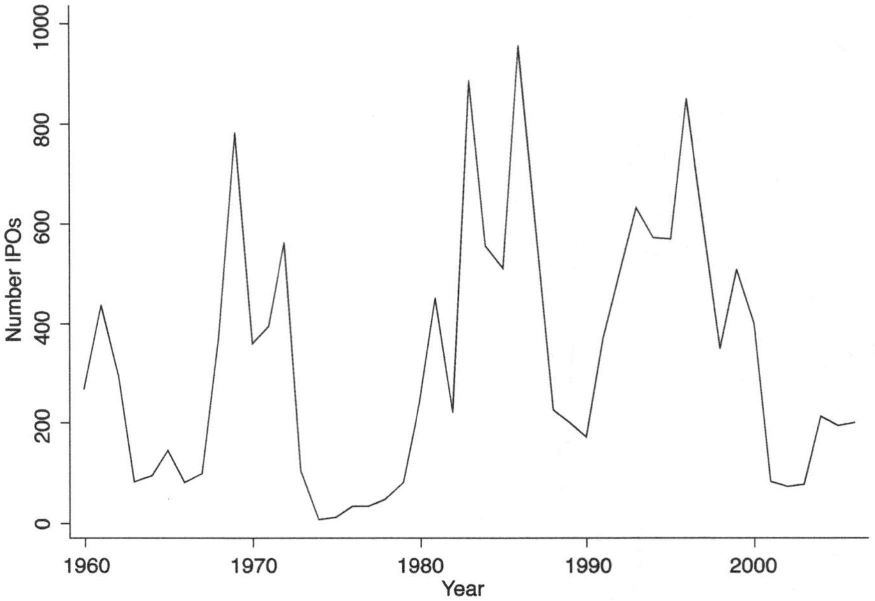

Figure 6.2 Annual data on Initial Public Offerings

Notes: This figure shows annual data for the US economy on the number of IPOs, using data collected by Jay Ritter and made available on his website in 2007. See Appendix 1 for more details.

supports the narrative in which entrepreneurs are willing to defer bringing an idea to the market.

Again, there are several possible explanations for the observed waves in IPOs. These include the possibility that entrepreneurs wish to take advantage of mispricing in equity markets. As Pástor and Veronesi (2005) argue, it is not clear why the mispricing is clear to entrepreneurs but less readily observable to other market participants. Their preferred explanation is that the decision to go public can be seen as exercising a real option. Entrepreneurs might wish to delay an IPO, exercising the option only when there is a favourable change in market conditions. They present evidence that movements in expected aggregate profitability, including revisions to analysts' earnings forecasts, are one determinant of the timing of IPOs. This endogeneity in the timing of IPOs. This endogeneity in the timing of investment can be seen as a specific instance of the general argument in Shleifer (1986).

If we see MFP growth and the number of IPOs as two alternative measures of the level of innovative activity in the economy, it is natural to ask whether the two move together in similar ways. We are not primarily interested in whether there is a causal relationship, but in whether there is a tendency for these two series to move in the same way over time. If they move together, that will support our use of the two measures as proxies for innovation. To the extent that each series indicates a 'bunching' of innovations, with similar timing, that will support our emphasis on cycles in implementation.

Figure 6.3 combines the annual data on IPOs with that on MFP growth. The correspondence between the two is weak for the 1960s and 1970s, but strengthens greatly thereafter, with a slight tendency for IPOs to anticipate movements in MFP growth. This relationship is stronger when we restrict attention to MFP growth in the manufacturing sector, disaggregated into durables and non-durables. Figures 6.4 (for non-durables) and 6.5 (for durables) again reveal the tendency for IPOs (for the whole economy) and MFP growth to move together after 1980.

The visual impression is confirmed by two further ways of looking at the data. First, we report simple correlations between MFP growth and the (contemporaneous and lagged) number of IPOs. Second, we show that the number of IPOs helps to forecast MFP growth, even when conditioning on past MFP growth rates.

Table 6.1 shows the correlations between MFP growth and lagged IPOs for the whole period (1963–2001) and for the sub-period 1980–2001. Given the likely measurement error in MFP growth, and the various influences on decisions to go public, the contemporaneous correlation for the post-1980 data is surprisingly high at 0.63. There is also some evidence that MFP growth is correlated with past numbers of IPOs, especially for the post-1980 period.

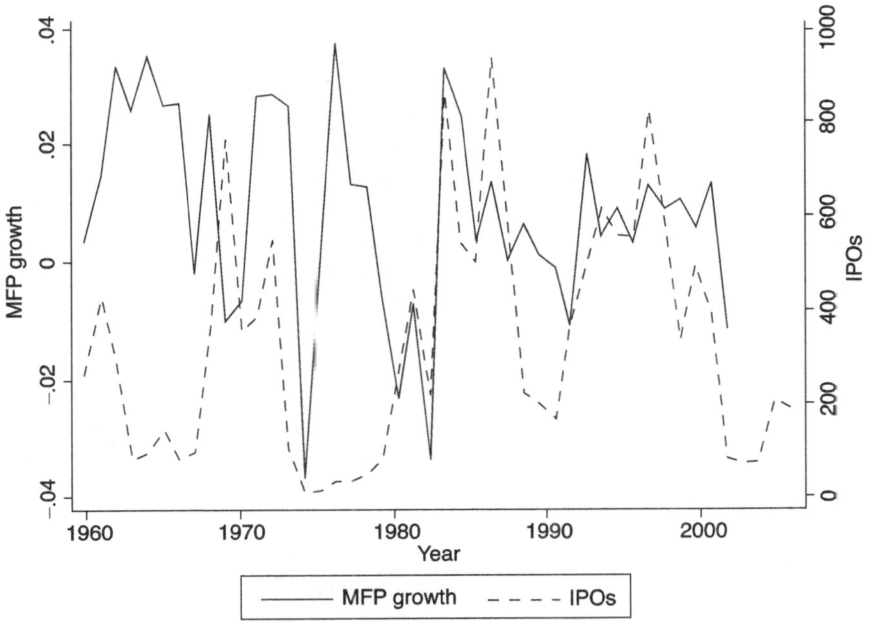

Figure 6.3 The co-movement of IPOs and MFP growth

Notes: This figure shows the co-movements of annual data on MFP growth for the private non-farm business sector, and annual IPOs. These co-movements are stronger after 1980. See Appendix 1 for data sources.

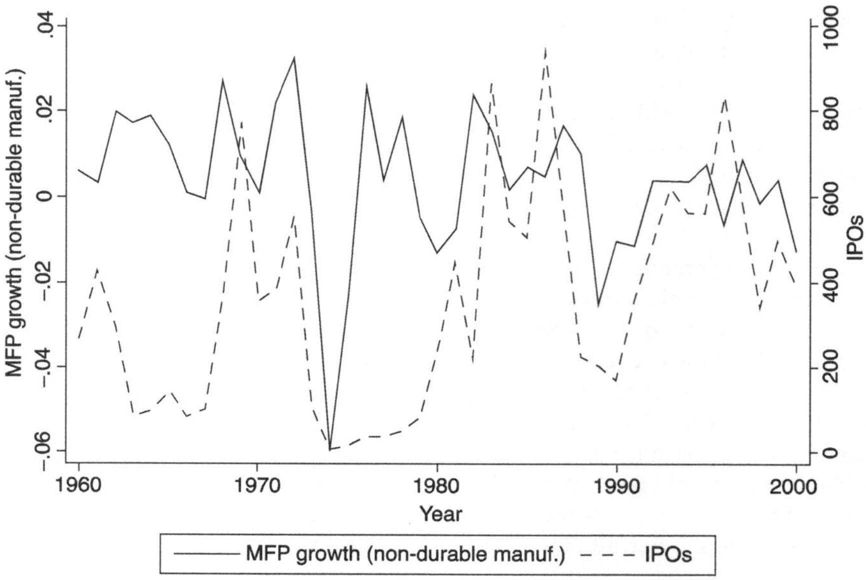

Figure 6.4 Annual MFP growth (manufacturing, non-durables) and IPOs

Notes: This figure shows the co-movements of annual data on MFP growth for the non-durables manufacturing sector, and annual IPOs. See Appendix 1 for data sources.

We now carry out simple Granger-causality tests, by regressing annual MFP growth on two lags of MFP growth and one lag of the number of IPOs. The intention here is not to make statements about causality in any structural sense, but to examine the extent of co-movement between the productivity growth and IPO series. To do this, we test the null hypothesis that the coefficient on lagged IPOs is equal to zero, using Newey-West standard errors to construct our test statistics.[3] We also report the incremental R^2, the increase seen in the R^2 of the model for productivity growth by adding the lagged IPO variable. The results are shown in Table 6.2. For the period beginning in 1962, lagged IPOs help to forecast MFP growth only in the durables manufacturing sector (the zero restriction is not rejected in the other cases). For the period after 1980, however, the IPO series helps to forecast all three MFP growth series (business, durables manufacturing, and non-durables manufacturing) and the effect on the R^2 of adding lagged IPOs is substantial.

This result is not necessarily surprising, given that IPOs are inherently forward-looking, and we are not claiming to have identified a genuine causal effect. We are interested in these correlations for the more general relationship that is revealed: the extent to which IPOs and MFP growth fluctuate in similar ways over the business cycle. The tendency for these series to move jointly, with periods of high IPO activity preceding high MFP growth rates, supports the idea of innovative activity that is 'bunched' in distinct periods or waves. The evidence we present

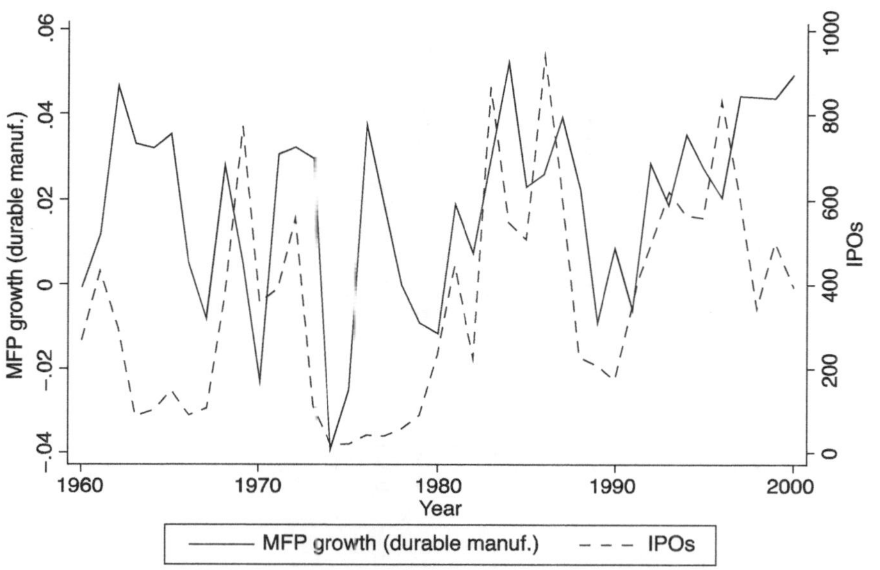

Figure 6.5 Annual MFP growth (manufacturing, durables) and IPOs

Notes: This figure shows the co-movements of annual data on MFP growth for the durables manufacturing sector, and annual IPOs. See Appendix 1 for data sources.

Table 6.1 Correlations between MFP growth and lagged IPOs

1963–2001				
	IPO(t-3)	*IPO(t-2)*	*IPO(t-1)*	*IPO(t)*
Business MFP(t)	0.14	–0.02	–0.00	0.10
	(0.39)	(0.89)	(0.97)	(0.53)
Durables MFP (t)	**0.30**	0.24	**0.45**	**0.40**
	(0.07)	(0.13)	(0.00)	(0.01)
Non-durables MFP(t)	–0.16	–0.08	0.14	0.24
	(0.34)	(0.62)	(0.39)	(0.13)
1980–2001				
Business MFP(t)	**0.37**	0.29	0.33	**0.63**
	(0.09)	(0.19)	(0.13)	(0.00)
Durables MFP (t)	0.28	0.28	**0.71**	**0.46**
	(0.21)	(0.20)	(0.00)	(0.03)
Non-durables MFP(t)	–0.18	0.20	**0.49**	**0.46**
	(0.42)	(0.37)	(0.02)	(0.03)

Notes: This table shows simple correlations between three MFP growth series (row) and IPOs in different periods (column) using annual data. Data sources are described in Appendix 1. Figures in parentheses are significance levels. Correlations significantly different from zero at the 10 per cent level are shown in bold.

Table 6.2 Do IPOs help to predict future MFP growth?

Regression	*(1)*	*(2)*	*(3)*	*(4)*	*(5)*	*(6)*	*(7)*
Sector	Business	Business	Non-durables	Durables	Business	Non-durables	Durables
Time period	1962–2006	1962–2001			1980–2001		
Observations	46	41	41	41	22	22	22
Constant	0.01	0.01	0.00	0.00	–0.01	–0.01	–0.01
	(0.01)	(0.01)	(0.00)	(0.01)	(0.01)	(0.01)	(0.01)
MFPG(t-1)	0.16	0.17	0.40**	0.22	-0.27	0.20	–0.03
	(0.14)	(0.15)	(0.11)	(0.12)	(0.21)	(0.21)	(0.20)
MFPG(t-2)	–0.03	–0.06	–0.41	–0.22	0.10	–0.38	0.18
	(0.17)	(0.19)	(0.22)	(0.15)	(0.25)	(0.22)	(0.13)
IPO(t-1)	–0.07	0.02	0.08	**0.36***	**0.33***	**0.27***	**0.58****
	(0.09)	(0.10)	(0.09)	(0.17)	(0.15)	(0.10)	(0.14)
R^2	0.03	0.03	0.24	0.27	0.16	0.37	0.53
LM(1)	0.81	0.85	0.43	0.49	0.66	0.70	0.24
LM(2)	0.53	0.53	0.69	0.45	0.50	0.26	0.46
Lagged IPO	0.44	0.84	0.36	0.04	0.05	0.01	0.00
Incremental R^2	0.01	0.00	0.01	0.13	0.16	0.22	0.35

Notes: Dependent variable: MFP growth, MFPG. * significant at 5 per cent; ** significant at 1 per cent. Newey-West standard errors in parentheses, corrected for heteroscedasticity and serial correlation up to two lags. LM(*n*) is the *p*-value for a Breusch-Godfrey LM test for serial correlation, where the null hypothesis is no serial correlation of order *n*. 'Lagged IPO' is the *p*-value for the null that *IPO(t-1)* has a zero coefficient, based on Newey-West standard errors. 'Incremental R^2' is the increase in the R^2 achieved by including *IPO(t-1)* in the model. In presenting the results, the IPO series has been rescaled by dividing by 10,000.

suggests that bunching of innovative activity is initially reflected in the observed timing of IPOs and subsequently in MFP growth.

We now turn to a further set of stylised facts, related to a central argument of our paper. We will argue that implementation cycles were weakened in the 1990s, consistent with a tendency for innovations to be implemented rapidly rather than deferred to better times. This shows how Shleifer's model might be used to interpret the stylised facts of the New Economy period. Although a direct test of this hypothesis is hard to implement, we can at least examine whether the aggregate data are consistent with weaker implementation cycles.[4]

It is well known that the 1990s were a period of unusual stability for the US economy: the 'Great Moderation' that is documented in McConnell and Perez-Quiros (2000), Blanchard and Simon (2001), and Stock and Watson (2002), among others. To the extent that implementation has been smoother and less subject to distinct waves, we would expect to see reduced volatility in our proxies for implementation, namely MFP growth and IPOs. We now examine whether volatility in these measures declined over the course of the 1990s, a necessary condition for claiming that implementation cycles have weakened.[5]

We will use as our measure of volatility the (scaled) median absolute deviation: 1.484 $median\ |x_i - median\ (x_i)|$ or, in other words, the median of the absolute deviations from the median. This is a standard robust estimator of the scale of a distribution, which is less influenced by any single observation than the standard deviation. The scaling factor 1.484 ensures that the statistic will be a consistent estimator of the standard deviation of a normally distributed variable.[6]

First of all, Figure 6.6 plots a nine-year rolling median absolute deviation of MFP growth in the private non-farm business sector. An especially noteworthy aspect of this figure is that MFP growth was stable, by historical standards, even through the recession of 2001. This supports our argument that implementation cycles were temporarily in abeyance.

Figure 6.7 restricts attention to the volatility of MFP growth in the manufacturing sector, disaggregated into durables and non-durables, again using a nine-year rolling median absolute deviation for each series. Figure 6.8 examines the volatility of the annual number of IPOs. All of the figures reveal the same pattern, namely that volatility was noticeably lower in the 1990s than previously, although the volatility of IPOs shows an increase as the boom finally draws to a close.

These findings are consistent with other research; for example, Leduc and Sill (2007) present evidence that reduced output volatility in the US should be attributed to smaller shocks to total factor productivity. In the remainder of the paper, we will present a theoretical argument which could explain this reduced volatility. The argument relies on uncertainty over the underlying rate of productivity growth, which can eliminate the multiplicity of equilibria obtained by Shleifer. At first sight, our argument might appear to be on unsafe ground, because superficially the 1990s were a period of stability rather than uncertainty. Here, however, the distinction between volatility and predictability is crucial. It is well known that a series can be volatile but predictable, but in the 1990s the reverse was obtained. The New

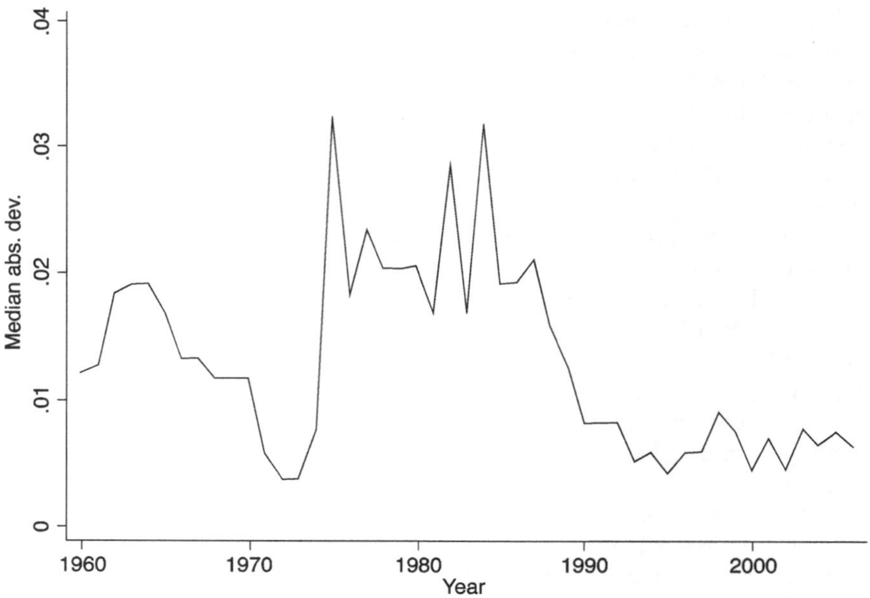

Figure 6.6 The declining volatility of MFP growth

Notes: The plotted value at date T is the nine year rolling median absolute deviation of annual MFP growth in the private non-farm business sector, using data from year T–8 to year T. See Appendix 1 for data sources.

Economy period was one in which major macroeconomic variables were unusually stable, to an extent that caught out many observers. As stated by Robert Hall in his comments on Blanchard and Simon (2001), five-year and ten-year forecast errors for the US economy were unusually large in the 1990s.

Much the same point is made in Jorgenson and Stiroh (2000). They note that forecasters repeatedly had to raise their growth projections, and that the Congressional Budget Office revised forecasts of TFP growth upwards on a number of occasions. The uncertainty arose partly because the 1990s expansion was sustained to an unusual extent, making it harder to rely on past cycles for guidance. Moreover, since growth consistently exceeded expectations, there was speculation that trend growth had increased, and disagreement over the extent to which this had happened.

The combination of a sustained expansion and a massive stock market boom led to wide discussion of the possibility that trend growth had increased, in both the business press and in more academic commentary. Views differed, indicating the uncertainty even among close observers. In reviewing productivity growth in the 1990s, Jorgenson and Stiroh (2000) argued that there was a case for an upwards revision of medium-term growth forecasts. In contrast, Gordon, in his comments on the same paper, argued that some of the productivity gains of

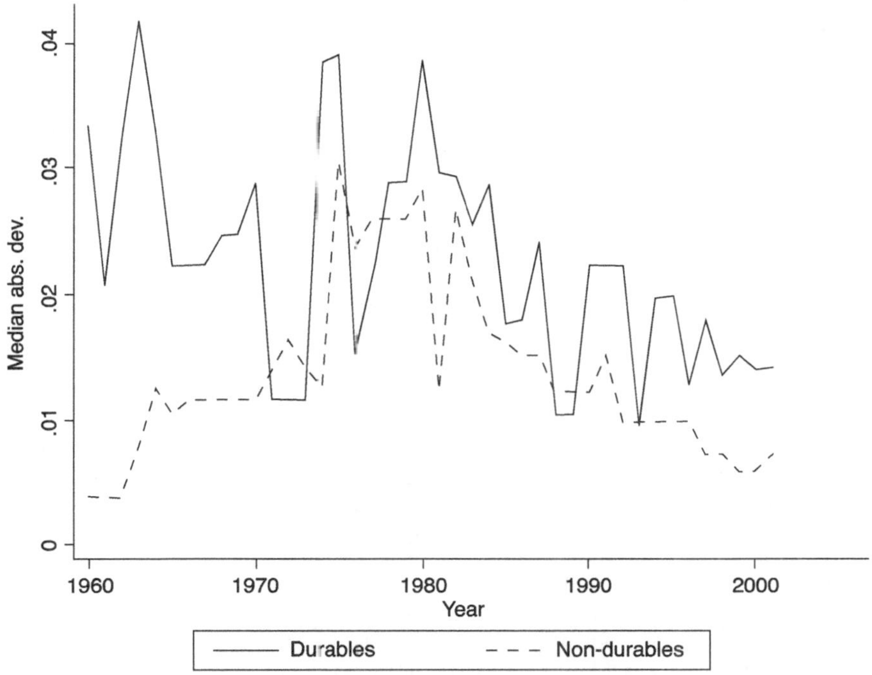

Figure 6.7 The declining volatility of MFP growth in manufacturing

Notes: The plotted values at date T are the nine year rolling median absolute deviation of the two MFP growth series using data from year T–8 to year T. See Appendix 1 for data sources.

1995–99 were likely to prove transient, and that the reputation of the New Economy had been inflated by cyclical factors. More recently, it is clear from Figure 6.1 that productivity growth continued its strong performance despite the 2001 recession, another departure from previous cyclical patterns.

As noted by Sichel in his commentary on Jorgenson and Stiroh (2000), the decomposition of output growth into trends and cyclical effects is particularly difficult when the length and nature of an expansion has departed so sharply from previous norms. Stiroh (1999), in discussing the possibility of a rise in trend growth, argued that conclusions would have to await new evidence. Overall, the lack of consensus illustrates the uncertainty about the trend growth rate, which was an important feature of the late 1990s.

The theoretical analysis in the remainder of the paper will explain why uncertainty of this kind could have implications for Shleifer's explanation of business cycles. Although it may seem paradoxical at first sight, the Shleifer model can explain the unusual stability of the 1990s, if we appeal to contemporaneous uncertainty about the underlying trend growth rate. This apparent paradox arises because the uncertainty about trend growth can potentially eliminate cyclical equilibria.

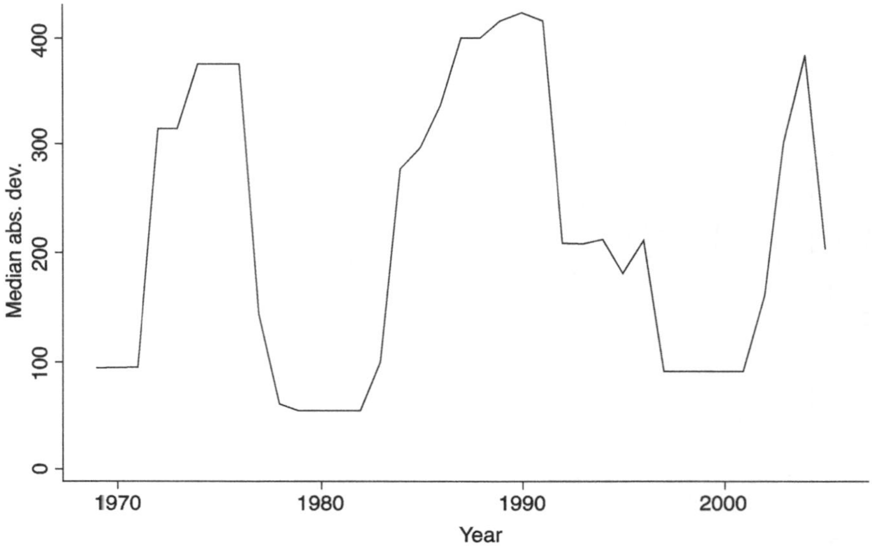

Figure 6.8 The declining volatility of IPOs

Notes: The plotted value at date T is the nine year median absolute deviation of the annual IPO series using data from year T–8 to year T. By the year 2000, volatility on this measure is clearly much lower than 10 years previously, although volatility increases again as the boom draws to a close.

6.3 Implementation cycles

In this section, we provide an overview of the arguments in the remainder of the paper. The arguments build on a long tradition in macroeconomics, emphasizing the importance of expectations and beliefs for macroeconomic behaviour. This has been stressed at least since Keynes (1936) argued that 'animal spirits' may give rise to instability. Expectations of booms and recessions can be self-fulfilling, as agents bring forward or postpone their investment decisions depending upon their perceptions of how the economy will evolve in the future. If some firms anticipate an increase in aggregate demand, they may decide not to invest in the present period and delay their investment to some future date. This will enable those firms to maximize the revenue from their sales during a boom. If other firms in the economy share the same expectations about future demand, they will also postpone their investment. This will bring about a recession in the current period and a boom at a later date.

Based upon this kind of intuition, there is now a large literature on self-fulfilling prophecies, stemming from the theoretical analyses of Azariadis (1981) and Cass and Shell (1983), and surveyed by Farmer (1993), Silvestre (1993) and Matsuyama (1995). Many of these models imply that, under some conditions, there are several possible outcomes or even a continuum of equilibria.

The practical relevance of multiplicity has been questioned by examining the role of higher order beliefs (beliefs about beliefs). Recent contributions emphasize that certain equilibria will be observed only under restrictive assumptions on the informational structure of the economy.[7] Coordination on certain equilibria often requires an assumption that agents have common knowledge about the fundamentals of the economy and about the beliefs (of all orders) of the other agents. In particular, the expectations of all the agents in the economy should be common knowledge, in the technical sense of that term.

This is clearly an unrealistic assumption to make in macroeconomic models. A more satisfactory assumption is that agents have imperfect knowledge of the fundamentals of the economy and of the beliefs held by everybody else. Their beliefs may still be related to those of other agents: individuals can learn about the information and beliefs of others, simply by observing their actions. Furthermore, they share access to public information. The key point, however, is that the beliefs of all agents are unlikely to be common knowledge.

This apparently minor change in assumptions has dramatic implications. Imagine that agents receive noisy signals about the same key parameter, and the noise affecting the signal is idiosyncratic, so that agents' signals may be different. In this case, and under quite general conditions, agents will select what they perceive to be their least risky course of action. As a consequence, some of the equilibria in the economy can be ruled out.

In the analysis that follows, we apply these ideas to the multiplicity of equilibria in Shleifer's model of implementation cycles. His framework is particularly appropriate for looking at the role of information assumptions in macroeconomics, since the cyclical equilibria rely on expectations about expectations.

In Shleifer's model, the rate of technological progress is a known constant. In the analysis that follows we demonstrate that, if there is uncertainty about the rate of technological progress, and if signals about this variable are correlated across agents, then agents will coordinate on a single equilibrium. Under some conditions, immediate implementation is the only undominated strategy for firms. According to this result, it would not be profitable for firms to delay the implementation of their innovations. The potential relevance to the New Economy period should be clear. Uncertainty about the trend growth rate, which encourages firms to implement immediately rather than delay, eliminates implementation cycles and can be associated with an unusually long expansion.

The intuition for our results can be summarized as follows. Suppose that we are in a situation where the fundamentals of the economy are only consistent with immediate implementation, and this is the dominant strategy for firms. Suppose now that the fundamentals change slightly, and that immediate implementation is only 'almost' dominant. Firms might choose to delay the implementation of their innovations. Yet, if there is some noise about the fundamentals, and if agents are uncertain regarding the beliefs of the other agents in the economy, delaying the implementation is a riskier strategy than immediate implementation. Firms will therefore tend to implement immediately.

More generally, the optimal strategy depends upon what other firms will do in nearby states of the world, including those in which immediate implementation is a dominant strategy. Taking these into account, the *ex ante* dominant strategy is not to wait for a boom. This logic applies even to circumstances in which the fundamentals of the economy are not close to making immediate implementation 'almost' dominant, as we clarify below.

6.4 The basic setup

The basic structure of the model is identical to Shleifer (1986), and we refer the reader to that paper for full details. Briefly, an infinitely-lived representative consumer maximizes utility:

$$\sum_{t=1}^{\infty} \rho^{t-1} \frac{\left(\prod_{j=1}^{N} x_{tj}^{\lambda}\right)^{1-\gamma}}{1-\gamma} \tag{6.1}$$

where $0 < \rho < 1$ is the subjective discount factor, $0 \leq \gamma < 1$ indexes the extent of relative risk aversion, x_{tj} is the consumption of good j in period t, N is the number of commodities, and $\lambda \equiv 1/N$, where N is a large number. Preferences are assumed to be Cobb-Douglas to ensure that equilibrium in each sector is determined by aggregate demand. There are perfect capital markets. The lifetime budget constraint of the representative agent is:

$$\sum_{t=1}^{\infty} \frac{y_t - \sum_{j=1}^{N} p_{tj} x_{tj}}{D_t - 1} = 0 \tag{6.2}$$

where p_{tj} is the price of commodity j in period t, y_t is income, and $D_t = (1 + r_1) \ldots (1 + r_t)$ is the inverse of the discount factor, where $1 + r_t$ is the rate of interest paid in period $t + 1$ and where D_0 is set equal to unity. Consumption at time t is given by $c_t = \sum_{j=1}^{N} p_{tj} x_{tj}$. The structure of preferences implies constant expenditure shares:

$$p_{tj} x_{tj} = \lambda c_t. \tag{6.3}$$

No storage technology is assumed to exist: hence, $c_t = y_t$ and the consumer is neither a borrower nor a saver. As in Shleifer (1986), the equilibrium interest rate is thus:

$$1 + r_t = \frac{1}{\rho} \cdot \left(\frac{y_{t+1}}{y_t}\right)^{\gamma} \cdot \left(\frac{\prod_{j=1}^{N} p_{j,t+1}^{\lambda}}{\prod_{j=1}^{N} p_{j,t}^{\lambda}}\right)^{1-\gamma}. \tag{6.4}$$

Let Π_t be aggregate profits. Labour is inelastically supplied at L. A unit of labour is the numeraire and so the wage rate is normalized to unity. The income identity is then given by:

$$y_t = \Pi_t + L. \tag{6.5}$$

There are N ordered sectors in the economy. In the first period, one firm in each of the sectors *1, 2, . . ., n* generates an invention (so there are n inventions in the first period). In the second period, one firm in each of the sectors $n + 1, n + 2, \ldots, 2n$ generates an invention. In period $T^* = N/n$ one firm in each of the sectors $(T^* - 1) \cdot n + 1, (T^* - 1) \cdot n + 2, \ldots, T^* \cdot n$ generates an invention, before the cycle repeats starting in period $T^* + 1$. An invention in period t enables firms to produce output using a fraction $1/\mu$ of the labour input which was previously required, where $\mu > 1$ is the rate of technical progress. It is this rate that we will model as uncertain in the analysis of the next section.

Firms that invent can implement immediately or delay. When a firm implements its invention, it becomes a monopolistic supplier in its sector. Its profits are

$$\pi_t = m \cdot y_t \tag{6.6}$$

where $m \equiv \lambda(1 - 1/\mu)$. In the period following the implementation, imitators enter the market and drive the profits of the innovating firm down to zero. Hence, firms have an incentive to maximize the short-run returns from implementing the innovation. They will trade off the opportunity cost of delaying the innovation to the future against the potential gain from implementing during a period of high aggregate demand.

Let $0 \leq \alpha_t \leq 1$ be the fraction of the n firms receiving an invention at time $t = 1, \ldots, T - 1$ that implement immediately. Let $\beta_T = \left(T - \sum_{t=1}^{T-1} \alpha_t\right)$ so that $\beta_T \cdot n$ denotes the number of firms that implement at time T: those who received an invention during the cycle and waited, and those who received an invention at time T. Note that $\beta_T = 1$ when all firms implement immediately and $\beta_T = T$ when they all wait until time T.

Cycles of period $T \leq T^*$ are an equilibrium if and only if $\pi_t/D_{T-1} > \pi_1$, or

$$\rho^{T-1}\left(\frac{1 - nm\beta_T}{1 - nm\alpha_1}\right)^{\gamma-1} \mu^{\lambda n(1-\gamma)\cdot\sum_{t-1}^{r-1}\alpha_t} > 1. \tag{6.7}$$

Equation (6.7) simply requires that, prior to the boom, interest rates are not so high as to offset the incentive for firms to wait until the boom to receive their profits (Shleifer, 1986) By investigating the left-hand-side of equation (6.7) we can make a number of useful observations about the degree of coordination required to sustain a T-boom equilibrium.

First, if everyone waits until time T, i.e. $\alpha_1 = \alpha_2 = \ldots = \alpha_{T-1} = 0$ and $\beta_T = T$, then equation (6.7) collapses to Shleifer's equation (12): $f(T) \equiv \rho^{T-1}(1 - nTm)^{\gamma-1} > 1$. Second, if no one waits or $\alpha_1 = \alpha_2 = \ldots = \alpha_{T-1} = \beta_T = 1$ then the LHS < 1.

Since the LHS is continuous in $\sum_{t=1}^{T-1}\alpha_t$ then there exists a $0 \leq k \leq 1$ such that LHS < 1 if and only if $\frac{\sum\alpha_t}{T-1} \leq k$. In other words, a *T*-boom can be supported as a Nash equilibrium if and only if at least a fraction *k* of the $n(T-1)$ firms receiving an innovation at periods 1, . . ., $T-1$ wait. The precise value of *k* will depend upon the parameters of the model, and from now on we will restrict attention to the case where *k* is greater than 0.5.[8]

On the other hand, immediate implementation ('cycles' of length 1) always forms an equilibrium in Shleifer's model. The pay-offs for a firm that chooses to implement immediately do not depend on the behaviour of any of the firms receiving an innovation in the same period or in the future.[9]

An important point is that cycles of size $T \geq 2$ require a much greater degree of coordination than an equilibrium with immediate implementation. It is this need for coordination that makes the cyclical equilibria potentially fragile. In the next section, we will argue that when there is uncertainty about the rate of technical progress, only the equilibrium where firms implement immediately is robust.

6.5 Extending the basic model

This section reconsiders Shleifer's cyclical equilibria, from the viewpoint of the literature on global games. Our departure from Shleifer (1986) is that we assume that the rate of technical progress μ is not known to firms. Instead, they receive a noisy signal about μ. Given equation (6.6), this corresponds to a noisy signal on profits from implementation, and our discussion will proceed in terms of the parameter *m*.

Formally, we assume that nature draws *m* once at the beginning of each cycle of inventions – that is at time 1, $T^* + 1$, $2T^* + 1$ and so on – according to a uniform distribution over $\lfloor \underline{m}, \overline{m} \rfloor$. A firm that receives an innovation also receives a signal on *m* and then chooses whether or not to delay implementing. Let *M* be a one-dimensional random variable and let $\{E_i\}_{i=1}^{n(T-1)}$ be an $n(T-1)$-tuple of i.i.d. random variables, each having zero mean. Each E_i is independent of *M*, with a continuous density and a support within $[-1,1]$. For $\varepsilon > 0$ we write:

$$M_i^{\varepsilon} = M + \varepsilon E_i. \tag{6.8}$$

If $\varepsilon = 0$ then *m* is common knowledge and we are back to Shleifer's model. We are interested in what happens when ε is arbitrarily small – that is, under almost common knowledge.

Denote by Ω the set of all values of *m* for which $f(T) < 1$ for $T = 2$ . . ., T^*.[10] Let m^* be the infimum of the set Ω. We shall now assume that $m^* \in (\underline{m}, \overline{m})$. Notice that for $m < m^*$, immediate implementation is the dominant strategy in Shleifer's model: $f(T) < 1$ for all $T \geq 2$ means that $\pi_T/D_{T-1} < \pi_1$ or that implementing immediately yields a higher pay-off than waiting until period *T even if everyone* *else is waiting*. We have seen from equation (6.7) that if some firms wait, the pay-off from waiting will increase and the pay-off from implementing immediately either does not change or decreases. But if $m < m^*$ the strategy 'implement immediately' yields a higher pay-off than the strategy 'wait until period T' for all T, regardless of the behaviour of the other firms – in other words, it is the dominant strategy.

In our extended model, this is still true: if a firm i receives a signal $m_i < m^*$ then it expects that the true value of m is smaller than m^*, because $E(M / m_i^{\varepsilon} = m_i) = m_i$. The firm will therefore implement immediately.

In Shleifer's model, for $m \geq m^*$ longer cycles, or T-booms ($T = 2, 3, \ldots, T^*$) can also be sustained as a Nash equilibrium. As we now show, this is no longer the case in our extension, when there is noise.

In deriving this result, it is important to note that we make a crucial, but artificial, simplifying assumption. We shall assume that firms select their strategies (that is, mappings from signals to behaviour) at the beginning of each cycle. This can be seen as consistent with the spirit of Shleifer's original model, but we rule out the more realistic case in which firms make their choices after receiving their signal, and in particular, after observing the actions of firms receiving innovations in previous periods. The more realistic case would require a more complicated proof: for example, a firm which receives an innovation in a given period will know that if it waits, this will affect the behaviour of firms receiving innovations in later periods.

Under more general assumptions, the arguments we use below, and in particular the symmetry argument we use to justify equation (6.10), would have to be modified to take into account the posterior distributions of firms that receive innovations later in the T-boom. This would correspond to applying the global games arguments to a more complex case where groups of players move sequentially, rather than all moving simultaneously.

It is possible that the general line of argument can be applied in such a case, as in Chamley (2007), but any extension to sequential moves will complicate the analysis. Some researchers have extended global game results to repeated games, but in the context of problems that are (in some respects) simpler. References include Giannitsarou and Toxvaerd (2007) and the papers cited therein. These extensions often consider a continuum of agents making decisions in what is essentially a repeated game (plus noise). In this paper, we do not assume a continuum, or that the game is the same in each period, and the proof in our paper relies heavily on the fact that there is a finite number of participating agents at each stage. The alternative assumption of a continuum of agents does simplify the analysis in some respects, but also misses out on the type of reasoning used in this paper: that some actions are more risky based on what I think the others think of me, which depends on what I think of them, and so on.

Extending this type of reasoning to the full dynamic case, with updating of beliefs, is far from straightforward and has not yet been resolved. We therefore focus on the case in which firms choose their strategies in advance of receiving their signal, and do not take into account the observed actions of other firms. In terms of eliminating cycles, this may or may not be a conservative assumption: arguably, a firm that observes another firm implementing immediately is less likely to delay, but, at the same time, a firm choosing to delay knows that this could encourage firms in future periods to delay. By ruling out these considerations, our assumption about the timing of strategy choices will keep the logic closer to that in the existing global games literature.

We can now prove our main result:

Proposition. In the implementation cycle model with noisy signals, the only possible equilibrium is one with immediate implementation.

Proof. By contradiction: Assume that there exists $\tilde{m}_i \geq m^*$ and a symmetric Nash equilibrium S where any firm i receiving an invention at time $t(\neq T-1)$ and a signal $m_i \geq \tilde{m}$ delays its implementation until time $T > 1$.

Denote by $\phi_i\,(m_i, s_{-i})$ the probability firm i attaches to the event that more than $kn\,(T-1)$ firms that receive innovations at periods 1, . . ., $T-1$ wait for a T-boom, when its own signal is m_i and their equilibrium strategies are s_{-i}.

Lemma. $\phi_i(m_i, s_{-i}) > 0$.

Proof. Given the structure of Shleifer's model, this result is a corollary of the maintained assumption that S is an equilibrium. A more formal proof is provided in Appendix 2.

Suppose now that $m_i - m^* < 2\varepsilon$.

$$\phi_i(m_i, S_{-i}) \leq Pr\,(\text{at least } kn \cdot (T-1) \text{ receive a signal} < m_i \,|\, M_i^E \approx m^*) \qquad (6.9)$$

$$= \Pr\,(E_j > E_i \text{ for at least } kn \cdot (T-1) \text{ components of the vector E}) \qquad (6.10)$$

$$= \sum_{m=kn}^{n} \binom{n}{m} \frac{1}{2^n}. \qquad (6.11)$$

Equation (6.9) follows from the fact that firms which receive a signal $m_j < \tilde{m}$ implement immediately (because it is a dominant strategy). As long as m_i lies inside the support of M, this probability is independent of the exact value of m_i.[11]

The expression in (6.10) follows from the symmetry of E.

Finally we note that (6.11) $\sum_{m=kn}^{n} \binom{n}{m} \frac{1}{2^n}$ converges to 0 as n increases for a fixed $k > 0.5$. Formally, using the central limit theorem, this sum converges to: $\frac{1}{\sqrt{2\pi}} \int_{(2k-1)\sqrt{n}}^{\infty} e^{-\frac{1}{2}t^2}\, dt$ which essentially is a step function equal to 1 for $k < 0.5$ and 0 for $k > 0.5$. Throughout Shleifer's paper it is assumed that n is large and so the limit applies.

If $\tilde{m} - m^* < 2\varepsilon$ then we are done. Otherwise. we can continue the same argument inductively until $\tilde{m}$ is reached. That is, we have just shown that for all signals m_i that are greater than m^* by an amount 2ε, firms will never wait before implementing their innovations. We can then repeat the same proof for m_i that are greater than m^* by 4ε and then 6ε, etc., until we reach $\tilde{m}$. We then get a contradiction with our lemma 1, which completes the proof of the proposition.[12]

6.6 Conclusions

In this paper, we have drawn attention to the contrast between the New Economy boom of the 1990s and previous cyclical fluctuations. We argue that this contrast might be explained using Shleifer's model of implementation cycles. In the first part of the paper, we present some indirect evidence in support of Shleifer's model. For example, the time series patterns and co-movements of initial public offerings and MFP growth are consistent with the view that innovations are implemented in waves. The association between these two proxies for innovation is quite strong: lagged IPOs help to predict MFP growth, even conditional on lagged MFP growth. This suggests that both these series may be useful proxies for the extent of implementation, and its variation over the business cycle.

The 1990s, however, clearly saw a decline in the volatility of productivity growth. There was a corresponding decline in the volatility of IPOs. Given the similar patterns shown by the two series, we argue that implementation cycles may have weakened in the 1990s. Again, we interpret this in terms of Shleifer's model. The decline in volatility can be explained if, instead of strategic delays, immediate implementation emerged as the equilibrium outcome.

Our theoretical contribution, the second part of the paper, explains this development in the following terms. Recall that, in Shleifer's model, the timing of the implementation of innovations is related to firms' expectations about future aggregate income. These expectations are self-fulfilling, and business cycles are driven by strategic delays supported by particular expectations. But when we extend Shleifer's model to incorporate uncertainty about the trend growth rate of the economy, the equilibria with delayed implementation are eliminated, because delay becomes risky. Business cycles with delayed implementation therefore rely on a strong common knowledge assumption, one that may not have been satisfied in the unusual circumstances of the 1990s. We argue that this could explain the reduced volatility in MFP growth and IPOs of that period: uncertainty about the trend growth rate led to immediate implementation as the sole equilibrium outcome.

Although previous researchers have demonstrated the importance of informational assumptions for multiplicity, we have shown that similar arguments apply to a classic model of the business cycle. More ambitiously, we have used this analysis to shed new light on the dynamics of the New Economy in the USA during the 1990s. We argue that introducing uncertainty into Shleifer's model of business cycles could help to explain some of the most important and unusual features of that decade.

Appendix 1. Data sources

Number of IPOs in USA: Updated data on IPOs collected by Jay Ritter, downloaded from website http://bear.cba.ufl.edu/ritter/ipoisr.htm on 3 August 2007. Note that there are small differences between the results in the current paper and those in the working paper version of this research, because of some minor revisions to the IPO data since the previous version was written.

Note that Figure 6.1, Figure 6.6, and the first regression in Table 6.2, are all based on MFP data for 1962–2006. This uses a linked series for MFP, downloaded from the BLS website, 2 August 2007. This linked series is for the private non-farm business sector (excluding government enterprises) and links SIC data for 1948–87 to NAICS (North American Industry Classification System) data for 1987–2006.

Linked series pre-1987 are not available disaggregated into durables and non-durables sectors. The other MFP series used in the paper were downloaded 29 June 2004, and are no longer updated, because the BLS has switched to the North American Industry Classification System (NAICS). Consistent series on the NAICS basis are not currently available before 1987. We therefore use the following older series that cover the much longer period 1949–2001:

MFP for private non-farm business sector: series MPU750023(K).
MFP for manufacturing, durables (SIC codes 24–25, 32–39): series MPU-320003(B).
MFP for manufacturing, non-durables (SIC codes 20–23, 26–31): series MPU310003(B).

Appendix 2. Proof of Lemma

Let $0 \leq j \leq n \cdot (T-1)$ denote the total number of firms who receive innovations at periods $1, \ldots, T-1$ and wait until period T before implementing. Let $\Pi_T/D_{T-1}(j)$ and $\Pi_1(j)$ denote the pay-offs from delaying until time T and implementing immediately (respectively) given that exactly j firms wait.

Since (by assumption) S is an equilibrium then the ex-post pay-off to the firm from waiting must be higher than that of implementing immediately:

$$p(j=n\cdot(T-1))\frac{\Pi_T}{D_{T-1}}(n\cdot(T-1))+p(j=n\cdot(T-1)-1)\frac{\Pi_T}{D_{T-1}}(n\cdot(T-1)-1)$$

$$+\cdots+p(j=0)\frac{\Pi_T}{D_{T-1}}(0)>p(j=n\cdot(T-1))\,\Pi_1(n\cdot(T-1)(+p(j=n\cdot(T-1)-1)$$

$$\times\Pi_1(n\cdot(T-1)-1)+\cdots+p(j=0)\Pi_1(0). \tag{6.12}$$

Which we can re-arrange as follows:

$$p(j = n\cdot(T-1))\left[\frac{\Pi_T}{D_{T-1}}(n\cdot(T-1)) - \Pi_1(n\cdot(T-1))\right] + \dots + p(j = kn\cdot(T-1)+1)$$

$$\times\left[\frac{\Pi_T}{D_{T-1}}(kn\cdot(T-1)+1) - \Pi_1(kn\cdot(T-1)+1)\right] > p(j = kn\cdot(T-1))$$

$$\times\left[\Pi_1(kn\cdot(T-1)) - \frac{\Pi_T}{D_{T-1}})(kn\cdot(T-1))\right]$$

$$+\dots + p(j=0)\left[\Pi_1(0) - \frac{\Pi_T}{D_{T-1}}(0)\right]. \tag{6.13}$$

Note that all the expressions in the square brackets – on both sides of the equation – are positive because $\Pi_T/D_{T-1}(j) > \Pi_1(j)$ when $j > k \cdot n \cdot (T-1)$ and $\Pi_T/D_{T-1}(j) < \Pi_1(j)$ otherwise.

Furthermore, the quantity $\Pi_T/D_{T-1}(j) - \Pi_1(j)$ increases with j (and conversely $\Pi_1(j) - \Pi_T/D_{T-1}(j)$ decreases with j). The left-hand size of (6.13) is therefore smaller than $\phi\left[\frac{\Pi_T}{D_{T-1}}(n\cdot(T-1)) - \Pi_1(n\cdot(T-1))\right]$ (because the probabilities sum up to ϕ). The right-hand side of equation (6.13) is greater than

$$(1-\phi)\left[\Pi_1(kn\cdot(T-1)) - \frac{\Pi_T}{D_{T-1}}(kn\cdot(T-1))\right].$$

Using these and the inequality (6.13) we get:

$$\phi\left[\frac{\Pi_T}{D_{T-1}}(n\cdot(T-1)) - \Pi_1(n\cdot(T-1))\right]$$

$$> (1-\phi)\left[\Pi_1(kn\cdot(T-1)) - \frac{\Pi_T}{D_{T-1}}(kn\cdot(T-1))\right]$$

or $\phi A > (1-\phi)B$ where A and B are both positive. Solving for ϕ we get $\phi > \frac{B}{A+B} > 0$.

Notes

1 The MFP growth series is constructed from historical Bureau of Labor Statistics data on MFP levels. The MFP data we use are based on the SIC classification, and are no longer updated by the BLS. We use them because they cover a longer span than the current measures based on the North American Industry Classification System (NAICS), which are not available before 1987. See the data appendix for more details and the data sources.

2 The main problem for this view is that it does not explain the significant positive autocorrelation seen in MFP growth, unless there are major technological shocks that have an economy-wide impact sustained over several years. As is sometimes discussed in the real business cycle literature, it is not clear that innovations are sufficiently

pervasive to generate the cyclical patterns seen in the aggregate data (see, for example, Stadler, 1994).

3 For these (Wald) test statistics to have their standard limiting distributions, the series must be stationary. For the various MFP growth series, we can easily reject the null of a unit root under a range of assumptions, using augmented Dickey-Fuller tests. For the IPO series, the results are slightly less clear-cut, but DF-GLS tests reject the null at the 10 per cent level for a wide range of lag choices.

4 Direct evidence is hard to find. For example, Jorgenson and Stiroh (2000) note a shortening in the product cycle of microprocessors, with new processors brought to market more quickly in the 1990s than previously, but this may reflect an acceleration in technical change rather than an elimination of implementation lags.

5 Note that the hypothesis of declining volatility is conceptually separate from our earlier hypothesis that IPOs and MFP growth move together. It is possible for the covariance of these two series to increase (as appears to have happened in 1980–2001, compared to the earlier 20 year period) at the same time as the short-run volatility of each series shows a decline.

6 We use the median absolute deviation because of the small number of observations and the possibility that the year to year variation in the series, such as MFP growth rates, might contain significant measurement errors. The patterns we describe below are also apparent when we use the standard deviation as a measure of volatility (see the working paper version of this research).

7 The fragility of some equilibria in the presence of uncertainty and correlated signals has been analysed by several authors in different contexts. Shin (1995) considers a decentralized economy with search externalities. Morris and Shin (1998) look at the timing of speculative attacks against a currency. Scaramozzino and Vulkan (2004) examine a model of local oligopoly with correlated noise about the competitive advantage of firms. See Morris and Shin (2000, 2003) for details of more applications, especially to macroeconomic issues.

8 This is not unduly restrictive. Numerical simulations suggest that k is substantially higher, around 0.75 for many parameter values that satisfy Shleifer's parameter restrictions.

9 Of course, the pay-off from implementing immediately will be higher if firms that received innovations in the past waited until this period.

10 In fact, in many cases it is sufficient that $f(2) < 1$ for this to hold. See figure 2, page 1176 in Shleifer and the discussion found there.

11 Formally, we also require that the variance of m must be sufficiently small relative to the dispersion of the private signals (that is, relative to ε) to ensure this is still true even if m is close to the boundaries of $\lfloor \underline{m}, \bar{m} \rfloor$.

12 The arguments used in this proof are in the spirit of the global game literature. In particular, we have followed an approach used by Kim (1996). While Kim's approach is more general than ours, our proof can be seen as an extension of his Proposition 4, since the game in this paper does not fall directly in the class of games he studies (the game specified by Shleifer's model does not satisfy Kim's assumption 1 and exhibits more than two equilibria).

References

Azariadis, C. (1981) 'Self-fulfilling prophecies', *Journal of Economic Theory*, 25(3): 380–96.

Beaudry, P. and Portier, F. (2006) 'Stock prices, news, and economic fluctuations', *American Economic Review*, 96(4): 1293–1307.

Beaudry, P., Collard, F. and Portier, F. (2011) 'Gold rush fever in business cycles', *Journal of Monetary Economics*, 58(2): 84–97.

Blanchard, O. and Simon, J. (2001) 'The long and large decline in US output volatility', *Brookings Papers on Economic Activity*, 32(1): 135–74.

Cass, D. and Shell, K. (1983) 'Do sunspots matter?', *Journal of Political Economy*, 91(2): 193–227.

Chamley, C. (2007) 'Complementarities in information acquisition with short-term trades', *Theoretical Economics*, 2(4): 441–467.

Farmer, R.E.A. (1993) *The Macroeconomics of Self-Fulfilling Prophecies,* Cambridge (MA) and London, MIT Press.

Fernández-Villaverde, J., Guerron-Quintana, P.A. and Rubio-Ramirez, J.F. (2010a) 'Reading the recent monetary history of the US, 1959–2007', *Federal Reserve Bank of St. Louis Review,* 92(4): 311–38.

Fernández-Villaverde, J., Guerron-Quintana, P.A. and Rubio-Ramirez, J.F. (2010b) 'Fortune or virtue: time-variant volatilities versus parameter drifting in US data', *CEPR Discussion paper*, 7813.

Francois, P. and Lloyd-Ellis, H. (2008) 'Implementation cycles, investment and growth', *International Economic Review,* 49(3): 901–42.

Giannitsarou, C. and Toxvaerd, F. (2007) 'Recursive global games', *CEPR Discussion paper*, 6470.

Hobijn, B. and Jovanovic, B. (2001) 'The information technology revolution and the stock market: evidence', *American Economic Review*, 91(5): 1203–20.

Jaimovich, N. and Rebelo, S. (2009) 'Can news about the future drive the business cycle?', *American Economic Review*, 99(4): 1097–118.

Jenkinson, T. and Ljungqvist, A. (2001) *Going Public,* 2nd edn, Oxford, Oxford University Press.

Jorgenson, D.W. and Stiroh, K. (2000) 'Raising the speed limit: US economic growth in the information age', *Brookings Papers on Economic Activity*, 31(1): 125–235.

Keynes, J.M. (1936) *The General Theory of Employment, Interest and Money,* London, Macmillan.

Kim, Y. (1996) 'Equilibrium selection in n-person coordination games', *Games and Economic Behavior*, 15(2): 203–27.

Leduc, S. and Sill, K. (2007) 'Monetary policy, oil shocks, and TFP: Accounting for the decline in US volatility', *Review of Economic Dynamics*, 10(4): 595–614.

Matsuyama, K. (1995) 'Complementarities and cumulative processes in models of monopolistic competition', *Journal of Economic Literature*, 33(2): 701–29.

McConnell, M.M. and Perez-Quiros, G. (2000) 'Output fluctuations in the United States: what has changed since the early 1980s?', *American Economic Review*, 90(5): 1464–76.

Morris, S. and Shin, H.S. (1998) 'Unique equilibrium in a model of self fulfilling currency attacks', *American Economic Review*, 88(3): 587–97.

Morris, S. and Shin, H.S. (2000) 'Rethinking multiple equilibria in macroeconomics', *NBER Macroeconomics Annual*, 139–61.

Morris, S. and Shin, H.S. (2003) 'Global games: theory and applications', in M. Dewatripont, L. Hansen and S. Turnovsky (eds) *Advances in Economic and Econometrics (Proceedings of the Eighth World Congress of the Econometric Society),* Cambridge University Press, Cambridge, England.

Pástor, Ĺ. and Veronesi, P. (2005) 'Rational IPO Waves', *Journal of Finance*, 60(4): 1713–57.

Pástor, Ĺ. and Veronesi, P. (2009) 'Technological revolutions and stock prices', *American Economic Review*, 99(4): 1451–83.

Robson, M., Townsend, J. and Pavitt, K. (1988) 'Sectoral patterns of production and use of innovations in the UK: 1945–83', *Research Policy*, 17(1): 1–14.

Scaramozzino, P. and Vulkan, N. (2004) 'Uncertainty and endogenous selection of economic equilibria', *Metroeconomica*, 55(1): 22–40.

Shin, H.S. (1995) 'Defining a notion of confidence in an economy with differential information', *Discussion Paper Series in Economics and Econometrics 9526*, University of Southampton.

Shleifer, A. (1986) 'Implementation cycles', *Journal of Political Economy*, 94(6): 1163–90.

Silvestre, J. (1993) 'The market-power foundations of macroeconomic policy', *Journal of Economic Literature*, 31(1): 105–41.

Stadler, G. (1994) 'Real business cycles', *Journal of Economic Literature*, 32(4): 1750–83.

Stiroh, K. (1999) 'Is there a New Economy?', *Challenge*, 42(4): 82–101.

Stock, J.H. and Watson, M.W. (2002) 'Has the business cycle changed and why?', *NBER Macroeconomics Annual*: 159–218.

Van Reenen, J. (1996) 'The creation and capture of rents: wages and innovations in a panel of UK companies', *Quarterly Journal of Economics*, 111(1): 195–226.

7 The impact of technological change on employment and skills

Theory and empirical evidence

Mariacristina Piva and Marco Vivarelli

7.1 Introduction

The assessment of the possible effects of technological change on employment is an old[1] and controversial issue. In a nutshell, the classical debate was characterised by two opposing views. On the one hand, fear of technological unemployment as a *direct* consequence of labour-saving innovation was the source of social and political concern; on the other, economic theory pointed out the existence of *indirect* effects which could compensate for a reduction in employment due to technological progress (see Marx, 1961).

Indeed, over the last three decades the worldwide diffusion of a 'new economy' based on ICT technologies has led to a re-emergence of this classical debate.

Nevertheless, recent economic analyses have also emphasised two additional important aspects of the relationship between technological change and employment: the role of product innovation and the relevance of human capital. First, with respect to product innovation, it is possible to single out a labour-friendly impact for new products which require additional employees; second, looking at the 'quality' of workers, there is a possible 'qualitative' effect of innovation which might be complementary to skilled workers and might displace unskilled workers.

In this chapter, the theoretical and empirical literature on the quantitative and qualitative employment impact of technological change will be critically discussed, the comparative explanatory power of the competing theories will be compared and the available macro and micro evidence on the issue will be presented.

The next section will be devoted to the possible 'compensation mechanisms' and it will present a survey of the available empirical evidence on the relationship between innovation and employment at different levels of data aggregation. Section 3, focusing on the 'quality' of workers, will analyse the literature (mainly empirical) on the so called 'skill-biased technological change' (SBTC). Section 4 will briefly draw conclusions.

7.2 Innovation and employment

7.2.1 Technological innovation and compensation mechanisms

Technological change allows production of the same amount of goods with a lower amount of production factors, namely capital and labour. Interestingly enough, 'technological unemployment' occurs as a direct effect of innovation irrespective of its intrinsic nature (neutral, labour-saving or capital saving). Nevertheless, back in the first half of the nineteenth century – while the Luddites were destroying the new machines – economists put forward a theory which Marx later called the 'compensation theory' (see Marx, 1961 and 1969). This theory is made up of six different market compensation mechanisms which are triggered by technical change itself and which can counterbalance the initial labour-saving impact of process innovation, especially in case of perfect competitive markets (for an extensive analysis, see also Petit, 1995; Vivarelli, 1995; Vivarelli and Pianta, 2000; Pianta, 2005). Hereafter, these mechanisms are recalled and critically discussed.

a) The compensation mechanism 'via new machines'

The same process innovations which displace workers in the user industries create new jobs in the capital sectors where the new machines are produced (see, for instance, Say, 1964). However, it should be kept in mind that the new machines can be implemented either through additional investments (see point *c*) or simply by substitution of the obsolete ones (scrapping). In this latter case – which is indeed the most frequent one – there is substitution, but no compensation at all (see, for instance, Freeman *et al.*, 1982).

b) The compensation mechanism 'via decrease in prices'

On the one hand, process innovations involve the displacement of workers; on the other hand, these innovations themselves lead to a decrease in the costs of production per unit and – in a competitive market – this effect is transferred into decreasing prices. In turn, decreasing prices could stimulate a new demand for products, and thus for additional production and employment (Steuart, 1966).[2]

Even in this case, as originally noted by Malthus (1964) and Sismondi (1971), the very first effect of labour-saving process innovation is a decrease in aggregate demand due to the cessation of the demand previously associated with the dismissed workers. So, the mechanism 'via decrease in prices' deals with a decreased demand and has to more than counterbalance the initial decrease in the aggregate purchasing power.

Furthermore, the effectiveness of this mechanism depends upon the hypothesis of perfect competition. If an oligopolistic market structure is dominant, the compensation effect is considerably weakened since cost savings are not necessarily and entirely translated into a decrease in prices (see Sylos Labini, 1969).

c) The compensation mechanism 'via new investments'

If there is not a simultaneous transfer of the reduced costs of production (due to innovation) into lower prices, extra profits may be accumulated by the innovative firms. These profits might be invested in order to stimulate new production and create new jobs (Ricardo, 1951; Marshall, 1961). In any case, it cannot be granted the translation of additional profits into 'effective demand'; in other words, the theoretical analysis has to take into account the economic agents' expectations, which can delay this translation (see Pasinetti, 1981; Freeman and Soete, 1987; Appelbaum and Schettkat, 1995).

d) The compensation mechanism 'via decrease in wages'

As with other forms of unemployment, the direct effect of labour-saving technologies may be compensated within the labour market through a price adjustment. In a neoclassical framework, a decrease in wages leads to an increase in the demand for labour (see Wicksell, 1961; Neary, 1981; Sinclair, 1981). Nevertheless, the reduced aggregate demand – following the decrease in the number of employees – lowers the employers' expectations, who might then tend to hire fewer workers.

e) The compensation mechanism 'via increase in incomes'

In a Fordist mode of production, unions take part in the distribution of the fruits of technical progress; thus, it has to be taken into account that a portion of the cost savings which result from technical change could be translated into higher income and hence higher consumption. This increase in demand leads to an increase in employment, which may compensate the initial job losses due to process innovations (see Pasinetti, 1981; Boyer, 1988a, 1988b and 1990). Nowadays, the Fordist mode of production is almost over (see Boyer, 1990) and the distribution of income follows different rules. On the whole, this compensation mechanism has been considerably weakened in the new institutional context.

f) The compensation mechanism 'via new products'

Technical change is not equivalent to process innovation, but in a number of cases it might assume the form of creation of new products; in this case, new branches might develop, leading to the creation of additional jobs (see Say, 1964, and, in the current debate, Freeman *et al.*, 1982; Freeman and Soete, 1987; Freeman and Soete, 1994; Vivarelli and Pianta, 2000; Edquist *et al.*, 2001). New products are still the more capable way to counterbalance labour-saving process innovations. Yet, different 'technological paradigms' (see Dosi, 1982) are characterised by different clusters of new products which turn out to have very different impacts on employment.

In conclusion, the relationship between innovation and employment can be represented by a very complex picture wherein the direct labour-saving impact of process innovation, the compensation mechanisms, the drawbacks and hindrances which can severely weaken the effectiveness of such mechanisms, and the labour-friendly nature of product innovation can combine in diverse ways.

Indeed, economic theory does not have a clear-cut solution to the employment effect of innovation; thus, attention should be turned to aggregate, sectoral and microeconomic empirical analyses which take into account the different forms of technical change, their direct effects on labour, the various compensation mechanisms, and the possible hindrances to these mechanisms.

7.2.2 Innovation and employment: empirical studies

While theoretical economists develop models about the employment impact of process and product innovations, applied economists have 'to measure' process and product innovations, as well as the compensation mechanisms and the final employment impact of innovation. In pursuing this task, three main problems arise.

First, innovation is hard to measure. Traditional indicators such as R&D (input indicator) and patents (output indicators) are seldom fully available and are often an inadequate representation of technological progress. For instance, most of process innovation is implemented (especially in Small and Medium Enterprises – SMEs) through 'embodied technological change'.[3]

Second, the final employment impact of innovation depends upon institutional mechanisms which can be very different at the micro, meso and macro levels and can vary in different institutional and legal contexts (such as, for instance, the protection of intellectual property rights).

Third, it is difficult to distinguish the final impact of innovation on employment, since the latter is possibly influenced by other factors such as macroeconomic conditions, labour market dynamics, trends in working time, and so on.

Empirical studies are run at the aggregate, sectoral and micro levels. Different aggregation levels allow us to better disentangle the employment impact of innovation in terms of the specific advantages and disadvantages they present.

Macroeconomic studies are attractive since they can take into account all of the direct and indirect effects of innovation, but they are often severely limited by the difficulty of finding a representative aggregate proxy of technological change and by the fact that the final employment national trend is co-determined by institutional and macroeconomic determinants which are difficult to ascertain.

To a lesser extent, these considerations also apply to the empirical analysis run at the sectoral level. On the opposite side, microeconometric firm-level studies have the great advantage of allowing a direct and precise firm-level mapping of innovation variables, both in terms of inputs and outputs. Obviously, the significant limitations of firm-level data should not be forgotten. First, the microeconomic approach cannot take into account the indirect 'compensation effects' which operate at the level of the firm but also through intersectoral linkages and aggregate dynamics. Second, there might be a 'positive bias', in so far as innovative firms tend to be characterised by better employment performances since they gain market shares, not taking into account the important effect upon the rivals, which are crowded out by the innovative firms ('business stealing' effect).

As far as aggregate empirical studies are concerned, macroeconometric studies have tried to test the validity of the compensation mechanisms within a partial or general equilibrium framework.

For instance, for the US, Sinclair (1981) puts forward a macro IS/LM scheme and concludes that positive employment compensation can occur if the demand elasticity and the elasticity of factor substitution are sufficiently high. The author finds strong evidence in support of the mechanism 'via decrease in wages', but not for the mechanism 'via decrease in prices'.

In the case of the UK, Layard and Nickell (1985) derive a labour demand in a quasi-general equilibrium framework and state that the crucial parameter is the elasticity of the demand for labour in response to a variation in the ratio between real wages and labour productivity. The authors estimate an elasticity coefficient equal to 0.9 and this is sufficient – in their opinion – to rule out technical change from the possible causes of British unemployment. Using a similar approach, Nickell and Kong (1989) focus their attention on the compensation mechanism 'via decrease in prices' in nine UK two-digit industries. The authors found that in seven of the nine sectors a sufficiently high demand elasticity is able to imply an overall positive impact of technical change upon employment.

In Vivarelli (1995), the direct labour-saving effect of process innovations, the compensation mechanisms, and the job-creating impact of product innovations have been tentatively represented and estimated through a simultaneous equations model for the period 1960–88. Running 3 SLS regressions using Italian and US data, the author shows how – in both countries – the more effective compensation mechanism is 'via decrease in prices', while other mechanisms turn out to be less important.

Finally, Simonetti *et al.* (2000) apply the same simultaneous equations macroeconomic model as Vivarelli (1995), using US, Italian, French and Japanese data for the period 1965–93. The authors provide evidence that the more effective compensation mechanisms are 'via decrease in prices' and 'via increase in incomes', especially in European countries. The other mechanisms are revealed to be less significant and conditional upon the institutional structures in different countries.

Turning to the sectoral empirical studies, the comparison between manufacturing and services might play a role, assuming that the compensation mechanism 'via new product' (or 'services') accelerates the secular shift from manufacturing to services (see Evangelista and Perani, 1998; Evangelista and Savona, 1998).

In a study based upon Italian data, Vivarelli *et al.* (1996) shows that in Italian manufacturing sectors product and process innovations have opposite effects upon the demand for labour, in line with what we discussed in the previous sections.

As previously mentioned, this scenario may change if service sectors are taken into account: for instance, using standardised sectoral data derived from national CISs (Community Innovation Surveys), Pianta (2000) and Antonucci and Pianta (2002) find an overall negative impact of innovation upon employment in manufacturing industries across five European countries, while, in contrast, Evangelista (2000) and Evangelista and Savona (2002) find a positive employment effect from technological change in the most innovative and knowledge-intensive service sectors.[4]

Recently, considering both manufacturing and service industries and using CIS cross-sectional sectoral data on relevant innovations for different European countries, Bogliacino and Pianta (2010) find a positive employment impact of product innovation against a negative impact of process innovation. Finally, Bogliacino and Vivarelli (2011) – running GMM-SYS panel estimations covering 25 manufacturing and service sectors for 15 European countries over the period 1996–2005 – provide evidence that R&D expenditures which foster product innovation induce a job-creating effect.

Overall, sectoral analyses are more recent than aggregate ones and, due to a greater availability of firm-level data-sets in the last decade, microeconomic studies started later, but are now flourishing. Hereafter, firm-level studies are presented clustered on the basis of country of analysis.

Starting from the earlier contributions for Germany, Entorf and Pohlmeier (1990) find a positive employment impact of product innovation – measured with a dummy – in a cross section of 2,276 West German firms in 1984. Smolny (1998) confirms this positive result in West Germany, using a panel of 2,405 manufacturing firms over the 1980–92 period. More recently, Lachenmaier and Rottmann (2011) – using a very comprehensive data-set of German manufacturing firms over the period 1982–2002 – put forward a dynamic employment equation including wages, gross value added, years and industries controls and alternative proxies (dummies) of current and lagged product and process innovations. Their GMM-SYS estimates show a significantly positive impact of different innovation measures upon employment, but, partially in contrast with expectations and previous contributions, the authors find a higher positive impact for process rather than product innovation.

Turning to the UK, using 1984 firm-level data, both Machin and Wadhwani (1991) and Blanchflower *et al.* (1991) find a negative raw correlation between ICT adoption and employment; however, once controlled for workplace characteristics and fixed effects, the same correlation turns out to be positive. Van Reenen (1997) matches the London Stock Exchange database of manufacturing firms with the SPRU innovation database and obtains a panel (more informative than previous cross-sectional data) of 598 firms over the period 1976–82. Running GMM-DIF estimates, the author finds evidence of a positive employment impact of innovation; this result turns out to be robust after controlling for fixed effects, dynamics and endogeneity. By the same token, Blanchflower and Burgess (1998) confirm a positive link between innovation (roughly measured with a dummy) and employment using two different panels of British and Australian establishments; their results are robust after controlling for sectoral fixed effects, size of firm, and union density.

In France, an interesting panel analysis is conducted by Greenan and Guellec (2000) using microdata from 15,186 manufacturing firms over the 1986–90 period. According to this study, innovating firms create more jobs than non-innovating ones, but the reverse is true at the sectoral level, where the overall effect is negative and only product innovation is job-creating. Interestingly enough, the opposite employment impact of innovation at the firm and sectoral level may be due to the 'business stealing' effect previously discussed.

In the case of Italy, even when taking the 'business stealing' effect into account by controlling for lagged firms' employment and current sales, Piva and Vivarelli (2004 and 2005) find evidence in favour of a positive effect of innovation upon employment at the firm-level. In particular – applying a GMM-SYS methodology to a longitudinal dataset of 575 manufacturing firms over the period 1992–97 – the authors provide evidence of a significant (although small in magnitude) positive link between a firm's gross innovative investment and employment. Furthermore, Hall *et al.* (2008), on a panel of manufacturing firms over the period 1995–2003, find a positive employment contribution from product innovation and no evidence of employment displacement due to process innovation.

In the case of the US, Doms *et al.* (1997) find that advanced manufacturing technologies – measured by dummy variables – imply higher employment growth in US manufacturing plants over the period 1987–91. However, limiting the analysis to specific US industries, Coad and Rao (2011) focus on high-tech manufacturing industries over the period 1963–2002 and investigate the impact of a composite innovativeness index (comprising information on both R&D and patents) upon employment. The main outcome of their quantile regressions is that innovation and employment are positively linked and that innovation has a stronger impact for those firms that reveal the fastest employment growth.

More recent studies try to overcome the single-country dimension of the analyses.

Using firm-level data from the CISs in four European countries (Germany, France, UK, Spain), Harrison *et al.* (2008) put forward a testable model – based on Peters (2004) – which is able to distinguish the relative employment impact of process and product innovation (discrete variables). The authors conclude that process innovation tends to displace employment, while product innovation is basically labour-friendly. However, compensation mechanisms at work, especially in the service sectors, are particularly effective through the increase in the demand for the new products. Finally, Bogliacino *et al.* (2012) – using a longitudinal database covering 677 European manufacturing and service firms over the period 1990–2008 and considering the sectors firms belong to – find that a positive and significant employment impact of R&D expenditures is detectable in high-tech manufacturing and service sectors but not in the more traditional manufacturing sectors.

Overall, the empirical evidence does not find clear-cut answers on the impact of innovation over employment. On the whole, most of the recent investigations tend to support a positive link between innovation and employment, especially when R&D and/or product innovation are adopted as proxies of technological change in high-tech sectors and services.

7.3 Innovation and skills

Beyond the quantitative effect of new technologies upon the number of employees, it is also possible to single out and investigate the *qualitative* effect of technological change on the different categories of workers. Initially proposed by Griliches (1969) and Welch (1970), the SBTC hypothesis is based upon the idea that there is close complementarity between new technologies and skilled workers, given that only the latter are fully able to effectively and efficiently implement those technologies. Therefore a positive relationship between new technologies and skilled workers is expected; meanwhile, a substitution effect between new technologies and unskilled workers is likely to happen.

A direct consequence of this hypothesis is that an insufficient number of qualified workers may be considered as a kind of 'human resource constraint' (Amendola and Gaffard, 1988), limiting the adoption of new technologies on the one hand, and the pursuit of full employment on the other.

Besides the intuition provided, the literature on the SBTC is mainly empirical, starting from the consideration that during the last three decades – while the new ICT technologies were spreading rapidly – the main OECD countries have shown a significant change, both in the composition of the labour force and in the wage shares, in favour of the skilled component of the labour force (see, for instance, Nickell and Bell, 1995; OECD, 1996 and 1998).[5] In accordance with different institutional systems and different degree of flexibility in the labour markets, economies have shown a dominant role for either the 'employment effect' or the 'wage effect'. In particular, an increase in wage differentials between skilled and unskilled workers has been registered in countries where labour markets are more flexible, such as the UK (Haskel and Slaughter, 1998) and the US (Autor *et al.*, 1998); meanwhile, in the continental European countries – i.e. Germany (Abraham and Houseman, 1995), Italy (Casavola *et al.*, 1996), Sweden (Hansson, 1997) and France (Goux and Maurin, 2000) – the increase in wage differentials has been more limited. Symmetrically, in these latter countries, there has been a greater impact upon employment figures, with higher unemployment levels at least partly due to the reduction of the employment of unskilled workers, which is only partially compensated by an increase in the employment of skilled workers.

From a methodological point of view, the empirical studies to test the SBTC hypothesis are generally based on the estimation of a restricted function of total variable costs where capital and technology are assumed to be quasi-fixed factors. Then, by means of Shephard's lemma, either the wage bill shares of each class of workers or the employment shares are derived and estimated using various proxies for technology.

The empirical papers are presented, country by country, starting from articles devoted to North America (US – the first country where the SBTC has been studied – and Canada), then shifting to UK and other European countries.

The seminal paper by Berman *et al.* (1994) is considered as the start of this stream of empirical literature. The authors analyse the dynamics of 450 US manufacturing sectors during the 1980s which showing a shift in employment structure in favour of skilled workers depending upon the investments in computers and R&D. This result turns out to be statistically significant in almost all sectors. Autor *et al.* (1998) extended the previous study over a longer period (1950–90), and also included non-manufacturing sectors. Their results confirmed the complementarity relationship between investment in computers and the skill structure. Nonetheless, sectoral outcomes are also confirmed by studies using microdata.

For instance, Dunne *et al.* (1997), using US manufacturing firms data from the period 1972–88, show a positive, if not robust, relation between 17 advanced industrial technologies and skilled labour; however, when they use R&D as a proxy for innovation, they find a positive and significant relationship between R&D and skilled labour. Doms *et al.* (1997), following a similar approach, provide evidence that the use of the most advanced industrial technologies – in a cross-sectional framework – could lead to a larger utilisation of highly qualified workers. Nevertheless, once the time dimension (1987–92) is taken into account, they show that the firms which introduced these innovations have better-skilled workers both before and after the introduction. Finally, Morrison-Paul and Siegel (2001), using an iterative 3SLS method and data on US manufacturing sectors covering a long period (1959–89) and considering a more refined classification of workers (four educational classes), show that technology has the strongest positive impact on highly educated workers, and simultaneously reduces the demand for workers without a college degree.

In Canada, Betts (1997), analysing the dynamics of the Canadian manufacturing economy over a two-decade period (1962–86), confirm the role of the technological variable (roughly measured as a temporal trend) in generating a statistically significant skill-biased effect in most of the industrial sectors (11 out of 18). Gera *et al.* (2001) update the previous analysis up to 1994 and highlight how technological proxies, this time measured in four alternative dimensions, cause a significant skill-biased effect.

Besides North American countries, empirical results generally support the SBTC hypothesis.

In the case of the UK, Machin (1996) uses different data levels of analysis – sector-level data (1979–90) and firm-level data (1984–90) – and shows a positive relation between proxies for innovation and skilled labour, i.e. R&D intensity/ number of innovations and skilled labour in the sectoral analysis, and the use of computers and skilled employees in the case of firms. These results are reinforced by a firm-level study by Haskel and Heden (1999), which found that the positive relation between investment in new technologies/computers and the presence of skilled workers is confirmed.

In the case of Italy, Casavola *et al.* (1996) used data on 36,000 Italian firms between 1986 and 1990 in order to study the effect of technological change on labour demand and wages for skilled workers, reaching the conclusion that the wage differences between categories in Italy is lower than elsewhere, but that technological change has a positive effect upon the employment of skilled workers. However, Piva and Vivarelli (2001) found – based upon a sample of 412 manufacturing firms (1991–97) – that the supposed link between R&D and the skill bias is not confirmed. Nevertheless, they provide evidence that the reorganisation process of production has a significant impact upon workers, especially at the work floor level, where organisational changes seems to drive the upskilling of the labour force. This result does not mean that the Italian manufacturing sector, characterised by the presence of SMEs, is immune to the SBTC; however, this probably does not take place through internal channels – in-house innovation from intra-moenia R&D expenditure – but by means of organisational changes connected to the purchase and use of new technologies. Along this interpretative line, Piva *et al.* (2005), estimating a SUR model for a sample of Italian manufacturing firms, confirmed that upskilling is more a function of reorganisational strategies than a consequence of technological change alone.

In Germany, Falk and Seim (1999) study the impact of new technologies on educated workers at the firm-level (900 firms over the period 1994–96), in service sectors only. The firms which devoted more resources into ICT turned out to employ more educated workers, even if the dimension of the relationship is small. Turning to manufacturing, Falk (2001) – using a system of static and dynamic factor demand equations and considering three categories of workers in 35 sectors for the period 1978–94 – shows how the increase in office machinery and computer capital has accounted for at least 60 per cent of the expanding employment of university graduates.

Looking at French studies, Mairesse *et al.* (2001) use firm-level data for different years (1986, 1990 and 1994) and obtain cross-sectional results supporting the SBTC hypothesis (the proxy for technology is ICT capital); however, only the negative relation between ICT and less-qualified labour turns out to be robust in the time-series. Moreover, Goux and Maurin (2000), analysing 34 French manufacturing sectors between 1970 and 1993, showed how an increased adoption of new technology (both computers and new industrial technology) explains only 15 per cent of the overall change in labour demand; meanwhile, the upskilling would be caused by the movements of the demand towards skill-intensive productions.

Finally, in the case of Spain, Aguirregabiria and Alonso-Borrego (2001) test the relationship between new technologies and upskilling using a panel of approximately 1,000 manufacturing firms. Taking as a technological variable a dummy on the introduction of 'technological capital', econometric results support the SBTC hypothesis, while no significant effect emerges when referring to physical capital or R&D expenditure. More recently, Luque (2005) – using an unbalanced panel data of 1,000 firms over the 1990–98 period – shows, via a statistical decomposition methodology, that a raise in the skill mix comes mainly from surviving firms increasing their skill mixes in response to retooling or upgrades in technology; moreover, firms belonging to high-tech sectors account for the majority of the increase in the skill mix.

A few studies go beyond the country level. Machin and Van Reenen (1998) set up a panel at the manufacturing-sector level for seven countries (Denmark, France, Germany, Japan, Sweden, UK and US) and show that the relative demand for skilled workers is positively linked to R&D expenditure. The robustness of the results is confirmed both with reference to alternative econometric specifications and with consideration of the possible problems of endogeneity, i.e. the reverse causal relation between technological innovation and skilled labour.[6]

On the whole, the current empirical evidence – based upon firm-level and sectoral level studies – is in favour of the skill-biased nature of new technologies and it turns out to be large, robust and proved across different OECD countries,[7] different economic sectors, and different proxies of innovation and skilled/ unskilled workers.

7.4 Conclusions

In this final section, the main results from the previous sections on the innovation-employment (quantitative and qualitative) relationship will be summarised.

1 With reference to the quantitative effects of innovation upon employment, process innovation tends to be labour-saving, while product innovation shows a labour-friendly nature.
2 According to classical 'compensation theory' mechanisms, market forces should assure a complete and automatic compensation of the initial labour-saving impact of innovations. However, both theoretically and empirically, these mechanisms do not work perfectly; they might be partially at work, depending upon institutional settings, market structures, macroeconomic cycles, and expectations of economic subjects.
3 Empirical studies on the quantitative effect of innovation over employment are run at the aggregate, sectoral, and microeconomic levels. In particular, microeconometric studies have the great advantage of allowing a direct, firm-level mapping of innovation variables and the testing of the direct impact upon the number of employees. Overall, the microeconometric literature tends to support a positive link between technology and employment, especially when R&D and/or product innovations are adopted as proxies of technological change, and when high-tech and service sectors are analysed.
4 Turning attention toward the qualitative employment effect of technological change, i.e. the so called 'skill-bias technological change', the sectoral and firm-level empirical evidence in favour of the skill-biased nature of new technologies is large and statistically robust among North American and European countries in both manufacturing and service sectors.

Notes

1 Historically, the fear of technological unemployment has always emerged in ages characterised by radical technological changes. For instance, the striking response of the English workers to the first industrial revolution was the destruction of machines in industrial areas and in the countryside (see Hobsbawm, 1968; Hobsbawm and Rudé, 1969).
2 Indeed, the compensation mechanism 'via decrease in prices' has been re-proposed many times in the history of economic thought both by neo–classical economists at the beginning of this century (see Clark, 1907; Pigou, 1962) and by modern theorists (see Neary, 1981; Stoneman, 1983; Dobbs *et al.*, 1987; Nickell and Kong, 1989).
3 The embodied nature of technological progress and the effects related to its spread in the economy were originally discussed by Salter (1960) and Solow (1960). More recently, the role of capital accumulation in inducing productivity growth and economic development has been recognised by growth theorists (see Hulten, 1992; Greenwood *et al.*, 1997; Hercowitz, 1998; Wilson, 2009).

4 While a negative one in the case of finance-related sectors and most traditional services like trade and transport.

5 Workers are split into two categories on the basis of their tasks: skilled and unskilled. Skilled workers do not undertake manual work (white collars), while the unskilled are manual workers (blue collars). An alternative classification is based on educational levels. The indicator based upon task seems more suitable for verifying the presence of upskilling deriving from labour demand, while the education-based indicator partly reflects the continuous increase in the supply of skill.

6 The possible problem of endogeneity is based upon the consideration that only the firms which already have skilled workers see the potential for innovative investment; in the study under consideration, the robustness control is carried out – successfully – by instrumenting the R&D expenditure technological variable with government expenditure in R&D, a variable which is not correlated with the error term (see also Piva and Vivarelli, 2009).

7 See Piva (2003) and Vivarelli (2011) for a critical survey of studies testing the SBTC hypothesis in developing countries.

References

Abraham, K. and Houseman, S. (1995) 'Earnings Inequality in Germany', in R.B. Freeman and L.F. Katz (eds) *Differences and Changes in Wage Structures*, Chicago: NBER, University of Chicago Press.

Aguirregabiria, V. and Alonso-Borrego, C. (2001) 'Employment Occupational Structure, Technological Innovation, and Reorganization of Production', *Labor Economics*, 8(1): 43–73.

Amendola, M. and Gaffard, J.L. (1988) *The Innovative Choice*, Oxford: Basil Blackwell.

Antonucci, T. and Pianta, M. (2002) 'Employment Effects of Product and Process Innovation in Europe', *International Review of Applied Economics*, 16(3): 295–307.

Appelbaum, E. and Schettkat, R. (1995) 'Employment and Productivity in Industrialized Economies', *International Labor Review*, 134(4–5): 605–23.

Autor, D., Katz, L. and Krueger, A. (1998) 'Computing Inequality: Have Computers Changed the Labor Market?', *Quarterly Journal of Economics*, 113(4): 1169–1214.

Berman, E., Bound, J. and Griliches, Z. (1994) 'Changes in the Demand for Skilled Labor within U.S. Manufacturing Industries: Evidence from the Annual Survey of Manufacturing', *Quarterly Journal of Economics*, 109(2): 367–97.

Betts, J. (1997) 'The Skill Bias of Technological Change in Canadian Manufacturing Industries', *Review of Economics and Statistics*, 79(1): 146–50.

Blanchflower, D. and Burgess, S.M. (1998) 'New Technology and Jobs: Comparative Evidence from a Two-Country Study', *Economics of Innovation and New Technology*, 5(2–4): 109–38.

Blanchflower, D., Millward, N. and Oswald, A. (1991) 'Unionisation and Employment Behaviour', *Economic Journal*, 101(407): 815–34.

Bogliacino, F. and Pianta, M. (2010) 'Innovation and Employment: A Reinvestigation Using Revised Pavitt Classes', *Research Policy*, 39(6): 799–809.

Bogliacino, F. and Vivarelli, M. (2011) 'The Job Creation Effect of R&D Expenditures', IZA Discussion Papers, 4728, Bonn, Germany: IZA.

Bogliacino, F., Piva, M. and Vivarelli, M. (2012) 'R&D and Employment: An Application of LSDVC Estimator Using European Microdata', Economics Letters, 116(1): 56–9.

Boyer, R. (1988a) 'New Technologies and Employment in the 1980s: From Science and Technology to Macroeconomic Modeling', in J.A. Kregel, E. Matzner and A. Roncaglia (eds) *Barriers to Full Employment*, London: Macmillan.

Boyer, R. (1988b) 'Assessing the Impact of R&D on Employment: Puzzle or Consensus?', paper presented at the International Conference on New Technology: *Its Impacts on Labour Markets and the Employment System*, Berlin, December 5–7, 1988.

Boyer, R. (1990) 'The Capital Labor Relations in OECD Countries: From the Fordist 'Golden Age' to Contrasted National Trajectories', Working paper CEPREMAP, 9020, Paris, France: Centre pour la Recherche Economique et ses Applications.

Casavola, P., Gavosto, A. and Sestito, P. (1996) 'Technical Progress and Wage Dispersion in Italy: Evidence from Firms' Data', *Annales d'Economie et de Statistique*, 41/42: 387–412.

Clark, J.B. (1907) *Essentials of Economic Theory*, New York: Macmillan.

Coad, A. and Rao, R. (2011) 'The Firm-level Employment Effects of Innovations in High–Tech U.S. Manufacturing Industries', *Journal of Evolutionary Economics*, 21(2): 255–83.

Dobbs, I.M., Hill, M.B. and Waterson, M. (1987) 'Industrial Structure and the Employment Consequences of Technical Change', *Oxford Economic Papers*, 39(3): 552–67.

Doms, M., Dunne, T. and Troske, K. (1997) 'Workers, Wages, and Technology', *Quarterly Journal of Economics*, 112(1): 253–89.

Dosi, G. (1982) 'Technological Paradigms and Technological Trajectories', *Research Policy*, 11(3): 147–63.

Dunne, M., Haltiwanger, J. and Troske, K. (1997) 'Technology and Jobs: Secular Changes and Cyclical Dynamics', *The Carnegie–Rochester Conference Series on Public Policy*, 46: 107–78.

Edquist, C., Hommen, L. and McKelvey, M. (2001) *Innovation and Employment: Product Versus Process Innovation*, Cheltenham: Elgar.

Entorf, H. and Pohlmeier, W. (1990) 'Employment, Innovation and Export Activities', in J.P. Florens, M. Ivaldi, J.J. Laffont and F. Laisney (eds) *Microeconometrics: Surveys and Applications*, London: Basil Blackwell.

Evangelista, R. (2000) 'Innovation and Employment in Services', in M. Vivarelli and M. Pianta (eds) *The Employment Impact of Innovation: Evidence and Policy*, London: Routledge.

Evangelista, R. and Perani, G. (1998) 'Innovation and Employment in Services: Results from the Italian Innovation Survey', Paper presented at the EAEPE Conference, Lisbon, Portugal, 5–8 November, 1988.

Evangelista, R. and Savona, M. (1998) 'Patterns of Innovation in Services. The Results of the Italian Innovation Survey', Paper Presented at the VIII Annual RESER Conference, Berlin, Germany, 8–10 October, 1998.

Evangelista, R. and Savona, M. (2002) 'The Impact of Innovation on Employment in Services: Evidence from Italy', *International Review of Applied Economics*, 16(3): 309–18.

Falk, M. (2001) 'The Impact of Office Machinery and Computer Capital on the Demand for Heterogeneous Labor', ZEW Discussion Paper, 01–66, Mannheim, Germany: The Centre for European Economic Research.

Falk, M. and Seim, K. (1999) 'The Impact of Information Technology on High-skilled Labor in Services: Evidence from Firm Level Panel Data', ZEW Discussion Paper, 99–58, Mannheim, Germany: The Centre for European Economic Research.

Freeman, C. and Soete, L. (1987) *Technical Change and Full Employment*, Oxford: Basil Blackwell.

Freeman, C. and Soete, L. (1994) *Work for All or Mass Unemployment? Computerised Technical Change into the Twenty-first Century*, London–New York: Pinter.

Freeman, C., Clark, J. and Soete, L. (1982) *Unemployment and Technical Innovation*, London: Pinter.

Gera, S., Gu, W. and Lin, Z. (2001) 'Technology and the Demand for Skills: An Industry level Analysis', *The Canadian Journal of Economics*, 34(1): 132–48.

Goux, D. and Maurin, E. (2000) 'The Decline in Demand for Unskilled Labor: An Empirical Method and its Application to France', *Review of Economics and Statistics*, 82(4): 596–607.

Greenan, N. and Guellec, D. (2000) 'Technological Innovation and Employment Reallocation', *Labor*, 14(4): 547–90.

Greenwood, J., Hercowitz, Z. and Krusell, P. (1997) 'Long-run Implications of Investment-specific Technological Change', *American Economic Review*, 87(3): 342–62.

Griliches, Z. (1969) 'Capital-skill Complementarity', *Review of Economics and Statistics*, 51(4): 465–68.

Hall, B.H., Lotti, F. and Mairesse, J. (2008) 'Employment, Innovation, and Productivity: Evidence from Italian Microdata', *Industrial and Corporate Change*, 17(4): 813–39.

Hansson, P. (1997) 'Trade, Technology and Changes in Employment of Skilled Labor in Swedish Manufacturing', in J. Fagerberg, P. Hansson, L. Lundberg and A. Melchior (eds), *Technology and International Trade*, Aldershot: Elgar.

Harrison, R., Jaumandreu, J., Mairesse, J. and Peters, B. (2008) 'Does Innovation Stimulate Employment? A Firm-level Analysis Using Comparable Microdata from Four European Countries', NBER Working Papers, 14216, Cambridge MA: NBER.

Haskel, J. and Heden, Y. (1999) 'Computers and the Demand for Skilled Labor: Industry and Establishment-level Evidence for the UK', *Economic Journal*, 109: C68–C79.

Haskel, J. and Slaughter, M. (1998) 'Does the Sector Bias of Skill Biased Technical Change Explain Changing Wage Inequality?', NBER Working Paper, 6565, Cambridge MA: NBER.

Hercowitz, Z. (1998) 'The 'Embodiment' Controversy: A Review Essay', *Journal of Monetary Economics*, 41(1): 217–24.

Hobsbawm, E.J. (1968) *Industry and Empire: An Economic History of Britain Since 1750*, Harmondsworth Middlesex: Penguin Books.

Hobsbawm, E.J. and Rudé, G. (1969) *Captain Swing*, London: London Reader's Union.

Hulten, C.R. (1992) 'Growth Accounting when Technical Change is Embodied in Capital', *American Economic Review*, 82(4): 964–80.

Lachenmaier, S. and Rottmann, H. (2011) 'Effects of Innovation on Employment: A Dynamic Panel Analysis', *International Journal of Industrial Organization*, 29(2): 210–20.

Layard, R. and Nickell, S. (1985) 'The Causes of British Unemployment', *National Institute Economic Review*, 111(1): 62–85.

Luque, A. (2005) 'Skill Mix and Technology in Spain: Evidence from Firm-level Data', Documento de Trabajo, 0513, Madrid: Banco de Espana.

Machin, S. (1996) 'Changes in the Relative Demand for Skills', in A.L. Booth and D.J. Snower (eds) *Acquiring Skills: Market Failures, Their Symptoms and Policy Responses*, Cambridge: Cambridge University Press.

Machin, S. and Van Reenen, J. (1998) 'Technology and Changes in Skill Structure: Evidence from Seven OECD Countries', *Quarterly Journal of Economics*, 113(4): 1215–44.

Machin, S. and Wadhwani, S. (1991) 'The Effects of Unions on Organisational Change and Employment: Evidence from WIRS', *Economic Journal*, 101(405): 324–30.

Mairesse, J., Greenan, N. and Topiol-Bensaïd, A. (2001) 'Information Technology and Research and Development Impacts on Productivity and Skills: a Comparison on French Firm Level Data', NBER Working Paper, 8075, Cambridge MA: NBER.

Malthus, T.R. (1964) *Principles of Political Economy*, New York: M. Kelley.

Marshall, A. (1961) *Principles of Economics*, Cambridge: Macmillan.

Marx, K. (1961) *Capital*, Moscow: Foreign Languages Publishing House.

Marx, K. (1969) *Theories of Surplus Value*, London: Lawrence & Wishart.

Morrison-Paul, C.J. and Siegel, D.S. (2001) 'The Impacts of Technology, Trade and Outsourcing on Employment and Labor Composition', *Scandinavian Journal of Economics*, 103(2): 241–64.

Neary, J.P. (1981) 'On the Short-run Effects of Technological Progress', *Oxford Economic Papers*, 33(2): 224–33.

Nickell, S. and Bell, B. (1995) 'The Collapse in Demand for the Unskilled and Unemployment across the OECD', *Oxford Review of Economic Policy*, 11(1): 40–62.

Nickell, S. and Kong, P. (1989) 'Technical Progress and Jobs', Centre for Labor Economics, Discussion Paper 366, London: London School of Economics.

OECD (1996) 'The Evolution of Skills in OECD Countries and the Role of Technology', STI Working Paper 1996/8, Paris: OECD.

OECD (1998) 'OECD Data on Skills: Employment by Industry and Occupation', STI Working Paper 1998/4, Paris: OECD.

Pasinetti, L. (1981) *Structural Change and Economic Growth*, Cambridge: Cambridge University Press.

Peters, B. (2004) 'Employment Effects of Different Innovation Activities: Microeconometric Evidence', ZEW Discussion Papers, 0473, Bonn, Germany: ZEW.

Petit, P. (1995) 'Employment and Technological Change', in P. Stoneman (ed.) *Handbook of the Economics of Innovation and Technological Change*, Amsterdam: North Holland.

Pianta, M. (2000) 'The Employment Impact of Product and Process Innovations', in M. Vivarelli and M. Pianta (eds) *The Employment Impact of Innovation: Evidence and Policy*, London: Routledge.

Pianta, M. (2005) 'Innovation and Employment', in J. Fagerberg, D. Mowery and R.R. Nelson (eds) *Handbook of Innovation*, Oxford: Oxford University Press.

Pigou, A. (1962) *The Economics of Welfare*, London: Macmillan.

Piva, M. (2003) 'The Impact of Technology Transfer on Employment and Income Distribution in Developing Countries: A Survey of Theoretical Models and Empirical Studies', International Policy Group Working Paper 15, Geneva: International Labor Office (ILO).

Piva, M. and Vivarelli, M. (2001) 'The Skill Bias in Italy: A First Report' *Economics Bulletin*, 15(2): 1–8.

Piva, M. and Vivarelli, M. (2004) 'Technological Change and Employment: Some Micro Evidence from Italy', *Applied Economics Letters*, 11(6): 373–76.

Piva, M. and Vivarelli, M. (2005) 'Innovation and Employment: Evidence from Italian Microdata', *Journal of Economics*, 86(1): 65–83.

Piva, M. and Vivarelli, M. (2009) 'The Role of Skills as a Major Driver of Corporate R&D', *International Journal of Manpower*, 30(8): 835–52.

Piva, M., Santarelli, E. and Vivarelli, M. (2005) 'The Skill Bias Effect of Technological and Organisational Change: Evidence and Policy Implications', *Research Policy*, 34(2): 141–57.

Ricardo, D. (1951) 'Principles of Political Economy', in P. Sraffa (ed.) *The Works and Correspondence of David Ricardo*, Vol. 1, 3rd edn, Cambridge: Cambridge University Press.

Salter, W.E.G. (1960) *Productivity and Technical Change*, Cambridge: Cambridge University Press.

Say, J.B. (1964) *A Treatise on Political Economy or the Production, Distribution and Consumption of Wealth*, New York: M. Kelley.

Simonetti, R., Taylor, K. and Vivarelli, M. (2000) 'Modeling the Employment Impact of Innovation', in M. Vivarelli and M. Pianta (eds) *The Employment Impact of Innovation: Evidence and Policy*, London: Routledge.

Sinclair, P.J.N. (1981) 'When will Technical Progress Destroy Jobs?', *Oxford Economic Papers*, 31(1): 1–18.

Sismondi, J.C.L. (1971) *Nouveaux Principes d'Economie Politique ou de la Richesse dans ses Rapports avec la Population*, Paris: Calmann-Levy.

Smolny, W. (1998) 'Innovations, Prices and Employment: A Theoretical Model and an Empirical Application for West German Manufacturing Firms', *Journal of Industrial Economics*, 46(3): 359–81.

Solow, R.M. (1960) 'Investment and Technical Progress', in K.J. Arrow, S. Karlin and P. Suppes (eds) *Mathematical Methods in the Social Sciences*, Stanford: Stanford University Press.

Steuart, J. (1966) *An Inquiry into the Principles of Political Economy*, Chicago: Oliver and Boyd.

Stoneman, P. (1983) *The Economic Analysis of Technological Change*. Oxford: Oxford University Press.

Sylos Labini, P. (1969) *Oligopoly and Technical Progress*, Cambridge MA: Harvard University Press.

Van Reenen, J. (1997) 'Employment and Technological Innovation: Evidence from U.K. Manufacturing Firms', *Journal of Labor Economics*, 15(2): 255–84.

Vivarelli, M. (1995) *The Economics of Technology and Employment: Theory and Empirical Evidence*, Aldershot: Elgar.

Vivarelli, M. (2011) 'Innovation, Employment and Skills in Advanced and Developing Countries – A Survey of the Literature', Science and Technology Division, Social Sector, Technical Notes No. IDB-TN-351, Washington: Inter-American Development Bank.

Vivarelli, M. and Pianta, M. (2000) *The Employment Impact of Innovation: Evidence and Policy*, London: Routledge.

Vivarelli, M., Evangelista, R. and Pianta, M. (1996) 'Innovation and Employment in Italian Manufacturing Industry', *Research Policy*, 25(7): 1013–26.

Welch, F. (1970) 'Education in Production', *Journal of Political Economy*, 78(1): 35–59.

Wicksell, K. (1961) *Lectures on Political Economy*, London: Routledge & Kegan.

Wilson, D.J. (2009) 'IT and Beyond: The Contribution of Heterogeneous Capital to Productivity', *Journal of Business and Economic Statistics*, 27(1): 52–70.

8 Design and evaluation of fiscal incentives for business innovation in Latin America

Lessons learned and future developments

Gustavo A. Crespi

8.1 Introduction

Since the beginning of the 1990s, several Latin American countries have witnessed the systematic growth of public programs aimed at enhancing firm-level innovation and technological upgrading. The overarching justification for these programs is that the market has failed to provide the incentives needed to reach an optimal level of private investment in innovation activities,[1] and therefore Latin American firms have failed to adopt the modern technologies and business practices that would have helped them to improve their productivity and competitiveness.

In this context, several Latin American countries have introduced various types of fiscal incentives in order to stimulate innovation activities and to strengthen the linkages among firms and other agents in the National System of Innovation (NSI).[2] The first of these fiscal incentives program was started in Chile in 1991, and since then they have spread throughout the region to the extent that nowadays about 70 per cent of the countries in LAC have put in place some form of fiscal incentive program for innovation.

Almost 20 years have elapsed since then and, pari-passu with the experience accumulated from the actual implementation of these policies, a systematic collection of evidence and methodological learning has been produced with regards to the effectiveness of these incentives in order to alleviate the different market failures hindering innovation and productivity in the region. The aim of this paper is to carry out a qualitative 'meta analysis' of these programs (and their respective impact evaluations) in order to take stock of the learning achieved so far and provide specific recommendations for how public policy should be better designed in order maximise additionality and productivity impacts.

This chapter is structured as follows. Section 2 provides the context for this chapter by presenting a short overview of the regional innovation performances over the last 20 years. Section 3 summarises the different rationales that justify public policy intervention with regards to business innovation. Section 4 presents the landscape of different policy interventions in the region and describes their main characteristics with regards to how they have actually been implemented. Section 5 presents a 'meta analysis' by making a comparative analysis of the most recent evaluations done so far. Section 6 concludes with a summary of the emerging issues with regards to the design of future innovation policies.

8.2 Innovation at glance: the regional landscape

On average, Latin American and Caribbean countries underperform, as compared to other developed and developing countries, in terms of productivity growth. This poor productivity outcome explains why the region has shown only modest economic growth in the last 30 years and why, despite unusually favourable international conditions over the last half decade, the region still lags behind other regions in terms of economic growth. Indeed, the top left panel of Figure 8.1 shows the evolution of the productivity gap between the typical LAC country and the US. Taking into consideration that the productivity gap was normalised to one in 1960, Figure 8.1 clearly indicates an increase in this gap since then. In other words, the productivity gap for the region with regards to the US – which is normally taken as the proxy for the best practice frontier – is today about 60 per cent higher than it was in 1960 (IDB, 2010a). This situation is seemingly unique to the LAC region, as other regions in the world have managed to either keep track with the best practice frontier or to catch up with it.

Innovation, broadly defined as the introduction of new products or processes by firms, has been credited as the ultimate engine of productivity growth by a vast amount of economic literature.[3] The entrepreneurs, motivated by profit, look for better – meaning more efficient – ways of doing things that can be commercially viable, and in the process they come up with new routines, production arrangements, materials or machinery that saves costs and/or improves the quality of output. In fact, the top right panel of Figure 8.1 shows that that there is a positive and statistically significant correlation between productivity and investments in R&D across a large cross-section of countries, and the strength of this correlation suggests an investment with high social return rates – indeed, about 40 per cent of the variations in productivity in the sample are due to variations in investments in R&D.

Yet it is also a well established fact that firms in Latin America register what appear to be sub-optimal levels of investment in innovation (both the tangible and intangible components). Certainly, the bottom left panel of Figure 8.1 compares the business innovation investment rates across a sample of OECD and LAC countries where information is available and internationally comparable. The figure also shows the average investment rate for these two regions, suggesting two clear findings: (i) average innovation investment rates are clearly lower in LAC than in OECD countries; and (ii) that the investment gap between these two regions is particularly severe in the case of the intangible component of that investment, which is R&D.[4]

Figures available at the macro level not only indicate that the general level of expenditure in innovation is low by international standards, but also, notoriously,

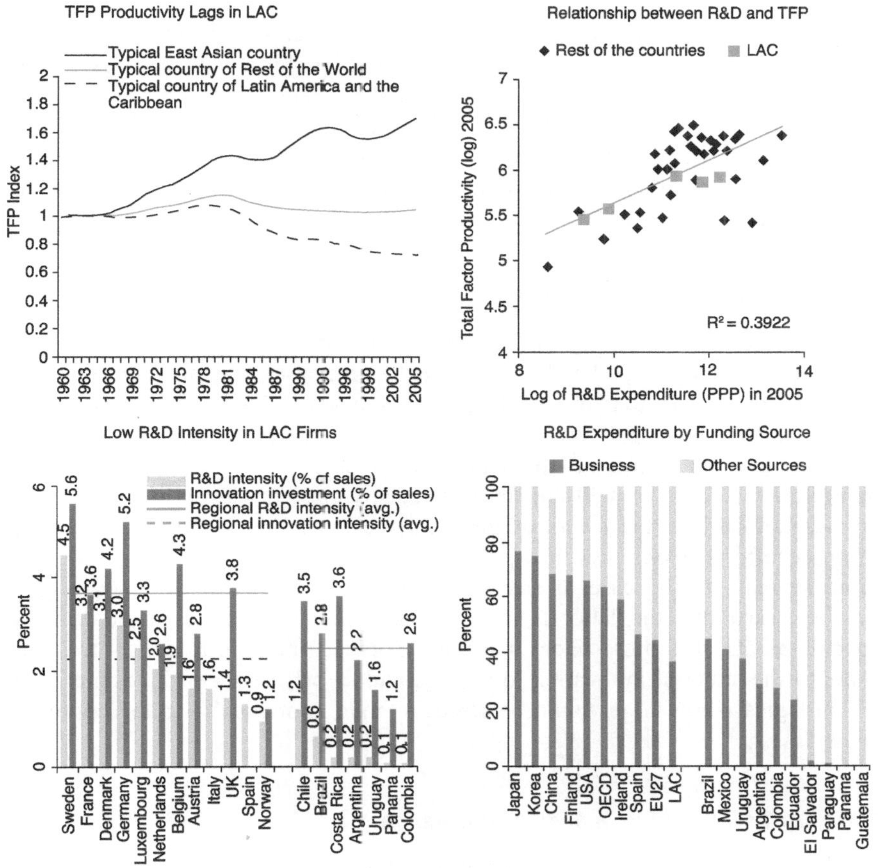

Figure 8.1 Latin America and the Caribbean: innovation at a glance

Source: IDB (2010a); IDB (2010b).

that most of the innovation investment in the region (about two-thirds of it) is directly performed by the public sector (see the bottom right of Figure 8.1). This is in stark contrast with OECD and successful catching-up economies, where about two-thirds of innovation investments come from private sources. This is also confirmed by micro-evidence which suggests that, more often than not, innovation is not a preferred path for many Latin American firms in their search for profitability, the opening of new markets, or the consolidation of advantages relative to competitors (IDB, 2010b). Such a weak propensity to engage in innovation activities by the private sector is all the more noteworthy given that studies which have looked closely at the profitability of innovation in Latin American firms show consistently positive returns (Crespi and Zuniga, 2011). Such reluctance of the private sector to invest in innovation has been traced to a set of ailments that hinder the private returns of these investments at the firm level.

8.3 The rationale for innovation policies

The fundamental premise for innovation policies is that government intervention would be necessary if profit driven actors underperform with regards to the production and/or exchange of technological knowledge from a social welfare perspective (Steinmueller, 2010). The literature on the economics of innovation has provided several rationales to justify that this is indeed the case. Broadly speaking, the rationale for public policy in this field can be articulated around the following considerations:

8.3.1 Spillovers and the 'public good' nature of knowledge

Since the seminal works by Nelson (1959) and Arrow (1962), scientific and technological knowledge has been regarded as a non-excludable and non-rival good. To the extent that private benefits associated with knowledge creation are not fully appropriable by the innovators, this creates a wedge between the private and social returns of knowledge investments, leading to a rate of investment in knowledge that will fall short of socially optimum levels. This rationale applies not only to the *levels* of innovation efforts, but also to the *direction* of these efforts. Certainly, the 'public good' rationale of knowledge applies more strongly in the case of *scientific* rather than *technological* knowledge.[5] To the extent that the latter is more applied, predictable and linked to firm specific assets, it is more likely that innovators will be able to collect a larger share of the value of innovation to society, and so private sector investments in technological knowledge would be closer – though not equal – to the optimum social levels.[6]

8.3.2 The problem of asymmetric information

The literature on economics of information (Stiglitz and Weiss, 1981) makes clear that asymmetric information in market transactions (due to the problems of adverse selection and moral hazard) can affect firm innovation from two different perspectives. From the perspective of investment theory, innovation projects have several peculiar characteristics which differentiate them from ordinary investments (Hall and Lerner, 2010). *First*, innovation projects are riskier than physical investment projects. Consequently, external investors might require a higher risk premium for the financing of innovation activities. *Second*, because of the problem of spillovers, innovators are themselves reluctant to share information about their projects with potential outside investors, further worsening the asymmetric information problem. *Third*, the difficulty of using intangible assets as collateral also leads to increased costs of external capital in the form of a higher risk premium. In summary, asymmetric information would lead to a wedge between the opportunity cost that private innovators require for their innovation investments and the capital cost that external investors are willing to charge to finance innovation projects, with the result that privately (and eventually socially) profitable innovation projects will not materialise due to the fact that financing costs are simply too high.

The second perspective on how asymmetric information affects innovation concerns knowledge dissemination, and relates to the fact that private actors do not have 'perfect information' on technology or production possibilities. In the same way, adverse selection and moral hazard problems also extend to the (imperfect) operation of technology markets. This claim is consistent with two empirical findings: (i) that persistent differences exist in the technological performance between countries, such that catching-up is far from being an automatic process consistent with the idea of knowledge as a global public good (Fagerberg and Verspagen, 2002); and (ii) that the process of technology diffusion, even within narrowly defined industries, is very sluggish, leading to a persistent firm heterogeneity in productive performance (Disney *et al.*, 2003).

8.3.3 The pervasiveness of coordination and institutional failures

A key contribution of the innovation systems literature is the fact that knowledge has non-negligible tacit components and, as such, innovation is the result of feedback and interaction involving numerous actors (Lundvall, 1992). Although many of these interactions are market-mediated, a large proportion of them are governed by non-market institutions: 'Because the efficiency of this process at the macro level depends on the behaviour of individual actors and the institutions that govern their interaction, coordination problems might arise' (Soete *et al.*, 2010). A nice example of where these coordination problems could emerge refers to the setting of standards that regulate producer–user interactions in the case of General Purpose Technologies (GPTs). GPTs are a set of technologies that spread out across different economic activities, leading them to innovate as well. Progress in the adopting sectors feeds back to the GPTs developers, generating a process of sustainable growth (Bresnahan and Trajtenberg, 1995; Aghion *et al.,* 2009). It is also clear that these technologies contribute to growth not only through the development of GPTs intensive sectors per se (the supply side), but also – and even more critically – through the development of the *complementary* innovations which facilitate their wider adoption across the *other* sectors of the economy, which then also start to innovate as a consequence of this. This requires a solution to coordination problems. The solving of the coordination problem requires not only putting attention to the linkages among the actors, but also to their absorptive capacities (Cohen and Levinthal, 1989). The concept of absorptive capacities is a key ingredient of the new literature on innovation, from the perspective of catching-up economies in particular. Following Steinmueller (2010), the concept of absorptive capacities refers to fact that new knowledge might not be employable without heavy co-investments in corresponding human capital and learning by the users; furthermore, it also implies that the new knowledge might not be reproducible without the direct assistance of the originator.

Due to the fact that human interactions are governed by institutions, the innovation systems literature places a strong emphasis on institutional governance and change. These refer to institutional design arrangements that foster public–private interactions and at the same time minimise problems of moral hazard. Institutional change interventions refer also to arrangements that build linkages between the different actors involved in the innovation process (such as universities, public research organisations, technology producers and users, consumers, etc.) either by defining new roles for existent institutions (such as allowing the patenting of university research in order to encourage technology transfer), or by creating clubs or consortiums that regulate interactions between the agents (Steinmueller, 2010). These sorts of arrangements may lead to a better equilibrium, either because innovation costs are not duplicated in separate efforts that lead to identical results, or because several externalities are internalised. In this case, public intervention is often required to reduce the transaction costs that may hamper the formation of the joint venture and to regulate their activities in order to achieve the desired balance between cooperation and competition.[7]

8.4 The implementation of innovation policies in lac

The previous analysis offers different conceptual frameworks that justify the implementation of innovation policies based on the idea that profit-seeking agents will produce both a level and a direction of knowledge investments which will fall short of socially desirable outcomes. Following David *et al.* (2000), broadly speaking, public policy has suggested two main approaches in order to solve the under-provision of innovation efforts by private firms: (i) direct production of knowledge in public institutions (laboratories and public research institutes); and (ii) fiscal incentives for a greater amount of private investment in knowledge generation. Without ignoring the importance of government investments into public research organisations, the issues related with the governance, funding incentives and productivity impacts related to the operations of these organisations are complex enough as to require a far more specific focus than the scope of this chapter. Despite this, in this chapter we will specifically focus on the second class of policy designs and their growing importance in the LAC region. In particular, we will focus on two particular classes of fiscal incentives – direct subsidies and tax incentives – without ignoring that other types of incentive designs are also available (such as adoption subsidies, technology acquisition policy, signalling strategies, information diffusion policies, thematic funding, entrepreneurship programs, etc.); however, the empirical evidence and learning accumulation of the impacts of these other designs in the region is far more limited.[8]

Both direct subsidies and tax incentive schemes have been in place in the LAC region since early 1990s, and more countries seem to be eager to adopt them. Figure 8.2 summarises the degree of penetration of fiscal incentives for innovation in LAC, comparing it against the OECD countries. In the OECD almost 80 per cent of the countries have implemented a matching grants system and 66 per cent have also implemented a tax incentive scheme, with 45 per cent of countries having both. The degree of penetration is somewhat lower in LAC: only

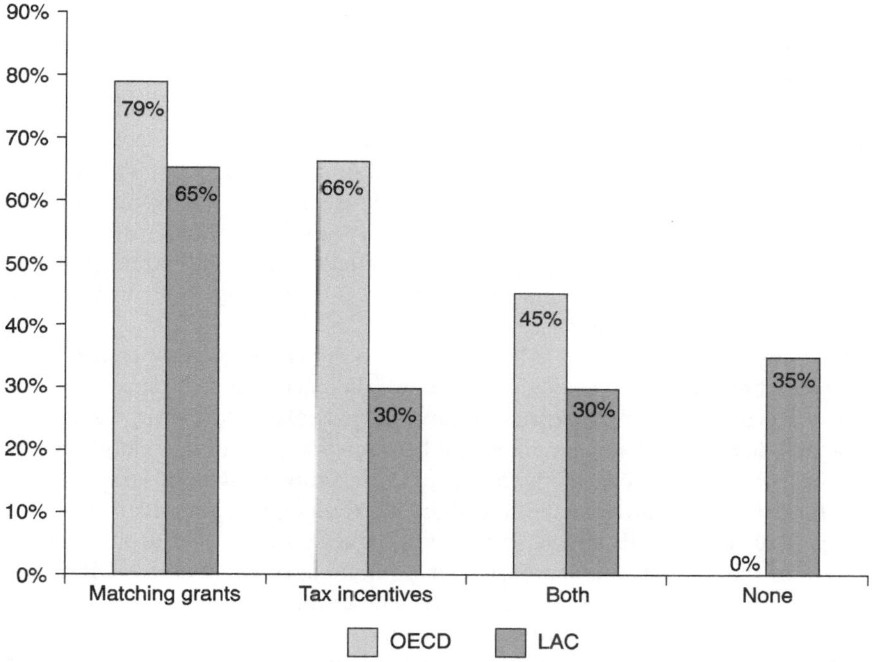

Figure 8.2 Percentage of countries with fiscal incentives for business innovation

Sources: Own elaboration using UNESCO and IDB database 'Politicas e Instrumentos en Cieencia, Tecnologia a Innovation en America Latin y el Caribe' and Table 2.6 of 'OECD Science, Technology and Industry Outlook 2010'.

Notes: OECD refers to 29 countries that outside LAC present either full members or in open discussions for accession. The list includes: Australia, Austria, Canada, Czech Republic, Denmark, Finland, France, Germany, Greece, Hungary, Ireland, Israel, Italy, Japan, Korea, Netherland, New Zealand, Norway, Poland, Portugal, Russian Federation, Slovenia, South Africa, Spain, Sweden, Turkey, United Kingdom, and United States. Russian Federation and South Africa are the only ones in the Open Discussion category. In the case of OECD, Matching grants comprises R&D subsidies and Competitive grants schemes. LAC comprises Argentina, Bolivia, Brazil, Chile, Colombia, Costa Rica, Dominican Republic, Ecuador, Guatemala, Honduras, Jamaica, Mexico, Nicaragua, Panama, Peru, Paraguay, Salvador, Trinidad and Tobago, Uruguay and Venezuela.

65 per cent of the countries have put in place a matching grants mechanism and just about 30 per cent have tax incentives. Moreover, only 30 per cent of the countries have both systems in place, while 35 per cent have neither of them. On the other hand, in the OECD there is no country without some sort of fiscal incentive for business innovation.

From a practical point of view it is worth emphasising the main differences between the direct subsidies and tax incentives and also their main pros and cons with regards to implementation:

Direct Subsidies: Subsidies are a type of *direct support* for business innovation which is *project-specific based*. They modify firms' marginal costs of capital and may raise the private marginal rate of return of the innovation investment by, for example, inducing collaboration with other actors with complementary assets. Because they are project based, subsidies allow public agencies to target projects with perceived high marginal social rates of return. Given the problem of asymmetric information between the public agency and the beneficiary, direct subsidies might suffer from opportunistic behaviour and moral hazard problems. Indeed, while the public agency might want to maximise a firm's innovation efforts, private entities might aim at maximising the size of the innovation project (and of the subsidy). However, although not fully eliminated, the moral hazard problem implicit in a direct subsidy could be controlled for by a design that considers the two following key attributes:

(a) Subsidies are normally allocated under a *matching grant* approach with *maximum* limits and a list of eligible expenses; in other words, the subsidy never covers the full costs of the supported project.[9] It is expected that by using this approach there will be a better alignment between the goals of the public agency and the firm, somehow controlling for the potential problem of moral hazard. If the beneficiary wants to increase the size of the innovation project in order to extract a higher subsidy, it will also have to pay a higher cost. Related to this, a nice feature of this cost-shared approach is that public agencies can also adjust the amount of the co-funding according to the main characteristics of the projects or the beneficiary. Indeed, in the case of innovation projects, the share of the subsidy might increase when the project implies the participation of more than one beneficiary, given that it is expected that projects which involve more than one firm (or a firm collaborating with a university) might lead to higher spillovers. Finally, the operation of the co-funding mechanisms is normally implemented through the ex-post reimbursement of the approved expenditures which qualify for the subsidy.

(b) Subsidies are also normally allocated using a competitive *call for proposal process*. This competitive process allows the public agency to identify the best proposals (selection based on merit) and to allocate resources among them according to some evaluation score normally set by external evaluators or peer reviewers. In some cases, on top of a technical evaluation, a social cost-benefit evaluation might be carried out in order to identify the economic relevance of the project. In order to allocate the subsidies, the public agencies normally put in place an adjudicatory commission formed by representatives from government, private, academic and civil society organisations, increasing transparency and reducing the risks of capture.

However, the above-mentioned counterbalances come at some cost. The main problem with direct subsidy schemes is that they need important institutional capacities in the executing agency, and when these capacities are not present the efficiency of the whole operation decreases dramatically.

8.4.1 Tax incentives

Different to direct subsidies, tax incentives are based on *firm-level* innovation activities rather than projects, thus allowing firms to obtain support for their whole portfolio of innovation activities without having to submit a project proposal for each one of them, which dramatically reduces firm compliance costs and agency administration costs. Strictly speaking, tax incentives operate through different approaches: tax credits, enhanced allowances, and accelerated depreciation of intangible investments, for example. Tax credits allow for a direct deduction from the payable tax, while enhanced allowances and accelerated depreciation represent a deduction (above the normal deduction rate of 100 per cent) from the taxable income of the company. The main difference between the two mechanisms is that the former directly reduces the tax liability, while in the latter the reduction of the tax liability depends upon the effective tax rates.

In developed countries, tax incentives normally apply to corporate income tax; however some countries (developing countries, in particular) have also experimented with other variations, such as reductions in tariffs for imported research machinery and equipment, deductions in the value added tax, and discounts on the social security and employers' contributions on the payroll of researchers' salaries. As in the case of direct subsidies, tax incentives are also subject to similar problems of moral hazard, such that firms might claim activities that they would otherwise conduct or have been conducting as innovation expenditures.[10] Averting this hazard requires the establishment of nominal limits, an auditable definition of innovation activity, and active enforcement by both the innovation agency and the tax authority. So, in principle the higher administration and compliance costs of the direct subsidies need to be compared with the higher policing costs of the tax incentive.

However, when analysing a tax credit system several caveats need to be considered. *First*, the actual impact of the tax incentives on the marginal cost of capital of innovation activities depends upon the general fiscal environment of the economy as fiscal incentives are less effective in a country with low taxes. In fact, this is one of the main reasons why the empirical literature has normally found that tax incentives are less effective as a stimulus for innovation in SMES rather than in large companies (Harris *et al.*, 2009). SMEs simplified tax treatments normally imply that corporate tax rates are lower in the case of this group of firms.[11] *Second*, the impact of the scheme also strongly depends on the tax position of the firm and on its ability to make profits; this does not seem to be the case for start-up firms that have just entered the market, so a priori, the power of this policy tool to promote entrepreneurship is limited. This limitation can be ameliorated to some extent through the inclusion in the scheme of carry-over provisions that allow unused portions of the credit to be carried forward to subsequent fiscal years (Criscuolo, 2009). *Third*, and most importantly, under a tax incentive scheme it is the firm who chooses those innovation projects that will be implemented, so to some extent it is a mechanism that is more market friendly than a direct subsidy, and this rests under the assumption that a firm should have better information about what project should be pursued. An important contradiction within this rationale is that in a world with market failures, market-friendly mechanisms might not be the best channels to guide resource allocation. In other words, giving complete freedom of choice to a firm does not guarantee that the selected projects will be also those with higher *social* return rates or those with a higher risk. So, in the end, additionality could be lower as funded projects could be similar to those that would have been funded by the firm anyway.

As mentioned above, several LAC countries have established innovation tax incentives during the last 15 years (Argentina, Colombia, Brazil, Mexico – recently discontinued – and more recently Chile and Uruguay). The typical LAC tax incentive presents some important differences with regard to the standard approach of developed countries. Given their importance, indirect taxes are sometimes also included among the deductions (such as the value added tax or import tariffs), but even more important is that the implementation of the scheme is mostly project-based. That is, in order to qualify for the tax incentive, firms are asked to submit a project proposal to a public agency, which then reviews whether the project qualifies as an innovation project; if so, it will recommend the approval of the eligible expenditures to the tax authorities, who will issue a tax credit certificate. The rationale for this approach is to have tighter control of the fiscal costs of the scheme. In addition to this, in many cases the allocation of resources is done following a competitive process whereby resources are allocated according to the merit of the project, until the overall budget for the program is achieved. The trade-off is that many of the administration and compliance costs of the matching grants schemes reappear, sometimes without the benefits of the matching grants system, as the decision-making power mainly remains in the firm (if the firm has submitted a project proposal that qualifies under the law as an innovation project, the agency is obliged to issue the tax credit certificate, even if social returns of the project are low and not very different from the private ones. Merit-based competition and the setting of fiscal quotas can alleviate this problem).

The remaining section of this chapter will focus on the main results emerging from the evaluation of the different programs which support business sector innovation in LAC. However, before this, we will summarise the main issues with regard to the impact evaluation of fiscal incentives for innovation.

8.5 Empirical results emerging from impact evaluations of fiscal incentives in LAC

8.5.1 Some methodological considerations

As is clear from the above discussion, innovation policies might be justified due to the presence of many different market, coordination, and institutional failures; however, successful implementation makes strong demands on the ability of governments to design programs that would rectify these identified failures. In real life, governments face informational constraints that may be as or more severe than those of firms. Although there might be a strong case for innovation policies, actual implementation could easily lead to the wrong results or, in other words, public support could lead to the crowding out of private funding.

One of the first issues to be defined in an impact evaluation is how and when to measure the effects of the program, i.e. the outcomes of interest. In the spirit of the CDM model (Crépon *et al.*, 1998), a distinction can be made between innovation-input indicators and economic-performance indicators. Innovation-input indicators are those indicators more directly affected by the intervention. For instance, for a fiscal incentives program, an innovation-input indicator is total investment in innovation by the beneficiary. While the relationship between the subsidy and the total investment seems in principle almost tautological, our previous discussion clearly highlights that this is not necessarily true (see e.g. David *et al.*, 2000). In other words, to the extent that innovation policies are capable of changing the firm's marginal cost of capital, and to the extent that investment decisions react to this change in the cost of capital, we might be capable of identifying the extent to which innovation policies generate *input additionality*.

However, simply assessing whether innovation efforts increase as a consequence of a subsidy is not sufficient for policy evaluation purposes. The whole portfolio of innovation projects held by the firm is normally affected. As a result of this, projects with different productivity might be executed, while others might be postponed. Hence, assessing the outputs of innovation investments is also important (*output additionality*). Innovation outputs are variables where the concrete realisation of innovation activities is observed and their impacts on economic performance materialise. So, particularly in the case of business innovation programs, important output variables for the measurement of *output additionality* are, for example, productivity growth, employment, wages, and exports, to cite just a few.

Normally, input additionality is measured in the short term, i.e. while the innovation project is being implemented. However, in the case of output additionality a 'time to build' period is necessary so as to find impacts. More generally, the impacts of different programs may display very different patterns over time. An intervention may generate a one-shot increase in the outcome, or may have strong impacts that fade out progressively with time; the impact of a program may only appear after a certain period, or may even generate an initial drop in the outcome that is later overshot by increases in subsequent years. As a result, a proper consideration of the timing of the effects is crucial in an impact evaluation setting, and failures to account for these issues may result in misleading conclusions and policy recommendations. A clear distinction should be made between short-term and long-term effects in order to properly evaluate the costs and benefits of a public program.

Each one of the evaluations that we review in this chapter is based on a unique data-set wherein both primary sources of information on beneficiaries are linked to secondary sources of information such as innovation and industrial surveys. This procedure allowed us to identify exactly the specific firms that participated in each evaluated program. In addition, in all of the cases the control group was constructed using a sample of firms that did not receive any other comparable treatment in order to maximise the accuracy of the impact estimation. Moreover, all of the evaluations aimed to identify comparable treated and non-treated firms in order to minimise the effect of potential 'selection biases' on the evaluation results. A control group was identified using a number of different methods: propensity score matching procedures, difference-in-differences estimation, fixed effect panel data estimation, and instrumental variables methods.[12] All the evaluations are done at the beneficiary level rather than at the project level.

8.5.2 A meta-analysis of the evaluation results

We first summarise the results on input additionality by looking at the impacts of the different programs on firms' innovation investments, and we attempt to infer the extent to which there might be crowding-in or crowding-out effects on private investment. We also provide some information on the impact evaluation methodology being used. After this, we focus our attention on the impact of output additionality, and in particular the extent to which any impact on productivity can be observed over a longer time period. As a caveat, it is important to note that the heterogeneity of available information and data sources did not allow the studies reviewed here to adopt identical estimation techniques for all impact evaluations. As a consequence, the results are sometimes not fully comparable across different cases.

8.5.3 Input additionality

In common with other regions, the evaluation of input additionality has been the preferred approach for impact evaluation in LAC. Table 8.1 summarises the results of the 13 impact evaluations performed so far in the region. In seven of the studies (summarised in the top half of the table), the main impact indicator variable is the absolute value of firm's innovation or R&D investment (in log). In five of these cases the main dependent variable is private R&D or innovation investment – in other words, the impact indicator is investment net of the subsidy, while in the two remaining cases the studies looked at total investment in innovation or R&D.[13] The bottom half of the table, on the other hand, summarises six studies in which the main indicator of interest is innovation intensity – that is, innovation or R&D expenditures as a fraction of sales or total investment. In two of these cases the results refer to innovation intensity efforts net of subsidy. In the remaining cases the results look at the impacts of total innovation efforts.

Several clear results emerge from Table 8.1. First, the evidence across the different studies is that fiscal incentives clearly stimulate innovation or R&D investments in the LAC region. In all of the cases, and regardless of the main indicator variable, a positive and significant average treatment effect was obtained. In other words, those firms which received some sort of fiscal support did increase their innovation investments. Based on the arguments made earlier, this implies that, in general, the programs are well focalised, to the extent that they seem to be targeting firms with either serious appropriability or financial constraints problems, or both. Thus whenever these constraints are relaxed, firms react favourably by increasing their investment in innovation. Furthermore, there are seven evaluations where the main impact indicator is private investment in innovation or R&D, and the results for this variable are also positive and significant, suggesting that companies which receive fiscal support also react by increasing their own investment in innovation. As mentioned above, this result might be the consequence of the operation of many different transmission mechanisms. On the one hand, fiscal support might target riskier projects, and so this might induce private finance to follow up less risky investments. On the other hand, public support might target infrastructure projects, thus reducing the subsequent capital costs of any subsequent project. Finally, the fiscal incentive might have a signalling effect on the quality of the project and the research team, thus allowing the firm to leverage additional resources from the financial markets. Moreover, the qualitative interviews that were part of the evaluation of the Chilean FONTEC suggest that this signalling effect was indeed important (see Benavente *et al.*, 2007).

With regard to the differences between instruments, it is observed that the only three cases where there was no evidence of crowding-in (in two cases there was no evidence on crowding-in or crowding-out, while in one remaining case there was some weak evidence of partial crowding-out) correspond to variations in the matching grant scheme. Indeed, it seems that subsidised loans or tax credit schemes are clearly more capable of generating multiplier effects on private investments. Therefore there is some evidence that, to the extent that matching grant programs provide financing at zero cost, this might reduce the potential multiplier effect of the fiscal scheme. However, an important qualification is needed in this case. There are two cases where the matching grant designs have had the highest positive impact on investment – even in comparison with the other instruments – and also the largest multiplier effect with regards to leveraging private investment in innovation. These are the cases of Brazil's FNDCT and Colombia's COFINANCIACION programs, where the main characteristic is that in both cases the scheme provides conditional funding for firm–university collaboration. So, it seems that matching grants programs are particularly well suited to encouraging the building of linkages among the different actors in the innovation system. This particular feature of these two programs could have contributed to the addressing of both financial and technical constraints (lack of human resources, lack of own specialised research infrastructure, and lack of technical knowledge, among others). The relaxation of these other technical constraints might have led to a multiplier effect in private funding.

One important limitation identified during the earlier evaluations was that sample size did not allow for assessing the extent to which impacts were heterogeneous according to different sub-groups of firms. However, in the case of Argentina's FONTAR-ANR program the sample size was large enough to assess whether there was impact heterogeneity according to a firm's experience with managing innovation projects. The evaluators found that although the program does not provide evidence of a multiplier effect at the aggregate level, the data indicates that new innovators have seen a substantial increase in their private investment in R&D. On the other hand, the effect of the matching grants is smaller in the case of more experienced innovators, for which some evidence of displacement of resources was found. In other words, although very limited, the evidence seems to point towards a favourable impact of matching grants on firms with limited experience on the formulation and execution of innovation projects (Chudnovsky *et al.*, 2006).

Finally, the different evaluations do not find any systematic differential impacts between the other two instruments: tax credits and subsidised loans. In both cases there are crowding-in multiplier effects, and the total of the impacts upon firms' total innovation investments are rather similar. As can be inferred from Table 8.1, the majority of the studies make use of propensity score based techniques in order to find similar treatment control pairs and select a common support for the impact evaluations. This allows an assessment of which are the main determinants affecting the probability of being selected for any of these programs are. Across many of the studies, it seems that firms with higher levels of human capital or some previous experience in managing R&D and innovation programs are more likely to be selected. This is expected, to a degree, given that in all of the cases, agencies' technical evaluators highly weighted these two indicators at the moment of scoring each proposal. The problem of an excellence-based selection system is that it is highly meritocratic and while it might be seen to produce good results in the short term, it could trigger un-expected dynamic effects in the longer term. Indeed, a heavily meritocratic evaluation system might lead to the selection of only very good candidates, candidates that might be selected again in future competitions. This might trigger a sort of 'Matthew Effect' dynamics that might end up affecting competition and inequality through the formation of powerful elites. More research is needed in order to assess whether these endogamy effects are present or not, but this is an issue that requires some follow-up in the near future. In other words, it is important to maintain clear exit strategies in the different schemes and to make it transparent to the firms that public support has a ceiling. This ceiling should be located at just the right level so as to foster excellence, but without compromising variety. The matching grant instrument seems to be particularly well-suited to keeping the right balance between excellence and diversity.

Are the LAC results very different from the international evidence? Several reviews of the impacts of fiscal incentives on business innovation investments tend to also reject the full crowding-out hypothesis. *David et al.* (2000) and Klette *et al.* (1999) provide a comprehensive review of the main empirical studies, measuring the impact of public funding on firms' investments in innovation during the 1990s in developed countries. According to David *et al.* (2000), two-thirds of the studies report that public R&D funding did not substitute private R&D investments. In the last decade, this kind of analysis has proliferated, thanks to the increasing availability of data, providing additional insights into the effectiveness of public support for private innovation. Aschhoff (2009) provides an updated review of the most significant results. Most of them confirm the absence of full crowding-out effects, and some also show evidence of multiplier effects on private investments. In summary, the LAC region compares quite well with regard to the international evidence on the effectiveness of fiscal incentives on input additionality.

8.5.4 Output additionality

At the international level, there are fewer studies which analyse the effects of public support on innovative output (patents, numbers of new products, and sales of new products) and firm performance. Although some positive effects are detected, the results are less conclusive. The main difficulty in this case is that a longer time horizon is required to detect these effects. In fact, while crowding-out or multiplier effects can be detected almost in conjunction with the receipt of public financing, other effects are detectable only after the innovation, the learning process, and the intra-firm diffusion of the technology have come to an end. This implies that rigorous impact evaluations of these effects may require panel data for a minimum period of at least five years after the receipt of public financing.

LAC evaluations do not escape to this problem either. In many of the evaluations carried out by the IDB between 2005 and 2007 (IDB, 2007) and summarised by Hall and Maffioli (2008), the studies do not find consistent effects on patents or new product sales, and the evidence on firm performance is also mixed with positive results in terms of growth but little corresponding impact on measures of productivity. As previously mentioned, this could be due to the fact that in many of these evaluations output additionality is asked with reference to the same time period during which the innovation project is taking place. So the time span for evaluating output additionality might have been too short.

In order to fill this knowledge gap the IDB is currently reassessing many of the programs mentioned in Table 8.1 over a longer period and looking at what sorts of impacts these interventions have had upon labour productivity growth. The following programs have already been re-assessed: COFINANCIACION (Crespi *et al.*, 2010), FOMOTEC (Maffioli *et al.*, 2011), and two for Chile FONTEC and FONDEF (Alvarez *et al.*, 2011). The implementation of these longer-term evaluations required a close articulation with national offices of statistics. Indeed, one of the drawbacks of the previous evaluations was that they were based on linking beneficiary data with innovation surveys. However, even though innovation surveys are important tools for gathering information on innovation investments (and therefore to assess input additionality), sampling is normally based upon repeated cross-sections of randomly selected firms. This makes the following of firms over longer time periods almost impossible. In order to lift this data constraint problem, the evaluators have been working with national statistics offices and linking beneficiary data with business register data, a process that allows the following of firms over a long time period. One limitation of this approach is that data linking can only be done on already existing data registers; so, in this case, the results are somewhat restricted to the analysis of manufacturing firms and firms with more than ten employees.

The results for the four evaluated programs are summarised in Table 8.2. In this case, all of the programs were evaluated using the same methodological approach,

Table 8.1 Input additionality – testing for crowding in/crowding out effects

Country	*Evaluation period*	*Program name*	*Intervention*	*Beneficiaries*	*Indicator*	*Impact*	*Crowding in/out*	*Method*
Argentina (a)	1994–2001	FONTAR-TMP1	Subsidized Loan	Firms	ln (Total R&D)	0.15**	In	FE-IV
Argentina (b)	1998–2006	FONTAR-ANR	Matching Grants	Firms	ln (Private Innov Exp)	0.18*	In	FE-CS
Panama (c)	2000–2003	FOMOTEC	Matching Grants	Firms	ln (Total R&D)	0.15**	No evidence	FE-CS
Uruguay (d)	2000–2006	PDT-I	Matching Grants	Firms	ln (Private Innov Exp)	0.84**	In	FE-CS
Mexico (e)	2004–2007	EFIDT	R&D Tax Credit	Firms	ln (Private R&D)	0.25**	In	FE
Colombia (f)	2000–2002	Tax Incentives	R&D Tax Credit	Firms	ln (Private R&D)	0.06**	In	SM
Argentina (g)	1995–2001	FONTAR CFF	R&D Tax Credit	Firms	ln (Private R&D)	0.13***	In	FE
Argentina (h)	2001–2004	FONTAR-ANR	Matching Grants	Firms	(Total R&D intensity) %	0.18**	No evidence	DID-PSM
Brazil (i)	1996–2003	ADTN	Subsidized Loan	Firms	(Priv. R&D intensity) %	0.66**	In	PSM
Brazil (j)	1999–2003	FNDCT	Matching Grants	Firms & UNIV	(Priv. R&D intensity) %	1.63**	In	PSM
Chile (k)	1998–2002	FONTEC	Matching Grants	Firms	(Total R&D intensity) %	0.74*	Partial out	DID-PSM
Panama (l)	2006–2008	SENACYT	Matching Grants	Firms	(Total R&D intensity) %	0.13**	In	PSM
Colombia (m)	2002–2003	COFINANCIACION	Matching Grants	Firms & UNIV	(Total R&D Intensity) %	1.20*	In	PSM

Source: FE-IV (fixed-effects, instrumental variable), FE-CS (fixed effects and common support), FE (fixed effect), SM (structural modelling), DID-PSM (difference in difference, propensity score matching), PSM (propensity score matching). (a) Chudnovsky *et al.* (2006), (b) Lopez *et al.* (2010), (c) Maffioli *et al.* (2011), (d) Lopez and Rossi (2010), (e) Calderon-Madrid (2010), (f) Mercer-Blackman (2008), (g) Binelli and Maffioli (2007), (h) Chudnovsky *et al.* (2006), (i) De Negri *et al.* (2006a), (j) De Negri *et al.* (2006b), (k) Benavente *et al.* (2007), (l) Crespi *et al.* (2011a), and (m) Crespi *et al.* (2011a). In the case of the evaluation of SENACYT-Panama, total R&D intensity is computed as R&D as a fraction of total innovation sales. *** 1 per cent significance level, ** 5 per cent significance level and * 10 per cent significance level.

Table 8.2 Output additionality – testing for productivity impacts

Country	*Evaluation period*	*Program name*	*Intervention*	*Beneficiaries*	*Indicator*	*Impact*	*Method*
Colombia (a)	1995–2007	COFINANCIACION	Matching Grants	Firms & UNIV	Labour Productivity	0.15***	FE-CS
Chile (b)	1998–2006	FONTEC	Matching Grants	Firms	Labour Productivity	0.09***	FE
Chile (b)	1998–2006	FONDEF	Matching Grants	Firms & UNIV	Labour Productivity	0.12***	FE
Chile (b)	1998–2006	FONTEC only	Matching Grants	Firms	Labour Productivity	0.06	FE-CS
Chile (b)	1998–2006	FONDEF only	Matching Grants	Firms & UNIV	Labour Productivity	0.10***	FE-CS
Chile (b)	1998–2006	FONDEF+FONTEC	Matching Grants	Firms & UNIV	Labour Productivity	0.24***	FE-CS
Panama (c)	2000–2003	FOMOTEC	Matching Grants	Firms	Labour Productivity	0.13*	FE-CS

Source: FE-CS (fixed effects and common support) and FE (fixed-effects). (a) Crespi *et al.* (2011), (b) Alvarez *et al.* (2011) and (c) Maffioli *et al.* (2011). *** 1 per cent significance level, ** 5 per cent significance level and * 10 per cent significance level.

with the main output indicators being labour productivity. In all of the cases the main dependent variable is in log. The results suggest an important impact of the fiscal incentives on firm performance with increases in labour productivity between 13 per cent in the case of Panama (which corresponds to the shortest time period) and 15 per cent in the case of the Colombian program. The results are significant to the standard levels. The two Chilean programs fall in between, with productivity increases of between 9 per cent in the case of FONTEC and 12 per cent in the case of FONDEF. Having two programs in the same country is interesting because it allows for a closer comparison between them. In particular, we could explore whether there are important synergies between the two programs. In other words, in the sample there is a small – but still important – group of beneficiaries that make use of both programs. Thus, we re-explored the analysis for Chile, taking into consideration the existence of three treatments or interventions: (i) firms that applied only for FONTEC; (ii) firms that applied only for FONDEF; and (iii) firms that made simultaneous use of both programs (FONTEC + FONDEF). We found strong evidence to suggest that there are important synergies between the two programs. Indeed, companies that received the FONTEC only treatment showed productivity increases of 6 per cent, while firms that made use of the FONDEF only approach showed productivity increases of 10 per cent. On the other hand, firms that used both programs showed a productivity growth of 24 per cent.

In order to explain the differences between FONTEC and FONDEF, we need to consider the main characteristics of both programs. In the case of FONTEC it is the typical grant-matching scheme for business innovation projects and where projects are selected based upon an open window system. In the case of FONDEF, we are looking at a program that gives grants to support university–firm collaboration and where the selection is based upon competition.

We systematically found that FONDEF firms have had a higher impact upon productivity than FONDEF firms; this could be due to either the incentives for collaboration (which reduces other market failures, in addition to lack of finance) or the nature of the competitive process used for the allocation of the resources. We also found evidence that there are important complementary effects between both programs, and that FONTEC produces its highest return when it is implemented together with FONDEF. In other words, we found that an incentive for U–I collaboration (which might be targeting a coordination failure) combined with an incentive to the firm (which could be targeting an appropriability or financial constrain problem) seems to be most effective.

In the cases of COFINANCIACION and FOMOTEC it was also possible to track the impacts along time by looking at the time since exposure to the treatment. The results provide evidence that these effects remain and (in some cases) increase over time, with these effects becoming more significant between three to five years after the firms started being treated. These findings suggest that longer-term impact evaluations enable the detection of impacts upon some of the most relevant variables of interest. This does not necessarily mean that final impact evaluations should be carried out five years after the project's execution; evaluations could focus instead on the first cohorts of treated firms, so that by the end of a program some results on performance could also be assessed. This is precisely the approach taken by the US Congress for the evaluation of the Small Business Innovation Research (SBIR) program. The moment that the program was approved in the early 1980s, Congress asked the Small Business Administration (SBA) to ensure that the beneficiaries of the first three cohorts were followed up over the next decade (Lerner, 1999).

8.6 Conclusions

This chapter presents the most important recent trends with regard to the design and evaluation of public policies for the support of business innovation in LAC. Several countries in the region have been experimenting with these policies since the early 1990s. In contrast with the OECD countries, the LAC's innovation support framework is clearly biased towards direct transfers to the private sector. Only a few countries have recently started to experiment with tax incentives, and even in the case of the most important direct support instruments, such as the matching grants, the fiscal budgets allocated to these programs are rather meagre. To some extent, business innovation policy in the region is still in its infancy.

Many of these pilot programs have already been assessed and this chapter takes advantage of the existent wealth of studies in order to provide a qualitative meta-analysis of the most pioneering programs in operation since the early 1990s. The main conclusions are rather straightforward: *there is clear evidence of a positive impact on input additionality*. In other words, fiscal incentives have been effective at the moment of increasing a firm's investment in innovative projects, and they have been also effective in leveraging private resources for this investment. The studies also found that different financing mechanisms have varying impacts on different group of beneficiaries. Although it seems that the risks of crowding-out private investments are lower in the case of programs based on subsidised loans or tax incentives, grant-matching seems to be more effective in the case of new innovators or at the moment of fostering linkages between firms and universities, which suggests a need for focusing programs upon these two issues. With regards to output additionality and productivity, impacts also seem to be positive whenever enough time has elapsed since the grant was approved. Indeed, the studies which looked at output additionality suggest that positive impacts in labour productivity are apparent only three to five years after the start of an innovation project.

Notes

1 Investments in innovation activities include both tangible components (machinery and equipment, computer hardware, etc.) and intangible components (research and development, design, software development, etc.)
2 Metcalfe (1995, p. 1164) states that NSI is

that set of institutions that jointly and individually contribute to the development and diffusion of new technologies and which provides the framework within which governments form and implement policies to influence the innovation process. As such it is a system of interconnected institutions to create, store and transfer the knowledge, skills and artefacts, which define new technologies. The element of nationality follows not only from the domain of innovation policy but from elements of share language and culture that bind the system together, and form the national focus of other policies, laws and regulations that condition the innovative environment.

3 The OECD defines innovation as 'the implementation of a new or significantly improved product (good or service), or process, a new marketing method, or a new organisational method in business practices, workplace organisation or external relations' (OECD, 2005, p. 46).
4 Although it is true that R&D investments are normally more oriented toward the introduction of innovations with a high level of novelty – in other words they are targeted toward innovations that push the technological frontier – there is also an important consensus in the literature that a minimum level of R&D is necessary in order to create enough absorptive capacities to search for, adopt and adapt already existent technologies to local contexts. The concern is that not even this minimum level is reached in the region.
5 On the other hand, projects with a significant component of basic research are unlikely to produce results with commercial application in the short term. Although this may discourage private investments, the projects could still have a high social return because of the skills and knowledge produced during their development, aside from their final achievements.
6 The applied nature of technological knowledge also made it more likely to be protected by intellectual property rights. However, this by no means implies that firm investment in technological knowledge will be socially optimum; appropriability problems also exist in the case of this knowledge as the coverage offered by intellectual property rights protection is usually limited.
7 The regulation may allow and encourage firms to coordinate their R&D investment during the first stage of a project (e.g. the basic research stage) and then force them to engage in Cournot or Bertrand-type competition in the second stage (e.g. prototype development). On this topic, see, among others, Martin and Scott (2000).
8 For some references see Steinmueller (2010) and IDB (2010a).
9 Although there are important variations in the sorts of expenditures that are eligible for the subsidy, the typical matching grant program normally pays for research variable costs (researchers' salaries, research inputs and the costs of outsourced R&D); however, in some cases a fraction of the fixed costs of laboratory and testing equipment is also included. Some programs also include among eligible costs those expenses related to either the acquisition of intellectual property rights (such as the purchase of a license) or patent/trademarks application costs.
10 Tax incentives normally end up making the tax system more complex, leading to a higher degree of uncertainty with regards to the interpretation of legal regulation, which requires more auditing resources from the administration authority, and to greater opportunities to manipulate the tax system, thus generating more room for evasion and avoidance (De Luis, 2010).
11 However, this needs to be balanced against the fact that the final impact of the tax incentive on capital costs also depends upon how the investment is going to financed. If financing is done with debt, some schemes – in Latin America, in particular – allow for the deduction of interest paid for this, de facto reducing the firm's taxable base. Because large and established firms are more likely to obtain external financing, this

rebalance the expected impact of the tax incentive towards SMEs and new firms (if they actually make profits) (Roca, 2010).

12 For more details see Hall and Maffioli (2008) and Crespi *et al.* (2011b).

13 In these two cases the crowding-in/out analysis is less precise. In this case some assumptions need to be done with regards to the potential ways in which the subsidy might have been disbursed for a representative plant, and also with regards to the project duration. In these cases the assumed project duration relates to the one specified in the grant contract. This approach will surely underestimate the actual length of the typical innovation project.

References

Aghion, P., David P. and Foray, D. (2009) 'Science, Technology and Innovation for Economic Growth: Linking Policy Research and Practice in 'STIG systems'', *Research Policy*, 38(4): 681–93.

Alvarez, R., Benavente, J.M., Campusano, R. and Cuevas, C. (2011) 'Employment, Firm Size and Innovation in Chile: The Microeconometric Evidence', IDB-TN, forthcoming.

Arrow, K.J. (1962) 'The Economic Implications of Learning by Doing', *The Review of Economic Studies*, 29(3): 155–73.

Aschhoff, B. (2009) 'The Effect of Subsidies on R&D Investment and Success: Do Subsidy History and Size Matter?', ZEW Discussion Papers 09–032, ZEW – Zentrum für Europäische Wirtschaftsforschung.

Benavente, J.M., Crespi, G. and Maffioli, A. (2007) 'Public Support to Firm Innovation: The Chilean FONTEC Experience', OVE Working Papers 0407, Inter-American Development Bank, Office of Evaluation and Oversight (OVE).

Binelli, C. and Maffioli, A. (2007) 'A Micro-Econometric Analysis of Public Support to Private R&D in Argentina', *International Review of Applied Economics*, 21(3): 339–59.

Bresnahan, T. and Trajtenber, M. (1995) 'General Purpose Technologies: Engines of Growth?', *Journal of Econometrics*, 65(1): 83–108.

Calderon-Madrid, A. (2010) 'A Microeconometric Analysis of the Impact of Mexico's R&D Tax Credit Program on Private R&D Expenditure', unpublished manuscript, El Colegio de Mexico.

Chudnovsky, D., López, A., Rossi, M. and Ubfal, D. (2006) 'Evaluating a Program of Public Funding of Private Innovation Activities. An Econometric Study of FONTAR in Argentina', OVE Working Papers 1606, Inter-American Development Bank, Office of Evaluation and Oversight (OVE).

Cohen, W. and Levinthal, D.A. (1989) 'Innovation and Learning: The Two Faces of R&D – Implications for the Analysis of R&D Investment', *Economic Journal*, 99(397): 569–96.

Crepon, B., Duguet, E. and Mairesse, J. (1998) 'Research, Innovation, and Productivity: An Econometric Analysis at the Firm Level', *Economics of Innovation and New Technology*, 7(3): 115–56.

Crespi, G. and Zuniga, P. (2011) 'Innovation and Productivity – Evidence from Six Latin American Countries', World Development, forthcoming.

Crespi, G., Maffioli, A. and Melendez, M. (2010) 'Business Support to Innovation. The Colombian COLCIENCIAS' Experience', IDB–TN, 264.

Crespi, G., Solis, G. and Tacsir, E. (2011a) 'Evaluación del Impacto de Corto Plazo en la Innovación de las Empresas Panameñas', IDB-TN-264, Washington DC.

Crespi, G., Maffioli, A., Mohnen, P. and Vazquez, G. (2011b) *Evaluating the Impact of Science and Technology Projects: A Methodological Toolkit*, Impact Evaluation Guidelines, Technical Notes,IDB-TN-333, Washington DC.

Criscuolo, C. (2009) 'Direct and Indirect Effects of Innovation Policy', OECD Presentation, Paris.

David, P., Hall, B. and Toole, A. (2000) 'Is Public R&D a Complement or Substitute for Private R&D? A Review of the Econometric Evidence', *Research Policy*, 29(4–5): 497–529.

De Luis, M.J. (2010) 'Methodology for Measuring the Fiscal Effect of Regional Tax Expenditures in Colombia', Discussion Paper, IDB-DP, 139.

De Negri, J.A., Borges Lemos, M. and De Negri, F. (2006a) 'Impact of P&D Incentive Program on the Performance and Technological Efforts of Brazilian Industrial Firms', OVE Working Papers 1406, Inter-American Development Bank, Office of Evaluation and Oversight (OVE).

De Negri, J.A., Borges Lemos, M. and De Negri, F. (2006b) 'The Impact of University Enterprise Incentive Program on The Performance and Technological Efforts of Brazilian Industrial Firms', OVE Working Papers 1306, Inter-American Development Bank, Office of Evaluation and Oversight (OVE).

Disney, R., Haskel, J. and Heden, Y. (2003) 'Restructuring and Productivity Growth in UK Manufacturing', *The Economic Journal*, 113(489): 666–94.

Hall, B. and Lerner, J. (2010) 'The Financing of R&D and Innovation', in B.H. Hall and N. Rosenberg (eds) *The Economics of Innovation*, Amsterdam: North Holland.

Hall, B. and Maffioli, A. (2008) 'Evaluating the Impact of Technology Development Funds in Emerging Economies: Evidence from Latin America', NBER Working Paper Series 13835.

Harris, R., Li, Q.C. and Trainnor, M. (2009) 'Is a Higher Rate of R&D Tax Credit a Panacea for Low Levels of R&D in Disadvantage Regions?', *Research Policy*, 38(1): 192–205.

Inter-American Development Bank, IDB (2007) 'IDB's Science and Technology Programs: An Evaluation of the Technology Development Funds (TDF) and Competitive Research Grants (CRG)', OVE Evaluation Report, Inter-American Development Bank, Office of Evaluation and Oversight (OVE).

Inter-American Development Bank, IDB (2010a) *The Age of Productivity. Transforming Economies from the Bottom Up* (ed. C. Pages), New York: Palgrave Macmillan.

Inter-American Development Bank, IDB (2010b) 'Science, Technology and Innovation in Latin America and the Caribbean: A Statistical Compendium of Indicators', IDB, Washington, DC.

Klette, T.J., Møen, J. and Griliches, Z. (1999) 'Do Subsidies to Commercial R&D Reduce Market Failures – Microeconomic Evaluation Studies?', Harvard Institute of Economic Research Working Papers 1861, Harvard Institute of Economic Research.

Lerner, J. (1999) 'The Government as Venture Capitalist: The Long-Run Impact of the SBIR Program', *Journal of Business, University of Chicago Press*, 72(3): 285–318.

Lopez, A. and Rossi, M. (2010) 'Evaluación de Impacto de un Programa de Financiamiento Público a Actividades de Innovación en Uruguay – Programa de Desarrollo Tecnológico', unpublished manuscript.

Lopez, A., Reynoso, A.M. and Rossi, M. (2010) 'Impact Evaluation of a Program of Public Funding of Private Innovation Activities: An Econometric Study of FONTAR in Argentina', OVE Working Paper 03/10.

Lundvall, B.A. (1992) *National Systems of Innovation: Towards a Theory of Innovation and Interactive Learning*, Pinter, London.

Maffioli, A., Pusterla, F. and Ubfal, D. (2011) 'Public Support to Firm's Innovation: The FOMOTEC Experience in Panama', IDB-TN, forthcoming.

Martin, S. and Scott, J. (2000) 'The Nature of Innovation Market Failure and the Design of Public Support for Private Innovation', *Research Policy*, 29(4–5): 437–47.

Mercer-Blackman, V. (2008) 'The Impact of Research and Development Tax Incentives on Colombia's Manufacturing Sector: What Difference Do They Make?', IMF Working Paper, WP/08/178.

Metcalfe, J.S. (1995) 'Technology Systems and Technology Policy in an Evolutionary Framework', *Cambridge Journal of Economics*, 19(1): 25–46.

Nelson, R. (1959) 'The Simple Economics of Basic Scientific Research', *Journal of Political Economy*, 67(3): 297–306.

Nelson, R. (1993) *National Innovation Systems: A Comparative Study*, Oxford University Press, New York.

Organization for Economic Co-operation and Development and Eurostat, OECD (2005) *Oslo Manual. The Measurement of Scientific and Technological Activities. Guidelines for Collecting and Interpreting Innovation Data,* 3rd edn, Paris: OECD/Eurostat.

Roca, J. (2010) 'Evaluation of the Effectiveness and Efficiency of Tax Breaks', Inter-American Development Bank, Institutional Capacity and Finance Sector, Discussion Paper, IDB-DP, 136.

Soete, L., Verspagen, B. and Ter Weel, B. (2010) 'Systems of Innovation' in B.H. Hall and N. Rosenberg (eds) *The Economics of Innovation*, Amsterdam: North Holland.

Steinmueller, E. (2010) 'Economics of Technology Policy', in B.H. Hall and N. Rosenberg (eds) *The Economics of Innovation*, Amsterdam: North Holland.

Stiglitz, J.E. and Weiss, A. (1981) 'Credit Rationing in Markets with Imperfect Information', *American Economic Review*, 71(3): 393–410.

Part III
Internationalisation

9 Trade policy and internationalisation

Dominick Salvatore

9.1 Introduction

In this chapter, I will begin by examining the current thinking on trade policy and the attacks on the free multilateral trading system, as well as the degree of trade restrictions actually in operation in the world today. I will then examine the degree of internationalisation of the current world economy and its effects upon innovation and growth.

9.2 Trade policies – theory

Free trade has been a universally accepted economic tenet for almost two centuries. Free trade leads to the most efficient use of world resources and benefits all trading nations through free voluntary exchange (the gains from trade). More important than the static gains from trade are the crucial dynamic benefits from increased competition, economies of scale, stimulus to investments, and access to foreign technologies. These have been formally recognised at least since the time of Adam Smith. Certainly, there have been challenges to the free trade tenet from time to time, but Ricardo's theory of comparative advantage has proven to be one of the most enduring of all economic laws.

In recent decades, however, the liberal trading system has been subjected to increasing attacks in the form of (1) strategic trade and industrial policies; (2) the deindustrialisation of advanced countries; (3) rapid globalisation and outsourcing; (4) demands by advanced countries for strict labour and environmental standards in emerging markets; (5) the proliferation of regional trade agreements; and (6) rising protectionism as most countries try to combat the present serious global economic crisis. In this section, I will briefly review each of these attacks on the free trading system and show that they do not justify the call to move away from principle of a liberal multilateral trading system.

9.2.1 Strategic trade and industrial policies

The most powerful theoretical argument advanced against free trade during the 1980s was that Ricardo's traditional theory of comparative advantage was entirely static in nature and not very relevant in today's world, which is characterised by imperfect competition, technological breakthroughs, product cycles, intra-industry trade, multinational corporations, and integrated capital markets. This was the case for strategic trade and industrial policies to create comparative advantages in new technologies, as contrasted with traditional comparative advantages based upon inherited international differences in factor endowments across nations (see Krugman, 1987 and Tyson, 1992).

With the continued liberalisation of international trade and the rapid globalisation of the world economy during the 1990s, however, strategic trade theory became discredited. It became increasingly clear that a policy maker's ability to identify industries of potential comparative advantage and actually create them by strategic trade and industrial policies was so difficult and fraught with danger that even some of its staunchest original supporters (including Krugman) moved away from it as a practical, real-world policy. In other words, it was recognised that although free trade may be suboptimal in theory, it is in fact optimal in practice. Indeed, for every alleged success of strategic trade theory and industrial policy (such as semiconductors in Japan in the 1970s and 1980s), one could point to an equally spectacular failure (such as the Concorde – a technological feat, but a commercial disaster).

9.2.2 The deindustrialisation of advanced countries

Another argument advanced in favour of moving away from a free trading system was the counteraction of deindustrialisation in advanced countries, as reflected in their declining share of manufacturing employment since the 1970s. But this is also an argument that should have been put to rest soon after it was first advanced. Table 9.1 shows the relative importance of the different factors accounting for deindustrialisation in all industrial countries as a group, and in the United States, the European Union, and Japan. The table shows that from 1970 to 1994 manufacturing employment declined by about 10 percentage points in all industrial countries, as a group, as well as in the United States and in the European Union, and by about 4 percentage points in Japan.

However, Table 9.1 also shows that most of this decline resulted from the growth of labour productivity (which made possible higher levels of output with less labour), the decline in the rate investments, and other domestic forces. International trade actually resulted in an increase in industrial employment (the negative signs indicated the opposite of deindustrialisation), except in the United States (where it led to 9.6 percentage point decline in manufacturing employment). This, however, does not take into consideration the fact that the huge trade deficits in the United States led to many more job losses than gains.

But even if this were not the case, deindustrialisation only reflects the change in comparative advantage in the world economy, whereby advanced countries moved toward greater specialisation in services and left more of the manufacturing to emerging markets, in which the latter have become much more efficient. But this in no way justifies calling for trade restrictions and industrial protection on the part of advanced countries against emerging market exports.

Table 9.1 Factors responsible for deindustrialisation

	Industrial countries	*United States*	*European Union*	*Japan*
Share of Manufacturing Employment (%)				
1970	27.6	26.4	30.4	27.0
1994	18.0	16.0	20.2	23.2
Change	–9.6	–10.4	–10.2	–3.8
Percentage change due to:				
Productivity growth	65.6	65.4	59.8	157.9
Investment	18.8	3.8	20.6	71.1
Trade	(–)2.1	9.6	(–)2.9	(–)30.0
Other	17.7	21.2	22.5	(–)51.7
Total	100.0	100.0	100.0	100.0

Source: International Monetary Fund (1997, p. 68), Scott (2007), and 'Pain from Free Trade Spur Second Thoughts', *Wall Street Journal*, March 28, 2008, p. A1.

9.2.3 Globalisation and outsourcing

Since the turn of this century, the challenge to free trade seems to have re-emerged with a vengeance as a result of the alleged rapid exporting of jobs from rich countries to emerging nations via outsourcing. The acceleration of the globalisation of the world economy that has been taking place during the past decade is now leading to the transfer abroad not only of unskilled jobs, as part of the specialisation in production engendered by traditional free trade theory, but also of highly-skilled and high-paying jobs from the advanced sectors of rich nations. This problem has been brought to the forefront during the jobless recovery from the recessions of 2001 and 2008–2009, and the very high rates of unemployment which prevail in most advanced nations today (2011).

The reasoning goes as follows: if rapid globalisation and international competitiveness leads – or even requires – multinational corporations to send an increasing share of their investments for their most advanced technology to emerging markets, especially China, the comparative advantage (and with it many high-paying jobs) will also be transferred from the advanced nations to the leading emerging markets where capital and technology can earn higher returns. What, then, does remain to be done in the rich countries?

At one time, rich countries could rely upon the greater availability of capital, more advanced technology and higher skills in order to gain a comparative advantage in high value-added and high-wage industries. But as capital and technology is transferred abroad ever more rapidly under the pressure of intensifying international competition, an increasing share of the benefits of international trade accrues to the leading emerging markets in the form of rising wages and high-paying job, at the expense of today's rich countries. At least, so the argument goes.

In short, the world seems to have turned upside-down. In previous decades, it was developing countries that complained that the international economic system was biased against them, with most of the benefits from free international trade accruing to the rich countries. Now it is the rich countries that are complaining that an increasing share of the benefits from international trade and specialisation go to emerging market economies, especially China.

The only way for rich countries to now retain more of the benefits of international specialisation in production and trade is to speed up the process of technological progress and hope to retain more of its benefits, at least temporarily. Here, the old saying that nations need to run ever faster in order not to fall behind seems pertinent. The question being asked more vociferously today is 'Does getting cheaper and/or better products and services from abroad justify sacrificing so many domestic jobs?'

Table 9.2 shows that from 2003 to 2005 the percentage of jobs lost in US manufacturing was three times higher than in US service industries, but in all sectors (except professional and business services) job losses were much higher in the non-tradable sectors (and thus not caused by increased imports, outsourcing or off-shoring) than in the tradable sectors. As discussed above in the case of deindustrialisation, most job losses in the United States resulted from technological changes that raised labour productivity, rather than from international trade, and it mostly affected low-skilled industrial workers. The fear today, however, is that the revolution in telecommunications and transportation is making possible the export of an increasing number of high-skill and high-paying jobs from advanced countries to emerging markets, not only in manufacturing, but also in a growing range of services that until recently were regarded as secure.

What can be made of all of this? First of all, and as is pointed out above, only a very small portion of the millions of new, high-paying jobs created by the new economy in advanced countries during the past decade actually went abroad through outsourcing. Second, some of the outsourced jobs are complementary with, and hence create, more domestic jobs in advanced countries. Third, some companies from emerging markets also invest in and create jobs in rich countries. Fourth, as emerging markets grow and become increasingly similar to advanced countries in terms of their factor endowments, their trade will become more intra-industry, rather than inter-industry (i.e. like the trade among advanced countries), to the benefit of all countries involved.

Table 9.2 US job loss rates by industry (per cent)

Industry	*Overall*	*Tradable*	*Non tradable*
Manufacturing	12	12	17
Information	4	4	15
Financial Services	4	3	12
Professional and Business Services	4	6	3

Source: Bradford and Kletzer (2008).

The same arguments and reasons are valid against the claim that globalisation and outsourcing are putting a great deal of pressure on real wages and workers' income in advanced countries. A great deal of empirical evidence shows that stagnant wages for unskilled and semi-skilled workers in advanced countries is more the result of technological change than of trade and outsourcing (see Bradford and Kletzer, 2008).

Thus, even this attack on the free multilateral trading system seems inappropriate. Should advanced countries protect industries and jobs in sectors in which they have a comparative disadvantage? How would or could this help them? All of this means that in today's highly globalised world, more of the gains from trade accrue to the leading emerging markets, particularly China, and relatively less to advanced countries. To some extent, this is what Samuelson (2004) meant, rather than supporting a move away from a liberal trading system (as wrongly interpreted by some of his readers). In short, globalisation is operating more and more as the great equalizer (i.e. spreading the grains from trade more equitably) in the world economy today.

9.2.4 Labour and environmental standards

To put an end to the process of deindustrialisation, as well as to job losses and wage pressures in advanced countries which allegedly result from outsourcing and trading with emerging markets, the United States, France, and other advanced countries have been demanding labour and environmental standards in developing nations or emerging markets in order to create a level playing-field. Labour standard are advocated not only to prohibit child labour, but also to allow workers to receive 'decent' wages and have 'good' working conditions, thereby avoiding 'social dumping'. How can a US worker compete with a Chinese worker who receives a wage of only a few dollars a day? And how can a steel plant in the United States compete with a steel plant in India if the latter does not require anti-pollution equipment that often costs half as much as that of the steel plant itself?

Although these may seem legitimate demands to many people in advanced countries, the suspicion is that they are advocated more to protect domestic industries and labour than for altruistic purposes borne out of concern with the wellbeing of foreign labour and the environment. Even if legitimate, there is a real danger that labour and environmental standards arguments will be and can be easily captured by the protectionist elements in advanced countries and lead to greater trade restrictions and a less open international trading system.

There is a more strictly economic reason for not trying to push labour and environmental standards onto emerging markets: different nations have different tastes for environmental protection, and it is wrong to try to impose uniform international standards upon them. If emerging markets could afford the same labour and environmental standards as the advanced nations, they would not be emerging or developing markets! The premature imposition of these standards would also take away most of their current absolute trade advantage. As emerging markets grow richer and more developed, they will naturally begin to demand and impose stricter labour and environmental standards – exactly as occurred in today's advanced countries as they grew richer. Trying to impose stricter labour and environmental standards upon emerging markets today represents a form of protectionism and a serious attack on the multilateral trading system.

9.2.5 Regional trade arrangements

Another alleged attack on the liberal trading system is provided by the rapid proliferation of regional trade arrangements (RTAs). Bhagwati (2008) believes that RTAs are discriminatory that their proliferation represents a 'stumbling block', and that they detract from the effort toward free trade. He also points out that trade controversies are more dangerous in a world of RTAs, as they often involve continents rather than nations. Larger nations can bully smaller ones into forming an RTA and then extract from them most of the potential benefits from economic integration. Finally, Lee *et al.* (2008) found that the current proliferations of *overlapping* or 'spaghetti-bowl' RTAs may not lead to freer trade. Indeed, the United States only started to favour and move toward RTAs from the late 1980s as a result of frustration with the slow progress in negotiating the Uruguay Round.

The World Trade Organization (WTO), on the other hand, seems to believe (perhaps out of necessity) that RTAs represent 'building blocks' toward a liberal multilateral trading system. To the WTO, RTAs represent a second-best choice of partially moving toward free trade in a piecemeal fashion in view of the fact that the first-best choice of full world-wide free trade is presently impossible to achieve. The WTO advocates policies to make RTAs function more like building blocks than stumbling blocks by (1) the adoption of the lowest, rather than the average, tariff of the nations forming a RTA; (2) keeping membership open (as the European Union does, for the most part); and (3) negotiating multilateral reductions in averages rates of protection (as in the Uruguay Round). Globalisation also works as a force toward multilateralism and inclusion.

So far, RTAs seem to have been complimentary, rather than having hindered multilateralism, but as more overlapping RTAs come into existence the point may soon be reached where they will become stumbling blocks to a liberal multilateral trading system. As pointed out by Bhagwati (2008), even now, RTAs are basically discriminatory and detract from efforts toward multilateralism.

9.2.6 The present economic crisis and protectionism

In 2008 and 2009, the world faced the most serious financial and economic crisis since the Great Depression, with slow growth and high unemployment persisting into 2011. The current crisis started in the US subprime (high-risk) housing mortgage market in August 2007, then spread to the entire financial and real estate sectors of the US economy in 2008, and from there to the rest of the world. Advanced nations faced deep recession and emerging markets experienced significantly lower growth. The United States and other nations responded by rescuing banks and other financial institutions from bankruptcy, slashing interest rates, and introducing huge economic stimulus packages. These efforts, however, only succeeded in preventing the economic recession in advanced countries and the decline in economic growth in emerging markets from being deeper than it might otherwise have been.

The Organisation for Economic Cooperation and Development (OECD, 2010a) indicated that in 2009 the volume of world trade shrank by 13 per cent from 2008 levels. Despite warnings by most international institutions and pledges to avoid protectionist trade measures made at the G8 and G20 meetings in autumn 2008 and spring 2009, many countries have raised import duties or passed stimulus measures with 'buy American' clauses and other trade-distorting subsidies in a vain attempt to protect domestic jobs (these are examined in Section 3). These policies retarded recovery and led to trade disputes which represent a serious setback to the multilateral trading system.

9.2.7 Conclusions on the theory of trade policies

Free trade is under unprecedented attack, both theoretically and empirically. Protection has been justified on strategic trade and industrial policies, in order to reverse the process of deindustrialisation in advanced countries, to protect jobs and wages from globalisation and outsourcing, by demands by advanced countries for strict labour and environmental standards in emerging markets so as to establish a level-playing field in international trade and competition, from the proliferation of overlapping regional trade agreements, and in order to mitigate the effects of the world economic crisis. None of the above, however, justify changing trade policies and moving away from the goal of a liberal multilateral trade system.

9.3 Trade policies – evidence

This section presents data on the prevalence of trade restrictions in the world economy today. The section starts with tariff restrictions on various types of products imposed by the major advanced and developing countries and then goes on to examine the prevalence of non-tariff barriers. Tariffs have historically been the most important form of trade restriction, but as tariffs were negotiated down (from nearly 50 per cent on a trade-weighted basis in 1947 to less than 5 per cent in 2011) under various multilateral trade negotiations, non-tariff trade restrictions – in the form of 'voluntary' export restraints, antidumping actions, export subsidies, etc. – have greatly increased in relative importance, to the point where today they represent the greatest threat to the open multilateral trading system.

9.3.1 Tariff restrictions

Table 9.3 gives the average tariff imposed by the United States, the European Union, Japan and Canada (the leading developed countries and the European

Table 9.3 Tariffs on non-agricultural products in the United States, the European Union, Japan and Canada 2009 (percentages)

	United States	*European Union*	*Japan*	*Canada*
Fish and fish products	1.0	11.8	5.5	0.9
Minerals and metals	1.8	2.0	1.0	1.7
Petroleum	1.5	3.1	0.6	2.7
Chemicals	2.8	4.6	2.2	2.8
Wood, paper, etc.	0.5	0.9	0.8	1.1
Textiles	8.0	6.6	5.5	6.6
Clothing	12.1	11.5	9.2	16.9
Leather, footwear, etc.	4.0	4.2	9.7	5.3
Non-electric machinery	1.2	1.9	0.0	0.9
Electric machinery	1.7	2.8	0.2	1.7
Transport equipment	3.0	4.3	0.0	5.8
Other manufactures	2.6	2.7	1.2	3.0
Average	3.3	3.9	2.5	3.5

Source: WTO (2010b).

Union) on various non-agricultural products in 2009. The table shows that the highest tariff is invariably imposed on imports of clothing, textile and leather products (also on fish and fish products in the European Union and Japan), but the average tariff level on all non-agricultural products is less than 4 per cent.

Table 9.4 gives the tariff imposed by China, India, Russia, Brazil, Korea, and Mexico (the largest and most important developing countries) on various non-agricultural products in 2009. The table shows that the lowest tariffs are imposed by Korea and China, with the others having average tariffs of between 9.9 (Mexico) and 15.3 (Brazil).

9.3.2 Non-tariff restrictions

Table 9.5 gives the antidumping investigations initiated by G20 members (the most important developed and developing nations and the European Union) between January and September 2009 and 2010. It shows that the total number of antidumping investigations initiated declined from 119 in 2009 (at the height of the financial crisis) to 96 in 2010 (there were 105 in 2008). In 2010, India had the highest number (32), followed by Brazil with 24 and the European Union (EU – as a separate entity from its members) with 13. The other members of the G20 (Australia, France, Germany, Japan, United Kingdom and Saudi Arabia) had none in 2009 and 2010. The products on which most antidumping investigations were initiated were metals, chemicals, plastics, textiles and machinery.

Table 9.6 gives the financial assistance that the Organisation for Economic Cooperation and Development (OECD) countries provided to their agriculture,

Table 9.4 Tariffs on non-agricultural products in China, India, Russia, Brazil, Korea, and Mexico in 2009 (percentages)

	China	*India*	*Russia*	*Brazil*	*Korea*	*Mexico*
Fish and fish products	10.7	29.8	12.7	10.0	16.0	16.6
Minerals and metals	7.4	7.5	10.1	10.1	4.6	7.3
Petroleum	4.4	3.8	5.0	0.2	4.1	4.4
Chemicals	6.6	7.9	6.9	8.3	5.7	5.5
Wood, paper, etc.	4.4	9.1	14.2	10.7	2.2	9.2
Textiles	9.6	13.6	11.8	22.5	9.1	14.5
Clothing	16.0	16.1	19.7	35.0	12.6	30.0
Leather, footwear, etc.	13.4	10.2	9.8	15.7	7.8	11.1
Non-electric machinery	7.8	7.3	4.1	12.7	6.0	5.9
Electric machinery	8.0	7.2	8.7	14.2	6.2	7.7
Transport equipment	11.5	20.7	15.2	18.1	5.5	14.2
Other manufactures	11.9	8.9	13.0	15.3	6.7	10.5
Average	8.7	10.1	10.1	14.1	6.6	9.9

Source: WTO (2010b).

both in billions of US dollars and as a *producer subsidy estimate* (i.e. as a percentage of gross farm receipts) in 2005 and 2009. The table shows that in 2009, the European Union spent the most on agricultural subsidies ($120.8 billion), followed by Japan ($46.5 billion) and the United States ($30.6 billion). This was only a little less than in 2005. The producer subsidy estimate (PSE) in the European Union was more than 2.4 times and that of Japan and 4.8 times that of the United States. Norway, Switzerland and Korea provided the highest PSE. Agricultural subsidies were (and continue to be) responsible for some of the sharpest trade controversies in the world today, were responsible for the long delay in concluding

Table 9.5 Anti-dumping investigations initiated in 2009 and 2010 by G20 members

G20 Member	*Jan.–Sept. 2009*	*Jan.–Sept. 2010*	*G20 Member*	*Jan.–Sept. 2009*	*Jan.–Sept. 2010*
India	23	32	Canada	5	2
Brazil	4	24	United States	12	2
EU	10	13	Mexico	2	1
Argentina	23	7	Turkey	6	1
Australia	5	7	Indonesia	6	0
China	15	4	Russia	6	0
S. Korea	0	3	South Africa	2	0
			Total	119	96

Source: WTO (2010b), Table 5.

Table 9.6 Agricultural subsidies and producer-subsidy equivalent in various OECD countries in 2005 and 2009

Country	*Billions of US dollars*		*Subsidy as a percentage of agricultural output*	
	2005	*2009*	*2005*	*2009*
United States	41.0	30.6	15	10
European Union	130.8	120.8	32	24
Japan	44.6	46.5	54	48
Canada	6.5	7.8	22	20
Australia	1.4	0.9	4	3
Norway	3.1	3.7	67	66
Switzerland	5.6	6.2	68	63
Mexico	5.0	5.8	13	13
Korea	23.5	17.5	62	52
Turkey	12.6	22.6	25	37
All industrial countries	272.1	252.5	28	22

Source: OECD (2010b), Table 3.1.

the Uruguay Round, and were an important reason that the Doha Round (which started in 2001) has not yet been concluded (as of 2011).

During 2009, the worst year of the global financial crisis, world output declined by 2.5 per cent, but world trade fell by 12 per cent. Trade protectionism also increased as nations tried to protect domestic production and jobs, despite a pledge by the leaders of the G20 to avoid protectionist measures at their November 2008 summit meeting. Table 9.7 shows that China was the nation targeted for protectionist measures by most other nations (56), followed by the United States, Japan, Germany, France, the United Kingdom and Italy. The table also shows that the products most affected by protectionism measures were basic metals (with 65 such measures), followed by basic chemicals, special purpose machinery, transport equipment, fabricated metal products, agricultural products, and textile articles.

Developed nations generally impose higher tariffs the greater the degree of processing of the imported commodity in order to encourage domestic processing and employment. This cascading tariff structure, with *effective* tariff rates greatly exceeding corresponding *nominal* tariff rates on manufactured imports, is extremely detrimental and seriously discourages industrialization in developing countries; also, it clearly represents trade protectionism. Although average tariff rates on imports at all stages of processing have fallen during the past decade as a result of the implementation of the Uruguay Round, the cascading tariff structure, or the tariff escalation with the stage of processing, persists, especially in textiles and clothing, leather, rubber and travel goods – all products of crucial importance to many of the poorest developing countries.

Table 9.7 Nations and products most affected by protectionist measures, September 2008 and September 2009

Targeted nation	*Number of nations imposing protectionist measures*	*Product category*	*Number of protectionist measures*
China	56	Basic metals	65
United States	49	Basic chemicals	54
Japan	46	Special purpose machinery	52
Germany	30	Transport equipment	50
France	29	Fabricated metal products	49
United Kingdom	29	Agricultural products	40
Italy	25	Textiles articles	40

Source: Evenett (2009).

9.3.3 Conclusions on actual trade restrictions

Tariff restrictions have been negotiated down as a result of multilateral trade restrictions, but the cascading tariff structure, with the nominal tariff rate higher due to the degree of processing of the imported products, persists. Furthermore, while tariff restrictions have been reduced, non-tariff trade restrictions have increased significantly during recent decades. The number of antidumping investigations increased from an 11 year low of 163 in 2007 to 208 in 2008, and are expected to rise (they were 306 in 2001, after the burst of the dot.com bubble). The largest number of antidumping cases (about a quarter of the total) were imposed by India, followed by Brazil and Turkey. The majority were on steel exports from China. Russia raised tariffs on imported cars, China tightened import standards on food, India banned Chinese toys, Argentina tightened licensing requirements on autoparts and textiles, and a dozen or so countries are subsidising embattled automakers and car dealers. France's president, Nicolas Sarchozy, in one of his pronouncements, stated that Europe would turn into an industrial wasteland if it did not prop up its manufacturers. Clearly, protectionism is rearing its ugly head, especially during periods of economic crisis, such as the recent economic crisis. Recession must be fought with monetary, fiscal and competitive policies and not with trade protection, which would be both self-defeating and wrong.

9.4 Internationalisation

We now analyse the process of internationalisation or rapid globalisation of the world economy since the early 1980s. We will first examine the process itself and then provide evidence of the actual degree of internationalisation or globalisation and its effects upon growth across countries.

9.4.1 The process of internationalisation

We live in a globalised world. Tastes are converging (i.e. more and more people all over the world generally would like the same things) and production is rapidly globalising, with many goods that we consume either made abroad or having many imported parts and components. Many of the services we use are increasingly provided by foreigners, such as, for example, when a radiograph taken in a New York hospital is evaluated on the other side of the world in Bangalore (India). Even small companies that until a few decades ago faced only local or regional competition must now compete with firms from across the world.

The flow of technology from advanced to developing nations has also increased rapidly during the past two or three decades. Globalisation forces corporations from advanced nations to use their technology where it is most productive – which often means using it in developing countries. The largest and most advanced developing countries frequently demand technology transfer as a condition for allowing foreign corporations to set up production facilities in their country, or even for simply exporting high-tech goods to the nation. This form of protectionism is illegal under WTO rules, but it occurs frequently and openly.

Although not as free as the flow of international trade in goods, services, parts and components, or technology transfers, millions of workers at all skill levels have migrated and can migrate around the world from low-wage nations to higher wage areas which also offer more job opportunities, and thousands of jobs have moved from advanced countries to such emerging markets as India and China. Outsourcing, labour migration and technology transfers are often accused by domestic labour of causing job losses, and this has led to the birth of a strong anti-global movement.

Finance has also globalised. We can invest in companies anywhere in the world and purchase financial instruments (stocks and bonds) from any company located almost anywhere on the planet. Many pension funds are invested abroad, and a financial crisis in one financial centre quickly spreads across the world at the click of a mouse. We can exchange dollars for euros and most other currencies quickly and easily, but the rates at which we exchange our currency change frequently and drastically. In short, tastes, production, competition, labour markets, and financial markets have rapidly globalised. The process of internationalisation has slowed down somewhat during the past decade as a result of terrorism and financial crises, but it has not been reversed, and we can expect it to continue to expand.

A frequent rough measure of the economic relationship between nations, or their interdependence, is given simply by the ratio of the sum of their exports and imports to their GDP. In general, we would expect nations to trade more with larger nations (i.e. with nations with bigger GDPs) than with smaller ones, with nations that are geographically closer than with nations that are more distant (for which transportation costs are greater), with nations with more open economic systems than with nations with less open systems, and with nations with similar language and cultural background than with nations that are more different – as postulated by the gravity model.

9.4.2 The degree of internationalisation – trade

An important piece of evidence for the rapid internationalisation in the world economy is that international trade has generally grown much faster than GDP. Figure 9.1 shows that the growth in the volume of world merchandise trade (exports) has been much greater than the growth in world GDP over the 1998–2009 period, except during the recessionary years of 2001 and 2009.

Table 9.8 shows the average merchandise exports and imports as a per cent of GDP since the start of the recent period of rapid internationalisation or globalisation in the early 1980s in the G20 (the group of largest developed and developing countries), which replaced the G7 (largest developed nations) as the 'steering committee' for the world economy following the recent financial crisis. Table 9.8 shows that trade openness more than doubled in 10 countries (United States, Japan, Germany, France, Italy, Canada, Australia, Korea, China, and Brazil) and more than tripled in another four countries (India, Mexico, Argentina, and Turkey) over the past three decades. It should be pointed out that a great deal of the expansion of trade has taken the form of intra-industry trade (two-way trade in the same type of product based on product differentiation and economies of scale).

9.4.3 The degree of internationalisation – migration

Table 9.9 shows the international flow of labour to nations that received the largest numbers of migrants, the migrant stock as a percentage of the nation's population,

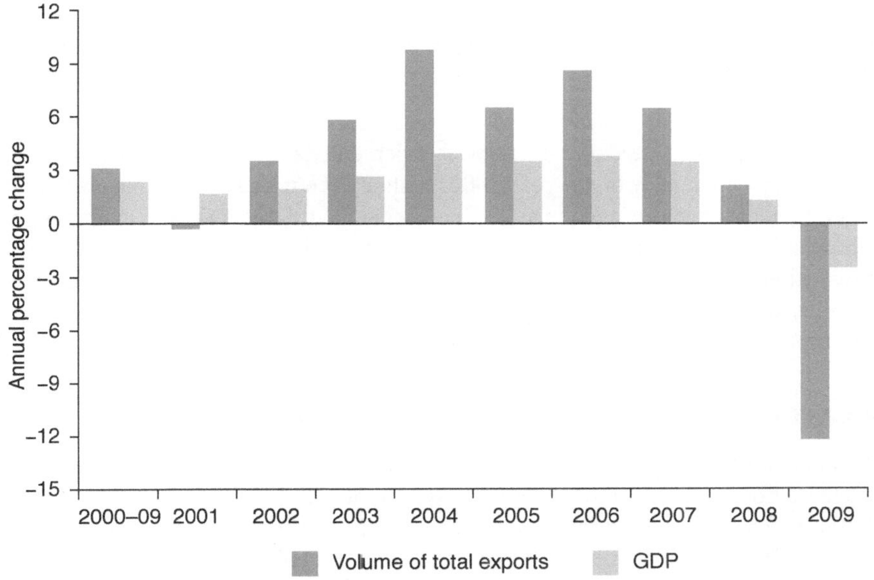

Figure 9.1 Growth of world trade and GDP, 2000–2009 (annual percentage changes)

Source: WTO (2010a, p. 3). (Figure: public domain, no permission required to reproduce).

Table 9.8 Trade openness in the G20 since 1980

Country or area	*1981–83*	*1990*	*2000*	*2009*
United States	8.1	15.8	20.7	18.8
EU	—	44.9	56.3	57.1
Japan	9.0	17.1	17.7	22.3
Germany	24.5	46.0	56.3	62.0
United Kingdom	23.5	41.3	43.9	38.4
France	19.3	37.1	46.6	39.4
Italy	17.3	32.0	44.2	38.7
Canada	22.1	43.8	75.8	48.4
Australia	15.7	26.4	34.7	34.6
Korea	31.9	53.4	72.8	82.5
China	16.7	32.5	43.9	44.3
India	7.4	13.1	20.3	29.9
Brazil	7.7	11.6	19.1	18.0
Russia	—	16.5	60.0	40.2
South Africa	28.3	37.5	47.4	47.6
Indonesia	26.9	41.5	62.4	39.1
Mexico	16.5	32.1	60.8	53.9
Argentina	9.0	11.6	18.1	30.7
Turkey	10.0	23.4	40.0	39.5
Saudi Arabia	49.5	65.4	66.0	76.6
World	27.3	32.4	40.0	42.8

Source: World Bank (1997, 2011), Table 6.1.

and immigrants with tertiary education as a percentage of the nation's population with tertiary education in the year 2000. It should be noted that, in general, it is much easier for a highly trained and skilled person to migrate than for an unskilled labourer. This represents a 'brain drain' from the point of view of the nation of emigration and a 'brain gain' for the nation of immigration. Although labour mobility is not as great as trade openness, it nevertheless represents a very important aspect of internationalisation and globalisation in the world economy today.

9.4.4 The degree of internationalisation – capital flows

Another aspect of rapid internationalisation in the world economy is indicated by the huge flows of international capital across national boundaries (OECD, 2011). Table 9.10 shows data on the major net exporters and importers of capital in the world in 2009. Most nations both export and import capital as their investors take advantage of foreign lending and investment opportunities, cover risk, and diversify their portfolio. Nations that export more capital than they import are the net capital exporters on the world scene, while those that import more capital than

Table 9.9 Movement of people

Country	*Net migration thousands 2005–2010*	*International migrant stock, % tot. pop. in 2010*	*Immigrants with tertiary educ. as % of pop. with tert. educ. in 2000*
Australia	500	21.1	2.7
Austria	160	15.6	13.5
Belgium	200	9.0	5.5
Canada	1,050	21.1	4.7
France	500	10.6	3.5
Germany	550	13.2	5.8
Greece	150	10.0	12.2
Ireland	200	20.2	33.7
Israel	85	38.8	7.8
Italy	1,650	7.4	9.7
Japan	150	1.7	1.2
Netherlands	100	10.5	9.6
Norway	135	9.9	6.2
Portugal	200	8.6	19.0
Russia	250	8.7	1.4
Saudi Arabia	150	28.0	0.9
Singapore	500	38.3	14.5
South Africa	700	3.7	7.4
Spain	1,750	13.8	4.2
Sweden	150	13.9	4.5
Switzerland	100	22.6	9.6
United Kingdom	948	10.4	17.1
United States	5,052	13.8	0.5
World	a	3.1	5.4

a = net migration = 0.

Source: World Bank (2011), Table 6.1.

they export are the net capital importers. From the table, we see that China is by far the largest net capital exporter, followed by Germany and Japan. The United States, on the other hand, is by far the largest net capital importer. The United States is simply spending too much and living beyond its means – a situation which it needs to correct.

9.4.5 Internationalisation, innovation and endogenous growth

We would expect internationalisation or the degree of openness of a nation to be an important determinant of its relative international competitiveness and growth. This is confirmed by Salvatore (2010), who found that the more internationalised

Table 9.10 Major world net exporters and importers of capital in 2009

Net exporters of capital	*Percent of world capital exports*	*Net importers of capital*	*Percent of world capital imports*
China	23.4	United States	41.7
Germany	13.3	Spain	7.3
Japan	11.7	Italy	7.0
Norway	4.8	France	3.9
Russia	4.7	Australia	3.7
Switzerland	3.6	Canada	3.6
Korea	3.5	Greece	3.5
Other	35.0	Other	29.3

Source: IMF (2010).

or globalised economies are usually more competitive than the less internationalised ones. He also found that that the more internationally competitive countries tend to grow faster than the less internationally competitive ones, but this is true only for large countries. Small, highly internationalised or globalised economies seemed to have reaped the benefits of openness from their early stages of development, so that more recently their growth has depended mostly on internal factors not directly related to internationalisation.

A theoretical explanation for expecting a positive relationship between internationalisation and growth is provided by endogenous growth theory, starting with Romer (1986) and Lucas (1988). Endogenous economic growth postulates that lowering trade barriers (i.e. opening up or internationalising the economy) will speed up the rate of economic growth and development in the long run by (1) allowing developing nations to absorb the technology developed in advanced nations at a faster rate than with a lower degree of openness; (2) increasing the actual benefits that flow from research and development (R&D); (3) promoting larger economies of scale in production; (4) reducing price distortions and leading to a more efficient use of domestic resources across sectors; (5) encouraging greater specialisation and more efficiency in the production of intermediate inputs; and (6) leading to the more rapid introduction of new products and services.

Certainly, many of these ways by which freer trade can stimulate growth and development had been recognised earlier (see, for example, Salvatore, 1983; Salvatore and Hatcher, 1991; Salvatore, 1996). Previous theorising, however, was much more casual and less rigorous. The new endogenous growth theory probes deeper and seeks to spell out more rigorously and in greater detail the actual channels or the ways by which lower trade barriers and internationalisation can stimulate growth in the long run. In particular, endogenous growth theory seeks to explain how endogenous technological change creates externalities that offset any propensity toward diminishing returns to capital accumulation (as postulated by neoclassical growth theory).

In spite of the progress made by the new endogenous growth theory in theoretically spelling out the channels through which freer trade leads to faster economic growth and development in the long run, it has been difficult to explicitly test these links in the real world because of a lack of more detailed data. In fact, as Edwards (1993) and Pack (1994) point out, most empirical tests have been based upon broad cross-sectional data for groups of countries and are not very different from the empirical studies conducted earlier. That is, new empirical studies have generally shown that openness leads to faster growth, but they have not been able to actually test in detail the specific channels by which trade is supposed to lead to faster growth in the long run, which is the major theoretical contribution of endogenous growth theory. For this, more specific country studies examining the relationship between innovation, trade, and growth are needed.

9.4.6 Conclusion on internationalisation

The rapid internationalisation or globalisation of the world economy is evidenced by the rapid convergence of tastes and international trade (especially intra-industry trade based on product differentiation and economies of scale), the rapid growth of outsourcing, and the establishment of production facilities abroad, labour migration, international capital flows, and technology transfers since the early 1980s. The rapid internationalisation or globalisation of the world economy over the past three decades has increased the international competitiveness of those countries that have opened themselves more widely to the world economy, and this has made a significant contribution to the growth of large nations. Small open economies captured the benefits of openness before the current period of rapid internationalisation and so their growth now depends more upon internal factors.

9.5 Summary and conclusions

This chapter examined (1) the current thinking on trade policies and the attacks on the free multilateral trading system, as well as the degree of trade restrictions actually in operation in the world today; and (2) the degree of internationalisation of the world economy today and its effects upon innovation and growth.

This paper showed that free trade is under unprecedented attack, both theoretically and empirically. Protection has been justified on strategic trade and industrial policies, in order to reverse the process of deindustrialisation in advanced countries, to protect jobs and wages from globalisation and outsourcing, by demands from advanced countries for strict labour and environmental standards in emerging markets so as to establish a level-playing field in international trade and competition, by the proliferation of overlapping regional trade agreements, and in order to mitigate the effects of the world economic crisis. None of these arguments, however, justify changing trade policies and moving away from the goal of a liberal multilateral trade system.

This chapter showed that tariff restrictions have been negotiated down as a result of multilateral trade restrictions, but the cascading tariff structure, with the nominal tariff rate higher due to the degree of processing of the imported products, persists. Furthermore, while tariff restrictions have been reduced, non-tariff trade restrictions have increased significantly during recent decades. Clearly, protectionism rose, especially during periods of economic crisis. Recession must be corrected with monetary, fiscal and competitive policies and not trade protection, which would be both self defeating and wrong.

The rapid internationalisation or globalisation of the world economy is evidenced by the rapid convergence of tastes and international trade, the rapid growth of outsourcing, and the establishment of production facilities abroad, labour migration, international capital flows, and technology transfer since the early 1980s. The rapid internationalisation or globalisation of the world economy during the past three decades has increased the international competitiveness of those countries that have opened themselves more widely to the world economy, and this has made a significant contribution to the growth of large nations. Small open economies captured the benefits of openness before the current period of rapid internationalisation and so their growth now depends more upon internal factors.

References

Bhagwati, J. (2008) *Termites in the Trading System*, New York: Oxford University Press.

Blinder, A. (2006) 'Offshoring: The Next Industrial Revolution', *Foreign Affairs*, 85(2): 113–28.

Bradford, J.J. and Kletzer, L.G. (2008) 'Fear of Offshoring: The Scope and Potential Impact of Imports and Exports of Services', *Policy Brief*, Petersen Institute.

Edwards, S. (1993) 'Openness, Trade Liberalization, and Growth in Developing Countries', *Journal of Economic Literature*, 31(3): 1358–93.

Evenett, S.G. (2009) *Broken Promises: A G20 Summit Report by Global Trade Alert*, London: Centre for Economic Policy Research.

IMF (1997) *Staff Studies for the World Economic Outlook*, Washington, DC: IMF.

IMF (2010) *International Financial Statistics*, Washington, DC: IMF.

Krugman, P.R. (1987) 'Is Free Trade 'Passe'?', *Journal of Economic Perspectives*, 1(2): 131–44.

Lee, J.W., Park, I. and Shin, K. (2008) 'Proliferating Regional Trade Arrangments', *The World Economy*, 31(12): 1525–57.

Lucas, R.E. (1988) 'On the Mechanics of Economic Development', *Journal of Monetary Economics*, 22(1): 3–42.

OECD (2010a) *International Trade: Free, Fair and Open?*, Paris: OECD.

OECD (2010b) *Agricultural Policies in OECD Countries: Monitoring and Evaluation*, Paris: OECD.

OECD (2011) *International Migration Outlook*, Paris: OECD.

Pack, H. (1994) 'Endogenous Growth Theory: Intellectual Appeal and Empirical Shortcomings', *Journal of Economic Perspectives*, 8(1): 55–72.

Romer, M.P. (1986) 'Increasing Returns and Long-Run Growth', *Journal of Political Economy*, 94(5): 1002–37.

Salvatore, D. (1983) 'A Simultaneous Equations Model of Trade and Development with Dynamic Policy Simulations', *Kyklos*, 36(1): 66–90.

Salvatore, D. (1996) 'International Trade Policies, Industrialization, and Economic Development', *International Trade Journal*, 10(1): 21–47.

Salvatore, D. (2010) 'Globalization, International Competitiveness, and Growth: Advanced and Emerging Markets, Large and Small Countries', *Journal of International Commerce, Economics and Policy*, 1(1): 21–32.

Salvatore, D. and Hatcher, T. (1991) 'Inward Oriented and Outward Oriented Trade Strategies', *Journal of Development Studies*, 27(3): 7–25.

Samuelson, P.A. (2004) 'Where Ricardo and Mills Rebut and Confirm Arguments Of Mainstream Economists Supporting Globalization', *Journal of Economic Perspectives*, 18(3): 135–46.

Scott, R.E. (2007) 'Costly Trade with China', Briefing Paper 188, Washington, DC: Economic Policy Institute.

Tyson, L. D'Andrea (1992) *Who's Bashing Whom?*, Washington, DC: Institute for International Economics.

World Bank (1997, 2011) *World Development Indicators*, Washington, DC: World Bank.

WTO (2010a) *World Trade Report*, Geneva: WTO.

WTO (2010b) *Report on G20 Trade Measures*, Geneva: WTO.

10 Models of speculative attacks and crashes in international capital markets

Giancarlo Marini and Giovanni Piersanti

Introduction

The past three decades of increased cross-border capital mobility and financial integration have seen the development of a variety of theoretical models aimed at understanding the causes and consequences of international capital market crises. These models can be arranged in two broad approaches, one based on the view that crises are the inevitable outcome of inconsistent macroeconomic policies and fundamentals imbalances, and the other based on the view that crises result from self-fulfilling expectations.

A more common classification scheme used in the literature categorises these two approaches as either first- or second-generation models of currency crises. Also, the financial and currency turmoils of the 1990s and early 2000s in Asia, Latin America and Russia have prompted many to include in this typology a third generation approach which emphasises the financial sector fragilities and contagion mechanisms in the onset of a crisis. Nevertheless, as it is not clear which new analytical features would organise the third approach into a distinct theoretical framework (see e.g. Flood and Marion, 1999; Jeanne, 2000; Piersanti, 2012), we will stick to the above two-approach-based classification scheme and think of the new models only as extensions of the existing setups that better evaluate the financial side of the economy.

This chapter is organised as follows. Sections 1 and 2 present the theoretical models for analysing crisis episodes in foreign exchange markets.[1] Section 3 deals with the issue of uniqueness versus multiplicity of equilibria in models of currency crises. Section 4 focuses on the relevant policy issue of crisis prevention and optimal foreign regime choice in a world of high cross-border capital mobility and full financial integration.

10.1 Fundamentals based crises

10.1.1 The basic model

The so-called first-generation or fundamentals based approach to currency crises follows from the seminal work of Krugman (1979) and Flood and Garber (1984). These models drew on earlier work by Salant and Henderson (1978) on the collapse of gold price-fixing policies.[2] The basic setup we focus on here shows that the long-run inconsistency between monetary, fiscal and exchange rate policies makes speculative attacks on fixed exchange rate regimes not only possible, but unavoidable.

Consider a small open economy facing a perfect world capital market and no barriers to trade, so that purchasing power parity (PPP) and uncovered interest parity (UIP) hold at all times. Agents have perfect foresight and consume a single tradable consumption good. The domestic supply of the good is exogenous, and its foreign currency price is fixed at unity. Financial wealth includes only two assets: domestic money and internationally traded bonds. Letting time run discretely, we can summarise the basic structure of the model in the following equations:

$$\frac{M_t}{P_t} = Y - \alpha i_t \tag{1.1}$$

$$P_t = S_t \tag{1.2}$$

$$i_t = i^* + E_t\left[\frac{S_{t+1}}{S_t} - 1\right] \tag{1.3}$$

$$M_t = M_t^H + M_t^F \tag{1.4}$$

$$M_t^H = M_{t-1}^H + \mu, \tag{1.5}$$

where M_t is the nominal money stock, Y exogenous output, i_t the domestic interest rate, i^* the foreign interest rate (assumed constant), P_t the domestic price level, S_t the nominal exchange rate, M_t^H the domestic component of money stock (domestic credit), M_t^F the domestic currency value of foreign reserves held by the central bank, and E_t the expectation operator conditional on information available at time t.

Equation (1.1) is a money market equilibrium condition, the right-hand side denoting the real money demand. Equations (1.2) and (1.3) reflect the PPP and UIP assumptions, respectively. Equation (1.4) is an accounting identity defining the money stock as domestic credit plus foreign reserves. Equation (1.5) states that domestic credit always increases at a positive constant μ. It reflects the higher priority given to other policy goals (e.g. the need to finance a fiscal deficit) which are inconsistent with the fixed exchange rate policy in the long run.

Using (1.2) and (1.3) to substitute for P_t and i_t in equation (1.1), and rearranging terms, yields

$$M_t = (Y - \alpha i^* + \alpha)\, S_t - \alpha S_{t+1} \tag{1.6}$$

where $E_t S_{t+1} = S_{t+1}$ under perfect foresight. This equation gives the dynamics of the exchange rate consistent with money market equilibrium in a flexible exchange rate regime.

Imagine now that the central bank fixes the nominal exchange rate at a level $\bar{S}$ and operates so as to accommodate any change in private demand for home currency through purchase or sale of foreign reserves to the public in order to maintain the money market equilibrium at the rate $\bar{S}$. As long as the exchange rate is fixed at $\bar{S}$ and expected to remain so for the next period, i.e. $S_t = S_{t+1} = \bar{S}$, the domestic interest rate i_t and the foreign rate i^* coincide and, by (1.6), the equilibrium level of money supply is

$$M_t = (Y - \alpha i^*)\bar{S}. \tag{1.7}$$

Equation (1.7) shows that in a fixed exchange rate regime, the nominal money demand is constant and the money supply is endogenously fixed at $(Y - \alpha i^*)\bar{S}$ in equilibrium.[3] Hence, in order to maintain money market equilibrium, reserves must adjust to compensate for the growth in domestic credit. Using (1.4) and (1.7), we obtain

$$M_t^F = (Y - \alpha\, i^*) - \bar{S} - M_t^H; \tag{1.8}$$

so that

$$\Delta M_t^F = -\Delta M_t^H = -\mu,$$

where Δ is the first difference operator. The increase in the domestic component of money supply (M_t^H) will cause foreign reserves (M_t^F) to fall steadily one-for-one with credit expansion, as the central bank intervenes in the foreign exchange market to hold the exchange rate fixed at $\bar{S}$. Any finite reserve stock committed to support the fixed rate will therefore be exhausted in a finite time, forcing the central bank to give way to a floating rate.

Letting $\bar{M}^F$ denote the lower bound on foreign reserves that the central bank is willing to accept to support the fixed rate, we can compute the date of collapse in the absence of a speculative attack as

$$\bar{t} = \frac{(Y - \alpha\, i^*)\bar{S} - \bar{M}^F - M_0^H}{\mu}$$

or

$$\bar{t} = \frac{M_0^F - \bar{M}^F}{\mu}, \tag{1.9}$$

since $M_0^F = (Y - \alpha i^*)\bar{S} - M_0^H$ by (1.8). Time $\bar{t}$ is called the point of 'natural collapse' (Grilli, 1986)[4].

The model, however, predicts that the breakdown of the fixed-rate regime comes before foreign reserves at the central bank approach the critical lower limit in the absence of speculation. This requires a speculative attack at some point $t^{\bullet} \in (0, \bar{t})$ in which private agents suddenly exchange domestic currency for foreign currency, driving the central bank's foreign reserves to their lower bound $\bar{M}^F$ and forcing a transition to a floating rate regime. The key to determining the exact point in time where the attack starts lies in the concept of the *shadow* exchange rate, which allows the fixed and the floating rate regime to be connected in a straightforward manner. According to Flood and Garber (1984), the shadow exchange rate is the floating rate that would prevail at any date t if the central bank abandoned the fixed rate regime, conditional on foreign reserves being at the critical level $\bar{M}^F$. Thus, if the exchange rate floats at any time t, the equilibrium rate S_t is again given by (1.6) with reserves fixed at their lower limit $\bar{M}^F$, that is,

$$(Y - \alpha i^* + \alpha) S_t = \left(M_t^H + \bar{M}^F\right) + \alpha S_{t+1}. \tag{1.10}$$

The bubble-free solution for S_t is[4]

$$S_t = \left(\frac{1}{Y - \alpha i^* + \alpha}\right) \sum_{j=t}^{\infty} \left(\frac{\alpha}{Y - \alpha i^* + \alpha}\right)^{j-t} \left(M_j^H + \bar{M}^F\right),$$

whence, using (1.5),

$$S_t = \frac{M_t^H + \bar{M}^F}{Y - \alpha i^*} + \frac{\alpha \mu}{(Y - \alpha i^*)^2}. \tag{1.11}$$

Agents anticipate the exchange rate that would prevail at any time t if the fixed rate regime collapsed. Therefore, if $S_t \geqq \tilde{m}$ is anticipated, a run on the central bank's foreign reserves will take place, as agents buy foreign reserves at price $\bar{S}$ and resell them at the post-collapse market price S_t, earning a profit of $(S_t - \bar{S}) \geqq 0$ per unit of reserves. On the other hand, if $S_t < \bar{S}$ is anticipated, no run will take place. As a result, the exchange rate will float whenever the condition $S_t \geqq \bar{S}$ holds.

Setting $S_t = \bar{S}$ in (1.11) and using (1.5), we find that the date of collapse $t^{\bullet}$ is

$$t^{\bullet} = \frac{(Y - \alpha i^*)\bar{S} - \bar{M}^F - M_0^H}{\mu} - \frac{\alpha}{(Y - \alpha i^*)}$$

or

$$t^{\bullet} = \frac{M_0^F - \bar{M}^F}{\mu} - \frac{\alpha}{(Y - \alpha i^*)}. \tag{1.12}$$

Substitution of (1.9) into (1.12) yields

$$\bar{t} - t^{\bullet} = \frac{\alpha}{(Y - \alpha i^*)},$$

which shows that the speculative attack will force the central bank to switch out of the fixed parity before the stock of official reserves sinks to the critical threshold level $\bar{M}^F$. Using (1.8) and (1.12), the size of attack can be computed as

$$M_{t^*}^F - \bar{M}^F = \frac{\alpha\mu}{(Y - \alpha i^*)}, \tag{1.13}$$

where $M_{t^*}^F$ denotes the stock of foreign currency reserves held by the central bank on the date of attack; so the crash of the exchange rate fixing policy materialises even when foreign reserves over the critical level $(M_{t^*}^F - \bar{M}^F)$ are still relatively high.

Note that $\bar{t} = t^*$ (and $M_{t^*}^F = \bar{M}^F$) under $\alpha = 0$. The reason is that under this restriction there would be no run on the central bank's reserves and the fixed parity would be abandoned only when the stock of official reserves comes to its lower bound. Therefore, the parameter measuring the interest sensitivity of money demand (α) is critical for determining the exact time of system's crash. It emphasises the role of speculation, by establishing the amount of contraction in money demand and reserves that occurs when the domestic interest rate jumps to reflect the expected depreciation of the exchange rate. As shown by (1.12) and (1.13), the higher α the earlier the time of attack and the higher the downward jump in foreign reserves and in the money stock will be. The higher the initial stock of foreign currency reserves held by the central bank (M_0^F) is, the longer the fixed exchange rate system will last, and the higher the growth of domestic credit (μ) is, the less chance there is that the system will survive.

10.1.2 Extensions

The basic Krugman-Flood-Garber model has been refined and extended in a variety of directions to include other important features of balance-of-payment crises. Flood, Garber and Kramer (1996), for example, have considered the effects of adding a risk premium to the simple interest parity condition and sterilisation policies at the time of speculative attack. They showed that a delayed collapse scenario is possible even when a full sterilisation policy is expected. The rationale is that the sterilisation of reserve losses spills over the bond market, shifting the adjustment mechanism from the money market to the bond market. Garber and Spencer (1995), Lahiri and Végh (2003), and Flood and Jeanne (2005) examined the effectiveness of interest rate defence policies and showed that they can be effective in postponing the date of a collapse. Obstfeld (1984), Dornbusch (1987), Djajic (1989), and Savastano (1992) analysed the effects upon the timing of a crisis of (perfectly anticipated) alternative post-collapse exchange rate arrangements. In the commonly observed case of a temporary period of floating followed by a new peg, they found that the crash date depends upon the size of expected devaluation and the length of the transition period of floating. Wyplosz (1986), Bacchetta (1990), Dellas and Stockman (1993), Agénor and Flood (1994), and

Daekeun and Sachs (1996) introduced capital controls. Wyplosz showed that imposing a ceiling on speculative transactions succeeds in maintaining a pegged-rate regime; Bacchetta, and Dellas and Stockman obtained that the anticipation of controls by private agents may advance the date of an attack; Agenor and Flood, and Daekeun and Sachs found that capital controls may be effective in delaying, but not in preventing, a collapse of a fixed rate regime. Buiter (1987) and van Wijnbergen (1991) analysed the effects of resorting to external borrowing. Buiter demonstrated that if borrowing occurs just before the fixed parity would have collapsed absent borrowing, then the crisis is likely delayed, but if borrowing occurs long enough before the fixed parity is abandoned, then the collapse occurs earlier. Finally, Connolly and Taylor (1984), Flood and Hodrick (1986), Connolly (1986), Blackburn (1988), Willman (1988), Goldberg (1991), Osakwe and Schembri (2002) and others extended the model to a world of sticky prices and analysed the real effects of speculative currency attacks.

Other research added the banking sector and focused on the role of banks in financing a speculative attack, and the interaction between bank solvency and currency stability. For example, Miller (1999) proved that the timing of an attack is brought forward when banks finance domestic currency speculation; Flood and Marion (2004) computed the different probability of banking and currency collapse when a real shock affects banks' assets returns. Gerlach and Smets (1995), Glick and Rose (1999), Corsetti *et al.* (2000), Forbes (2001, 2004), and Loisel and Martin (2001) focused on trade linkages and macroeconomic similarities in order to analyse contagion effects across countries. Using stochastic models, the consequences of uncertainty about government policies, reserves and other state variables have been investigated by, among others, Flood and Garber (1984), Obstfeld (1986b), Grilli (1986), Blanco and Garber (1986), Dornbusch (1987), Savastano (1992), and Goldberg (1994). The most important results obtained are that: (i) the transition to a floating rate is stochastic rather than certain; (ii) at any time *t* there always exists a non-zero probability of an attack in the next period, which causes the domestic interest rate to rise in the run-up to the regime crash; (iii) multiple equilibria and self-fulfilling speculative attacks may appear.

Buiter and Grilli (1991), Krugman and Rotemberg (1991), Flood and Garber (1991), Bertola and Caballero (1991), Delgado and Dumas (1993), Dumas and Svensson (1994), Werner (1995), Corrado and Holly (2000), and many others have extended the basic theory to the more empirically relevant case of exchange rate band regimes.

Corsetti, Pesenti and Roubini (1999) and Burnside, Eichenbaum and Rebelo (2001) showed that bad news about prospective budget deficits can trigger a currency crisis, and Annicchiarico, Marini and Piersanti (2011) demonstrated that a speculative attack against the home currency can start even when fiscal and monetary policy turn out to be consistent with the preservation of the fixed exchange rate regime.

Nevertheless, all of the extended models share the same primary view: crises occur as the result of an adverse trend in economic fundamentals and can be anticipated by looking carefully at some key economic variables that can serve as early warning signals of crises (e.g. Kaminsky, Lizondo and Reinhart, 1998; Kaminsky and Reinhart, 1999).

10.2 Self-fulfilling based crises

10.2.1 The basic model

The so-called 'second-generation' or endogenous-policy approach to currency crises follows from the influential work of De Kock and Grilli (1993), Obstfeld (1994, 1996, 1997), Isard (1994), Ozkan and Sutherland (1995, 1998), Sachs, Tornell and Velasco (1996), Jeanne (1997), and Jeanne and Masson (2000). The key point emphasised in this literature is that crises may happen even if the fundamentals are not inconsistent with the maintenance of a fixed exchange rate. The attack itself may induce an optimising regime-switching choice which makes the crisis self-validating.

The new approach is obtained by introducing into the basic framework an optimising policymaker committed to following a policy rule with an escape clause.[6] The policymaker minimises a loss function featuring a trade-off between the commitment to the fixed-exchange rate and other policy objectives, such as a high level of output, low unemployment, low domestic interest rates and low taxes. Conditional upon the realisation of exogenous shocks that cause deviations of the target variable from its desired level, the welfare losses of the policy authority may increase relative to that which could be obtained if the exchange rate were floating. Therefore, the problem facing the authority is to choose the value of the shock at which it is optimal to abandon the pegged rate regime. Private agents know the optimising problem of the policymaker and build this into expectations of a regime switch. This, in turn, can have an impact upon the policy decision and trigger a currency crisis even if economic fundamentals are not inconsistent with holding the currency parity unchanged. Private expectations are self-fulfilling and multiple equilibria arise for given fundamentals.

A simple model that allows an understanding of the role of fundamentals and self-fulfilling expectations in the onset of a crisis is found in Jeanne (1997). The model consists of the following three equations:

$$y_t = a_1(\Delta p_t - E_{t-1}\Delta p_t) + a_2 y_{t-1} \tag{2.1}$$

$$\Delta p_t = \Delta s_t \tag{2.2}$$

$$£_t = (y_t)^2 + \theta(\Delta p_t)^2 + \delta_t Z_t, \tag{2.3}$$

where all variables are measured in logs, y_t denotes the deviation of output from its full employment level, Δp_t is the rate of domestic inflation, and Δs_t the rate of change of the nominal exchange rate.

Equation (2.1) is the aggregate supply equation, where the lagged term $a_2 y_{t-1}$ sizes up the persistence in output fluctuations. Equation (2.2) states that domestic inflation equals the rate of depreciation of domestic currency under the PPP and zero foreign inflation assumptions. Equation (2.3) is an adaptation of the standard Kydland and Prescott (1977) and Barro and Gordon (1983) social loss function to include a fixed cost (Z_t) of switching regime, where $\theta > 0$ is a parameter reflecting the weight the government places on inflation stabilisation relative to that of output stabilisation and δ is a dummy variable which is equal to 1 if the central bank left the exchange rate to float and 0 if the official parity is left unchanged.[7]

Substitution of (2.2) into (2.1) and (2.3) reduces the model to

$$y_t = a_1(\Delta s_t - E_{t-1}\Delta s_t) + a_2 y_{t-1} \quad (2.4)$$

$$\pounds_t = (y_t)^2 + \theta(\Delta s_t)^2 + \delta Z_t. \quad (2.5)$$

These equations show that the policymaker's incentive to abandon the fixed-rate regime comes from the increase in output which can be obtained by a realignment of the exchange rate. As a result, the fixed parity commitment will almost never attain full credibility, and this, in turn, leads to a devaluation bias and the possibility of self-fulfilling speculative attacks.

Now let Δs be the amount of currency devaluation when the policymaker opts out of the fixed peg. Given uncertainty regarding the central bank's commitment to the official parity, in each period rational agents assess the policymaker's credibility by evaluating the probability that the exchange rate will be devalued next period. Therefore, if π_{t-1} denotes the perceived probability of devaluation at time t, estimated by the private sector at time $t-1$, $E_{t-1}\Delta s_t = \pi_{t-1}\Delta s$. Substitution into (2.4), shows that the output gap is

$$y_t^D = (1 - \pi_{t-1})a_1\Delta s + a_2 y_{t-1},$$

if the policymaker devalues at time t, and

$$y_t^F = a_2 y_{t-1} - a_1\pi_{t-1}\Delta s = y_t^D - a_1\Delta s,$$

if the fixed parity is maintained at time t, where the superscripts D and F stand for devaluing and fixing, respectively. Using these expressions to compute the policy losses under a flexible and a fixed exchange rate regime, we find that the *net benefit* of the fixed peg, B_t, is given by

$$B_t \equiv \pounds_t^D - \pounds_t^F = X_t - \alpha\pi_{t-1}, \quad (2.6)$$

where $X_t \equiv \mathbb{C}_t + 2a_1a_2y_{t-1}\Delta s + (a_1^2 + \theta)(\Delta s)^2$, and $\alpha \equiv 2(a_1\Delta s)^2$.

Equation (2.6) implies that the policymaker will choose to give up the fixed rate regime and devalue when $B_t < 0$. It also shows that the net benefit of staying in the pegged-rate system depends not only upon the objective economic conditions, or the 'fundamental' component, summarised in the variable X_t, but also upon the credibility of the policymaker's commitment captured by $a\pi_{t-1}$. Thus, (2.6) expresses, in the most simple way, the idea that both the state of fundamentals and self-fulfilling 'animal spirits' of the market (associated with the perceived probability of devaluation) have a role to play in the outbreak of the crisis.

Assume that the evolution of fundamentals is described by the following stochastic process

$$X_t = f_{t-1} + u_t, \tag{2.7}$$

where u_t is an i.i.d. random shock, and $f_t = E_t X_{t+1}$ is the expected value at time t of the fundamental at time $t+1$. Assume also that the density function $g(u_t)$ satisfies the following properties:

$$\begin{cases} g(-u_t) = g(u_t) \forall u_t \\ g'(u_t) > 0 \forall u_t \in (-\infty, 0) \\ g'(u_t) > 0 \forall u_t \in (0, \infty) \\ g'(0) = 0, g''(0) < 0 \end{cases},$$

which are met by a wide class of bell-shaped and symmetric density functions including the normal.

A rational expectations equilibrium is realised in this model when private agents' perceived probability of devaluation at time t equals the probability that the net benefit of maintaining the fixed parity is negative at time $t+1$, i.e. when the fixed point condition

$$\pi_t = \Pr_t \left[B_{t+1} < 0 \right],$$

is satisfied, where $\Pr_t$ denotes the probability conditioned on information available at time t. Using (2.6) and (2.7), this condition can be rewritten as

$$\pi_t = \Pr_t \left[u_{t+1} < \alpha\pi_t - f_t \right],$$

or

$$\pi_t = G(\alpha\pi_t - f_t),$$

where $G(u_t)$ is the cumulative distribution of the density function $g(u_t)$.

As both sides of equation (2.8) are increasing in π_t, there might be multiple solutions, which means that a multiplicity of devaluation probability levels may be associated with a given level of fundamentals. Nevertheless, the model implies that the multiplicity of solutions arises only when the economy finds itself in some critical zone where self-fulfilling speculative attacks are possible.

Figure 10.1 plots both sides of (2.8) for unique (a) and a multiple equilibria (b). The 45° line portrays the left-hand side of (2.8), whereas the G_f curve reproduces the right-hand side of that equation for a given value of the fundamentals, namely

$$G_{f_t} \equiv G(\chi_t),$$

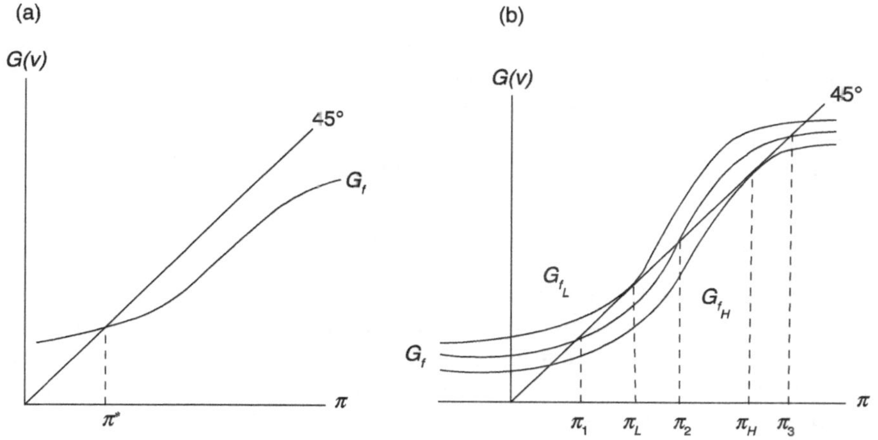

Figure 10.1 (a) Unique equilibrium; (b) multiple equilibria

where $\chi_t \equiv a\pi_t - f_t$, and f_t is the shifting factor. Fixed points, or equilibrium solutions of (2.8), are at the intersection between the 45° line and the G_f curve. The figure makes clear that the two curves may cross in one or several points depending on whether or not the slope of the G_f curve is higher than 1 at the fixed point. The slope of G_f is equal to $ag(\chi_t)$ and attains its maximum value at $\chi_t = 0$. Therefore, if $ag(0) < 1$, the slope of G_f is everywhere smaller than 1 and the expected probability of devaluation is uniquely determined by the fundamentals, as shown in panel a. Alternatively, if $ag(0) > 1$, the slope is greater than 1 at $\pi_t = f_t/a$ where $\chi_t = 0$, and the curve will cross the 45° line at more than one point (three points in panel b) if the fundamentals lie between the two threshold values f_L and f_H, as depicted in the figure.[5]

The relationship between the fundamentals and the devaluation expectations in this model can be summarised as follows:

- If $f_t < f_L$ – which means that the G_f curve is to the far right in figure 10.1b, above the G_{fL} curve – then the macroeconomic fundamentals are in such a bad state that the expected probability of being off the peg is uniquely determined and close to 1.
- If $f_t > f_H$ – which means that the G_f curve is to the far left in figure 10.1b, below the G_{fH} curve – then the macroeconomic fundamentals are in such a good state that the expected probability of switching to a floating rate is uniquely determined and close to 0.
- If $f_L < f_t < f_H$ – i.e. if the G_f curve lies between the two critical curves G_{fL} and G_{fH} – then macroeconomic fundamentals are such that the expected probability of devaluation may take multiple values (π_1, π_2, π_3 in figure 10.1b), which makes the fixed rate regime vulnerable to self-fulfilling attacks.

The model, therefore, predicts that multiple equilibria and self-fulfilling speculative attacks can occur only when the economy is in the 'grey' zone defined by the two critical values f_L and f_H. The reason for this is that when the economy enters into this range, the expected devaluation is not uniquely determined by fundamentals, so jumps across equilibria may be totally unrelated to the state of economic fundamentals and driven by a minor random event (a sunspot) that coordinates agents' expectations on one equilibrium point or the other.

10.2.2 Extensions

As with the first-generation research, the basic framework of the 'second-generation' approach has been refined and extended in a number of ways in order to capture the new features of the crises in the 1990s. This was usually worked out by incorporating into the objective function of the policymaker a number of other variables over and above the most standard ones, such as output, inflation, or unemployment. For example, De Kock and Grilli (1993), Obstfeld (1994), Sachs, Tornell and Velasco (1996), Velasco (1996), Cole and Kehoe (1996, 2000), and Benigno and Missale (2004) considered the level and the maturity structure of public debt. They showed that a government that faces a large and increasing debt, bad shocks, or high devaluation expectations will more likely choose to devalue and that, given the role that market expectations play in determining the size of total government commitments, multiple equilibria might arise for certain levels of public debt. Bensaid and Jeanne (1997), Drazen (2000), Christiano, Gust and Roldos (2004), and many others analysed the trade-off faced by a policymaker in implementing an interest rate defence policy. They found that this policy strategy is never at no cost and, in some circumstances, may lead the government into a vicious circle wherein defending the currency intensifies, rather than attenuates, speculation. Finally, Drazen and Masson (1994), Masson (1995), Agénor and Masson (1999), and Benigno and Missale (2004) examined the role of a government's credibility and reputation in determining the vulnerability of the economy to self-fulfilling speculation. The most striking result was that credibility and reputation may not increase over time, so that pursuing tough policies today may lower rather than raise the credibility of a no-parity-change promise in the future when policies have persistent effects upon the economy.

Other research shifted the focus onto spillover and contagion effects. Calvo (1999), Calvo and Mendoza (2000), Kodres and Pritsker (2002), Chari and Kehoe (2003, 2004) and others have stressed financial linkages, asymmetric information, and herding behaviour in transmitting crises across international markets. Calvo (1999), Masson (1999), Kodres and Pritsker (2002), and Berger and Wagner (2005) have examined the so-called pure contagion effect, whereby a crisis in one country can be transmitted in another country even if no change in its fundamentals has occurred.

The connections between financial fragility and currency instability, and the role of credit constraints and balance-sheet effects, have been examined in another set of papers incorporating the banking and financial sides of the economy. Goldfajn and Valdés (1997), Radelet and Sachs (1998), Chang and Velasco (2000a, 2000b, 2001),

Miller (2003), Burnside, Eichenbaum, and Rebelo (2004), Burnside (2004), Bernardo and Welch (2004), Rochet and Vives (2004), and Goldstein and Pauzner (2005), for instance, all focused on the links between bank solvency and currency stability. Calvo (1998, 2003), Krugman (1999), Aghion, Bacchetta, and Banerjee (2000, 2001, 2004), Caballero and Krishnamurthy (2001, 2003, 2004), Céspedes, Chang and Velasco (2004), Neut and Velasco (2004), and Schneider and Tornell (2004), explored the role played by capital flows volatility, financial sector weakness, credit constraints and balance-sheet mismatches in the build-up to crisis vulnerability.

Nevertheless, the fundamental implication of all these models is always the same: crises may be driven by self-fulfilling expectations with weak or no link to fundamentals, and therefore are essentially unpredictable events with some key information variables or early warning signals.

10.3 Uniqueness versus multiplicity of equilibria in models of currency crises

One problem with models of self-fulfilling crises is that they lack a theory of equilibrium selection, since the shift in beliefs which gives rise to jumps across equilibria appears to be quite arbitrary. Such an indeterminacy has made the multiple equilibrium framework open to a number of criticisms and has encouraged an intensive re-examination of its theoretical foundations (see, e.g., Morris and Shin, 1998, 1999, 2001, 2003, 2004; Heinemann and Illing, 2002; Metz, 2002; Rochet and Vives, 2004; Goldstein, 2005; Guimarães and Morris, 2007). This literature, which builds around the global game structure pioneered by Carlsson and van Damme (1993), argues that the multiplicity of equilibria in second-generation models follows from agents having common knowledge about the fundamentals of the economy. Removing this unrealistic assumption entirely eliminates the multiplicity of equilibria, thus raising a major challenge to the second-generation approach to regime change. In order to illustrate the main issues in a simple way, we discuss the basic structure of a global game model on currency attacks developed by Morris and Shin (1998, 2003, 2004).

Consider a small open economy with a pegged exchange rate. There is a government and a continuum of agents of measure one, indexed by i and uniformly distributed over $[0,1]$ Agents move simultaneously, choosing between two actions, either attack the peg or refrain from doing so. Let $a_i \in \{0,1\}$ denote the action of an agent, where $a_i = 0$ means 'not attack' and $a_i = 1$ means 'attack'. The payoff from attacking is $\Delta s = (S_t \bar{S}) > 0$, if the fixed peg is abandoned, and $-c$ if the peg is maintained, where ΔS is the size of devaluation and c is the cost of attacking. It is assumed that $\Delta S > c$, so that agents have a potential incentive to attack.

The policymaker's willingness to defend the current parity depends upon the state of economic fundamentals, as well as the proportion of agents attacking the currency. Let $f \in \mathbb{R}$ be an index of fundamentals state and let $q \equiv \int a_i \, di \in [0,1]$ be the mass of agents attacking. Assume also that the payoff for the policymaker is $f - q$ if the peg is maintained and 0 otherwise. Hence, the fixed-rate system will fail if and only if $q \geq f$.

The equilibria of this game when the state of fundamentals is common knowledge is as follows. Let $f \in (f_L, f_H) \equiv (0, 1)$ be the critical range of fundamentals over which the pegged rate is vulnerable to speculative attacks driven by self-fulfilling expectations. If $f \le f_L$, then 'attack' is a dominant strategy for each agent, since fundamentals are so weak that the peg is doomed to collapse even if no one attacks. Thus, the unique equilibrium in this case is every agent attacking. If $f \ge f_H$, then 'not attack' is a dominant strategy for each agent, since economic fundamentals are so strong that the peg will survive regardless of the size of the attack. Thus, the unique equilibrium in this case is every agent not attacking. If $f_L < f < f_H$ there are multiple equilibria sustained by self-fulfilling expectations where all speculators can either attack the currency and force a devaluation, if they expect everyone else to attack, or refrain from attacking and allow the fixed parity to be preserved, if they expect no one else to attack. This tripartition of fundamentals and equilibrium solutions occurs in a typical second-generation framework, as shown in section 2.

Following Morris and Shin, now let the common knowledge assumption about the true state of fundamentals f be dropped and imagine that each agent instead receives an idiosyncratic signal $x_i = f + \epsilon_i$ of the state f, where the noise ϵ_i is normally distributed with mean 0 and variance σ_x^2, and is i.i.d. across agents and independent of f. In addition, assume that agents observe an exogenous public signal $z = f + \xi$, where the common noise ξ is normally distributed with mean 0 and variance σ_z^2, and is independent of both f and ϵ.

In this game of incomplete information, the equilibrium solutions can be described in terms of switching strategies around threshold values $x^\blacktriangle$ and $f^\blacktriangle$ in which, for a given realisation of the public signal z, agents will attack the currency if $x \le x^\blacktriangle(z)$ and the fixed parity abandoned if $f \le f^\blacktriangle(z)$. If some conditions on the information context of private signals relative to the underlying distribution of fundamentals are satisfied, then the two thresholds $x^\blacktriangle$ and $f^\blacktriangle$ identify a unique solution for the game.

In order to compute these critical values, first note that, for given realisations of f and z, the incidence of attack q is given by the mass of agents with signals $x \le x^\blacktriangle$ (z), that is,

$$q(f, z) = \Pr(x \le x^\blacktriangle\ f, z) = \Phi\left[\sqrt{x_x}\,(v^\blacktriangle - f)\right],$$

where $\Phi(\cdot)$ is the cumulative distribution function of the standard normal, and $v_x \equiv \sigma_x^{-2}$ the precision of private signals. Notice that the size of attack q is decreasing in f, so that the threshold $f^\blacktriangle$ such that the fixed rate regime will always collapse can be computed as the solution to the policymaker's indifference condition $q(f^\blacktriangle, z), = f^\blacktriangle$, or equivalently,

$$\Phi\left[\sqrt{v_x}\left(x^\blacktriangle - f^\blacktriangle\right)\right] = f^\blacktriangle.$$

Rearranging, yields

$$x^\blacktriangle = f^\blacktriangle + \frac{1}{\sqrt{v_x}}\Phi^{-1}\,(f^\blacktriangle), \tag{3.1}$$

which expresses the first equilibrium condition of the model.

In order to find the second equation, note that when an agent observes a signal x_i, his posterior distribution of f is normal with mean

$$E(f \; x_i, z) = \gamma x_i + (1 - \gamma)z$$

and precision, $v \equiv v_x + v_z$, where $v_z \equiv \sigma_z^{-2}$ is the precision of public signal, v is the overall precision of information, and $\gamma \equiv \dfrac{v_x}{v_x + v_f}$ is the relative precision of private signal. Given that a devaluation occurs only when $f \leq f^{\blacktriangle}$, it then follows that the posterior probability of a regime switch is simply

$$\Pr(f \leq f^{\blacktriangle} \; x, z) = \Phi[\sqrt{v}\,(f^{\blacktriangle} - \gamma x - (1 - \gamma)\, z)].$$

Since $\Pr(f \leq f^{\blacktriangle} \; x, z)$ is decreasing in x, agents will find it optimal to attack the home currency up to the point where $x = x^{\blacktriangle}$, where $x^{\blacktriangle}$ is the 'optimal cut-off' condition solving the indifference condition

$$\Delta S \Pr(f \leq f^{\blacktriangle} \; x^{\blacktriangle}, z) = c,$$

or

$$\Delta S \Phi\, [\sqrt{v}(f^{\blacktriangle} - \gamma x^{\blacktriangle} - (1 - \gamma)\, z)] = c, \tag{3.2}$$

which gives the second equilibrium condition of the model. Solving (3.1) and (3.2) for the two unknown $f^{\blacktriangle}$ and $x^{\blacktriangle}$, yields the joint equilibrium values

$$f^{\blacktriangle} = \Phi\left[\frac{\sqrt{v_x}(1-\gamma)}{\gamma}\left(f^{\blacktriangle} - z - \frac{1}{(1-\gamma)\sqrt{v}}\Phi^{-1}\left(\frac{c}{\Delta S}\right)\right)\right]$$

$$x^{\blacktriangle} = \frac{1}{\gamma} f^{\blacktriangle} - \frac{(1-\gamma)}{\gamma} z - \frac{1}{\gamma\sqrt{v}}\Phi^{-1}\left(\frac{c}{\Delta S}\right)$$

or, using the definition of γ and v,

$$f^{\blacktriangle} = \Phi\left[\frac{v_z}{\sqrt{v_x}}\left(f^{\blacktriangle} - z - \frac{\sqrt{v_x + v_z}}{v}\Phi^{-1}\left(\frac{c}{\Delta S}\right)\right)\right] \tag{3.3}$$

$$x^{\blacktriangle} = \frac{v_x + v_z}{v_x} f^{\blacktriangle} - \frac{v_z}{v_x} z - \frac{\sqrt{v_x + v_z}}{v_x}\Phi^{-1}\left(\frac{c}{\Delta S}\right). \tag{3.4}$$

Such a solution always exists and is unique if and only if $v_x \geq v_z^2 / 2\pi$. Rewriting (3.3) as

$$\frac{v_z}{\sqrt{v_x}}(f^{\blacktriangle} - z) - \Phi^{-1}(f^{\blacktriangle}) = \sqrt{1 + \frac{v_z}{v_x}}\Phi^{-1}\left(\frac{c}{\Delta S}\right)$$

and letting

$$\psi(f, z) \equiv \frac{v_z}{\sqrt{v_x}}(f - z) - \Phi^{-1}(f),$$

for every $z \in \mathbb{R}$, the function $\psi(f, z)$ is continuous in f, with $\psi(f_L, z) = \psi(0, z) = \infty$ and $\psi(f_H, z) = \psi(1, z) = -\infty$, which implies that a solution necessarily exists and any solution $f^\blacktriangle \in (f_L, f_H)$. This proves existence. For uniqueness, it is required that the function $\psi(\cdot)$ be decreasing in f. Differentiating with respect to f,

$$\frac{\partial \Psi(f, z)}{\partial f} = \frac{v_z}{\sqrt{v_x}} - \frac{1}{\phi[\Phi^{-1}(f)]}.$$

Since $\max_{f \in \mathbb{R}} \phi\,[\Phi^{-1}(f)] = 1/\sqrt{2\pi}$, where $\phi\,(\cdot)$ is the probability density function of signals, a sufficient condition for $\partial\psi(\cdot) / \partial f \leq 0$ is

$$\frac{v_z}{\sqrt{v_x}} \leq \sqrt{2\pi} \Rightarrow v_x \geq \frac{v_z^2}{2\pi}. \tag{3.5}$$

Conversely, if

$$v_x < \frac{v_z^2}{2\pi},$$

the function $\psi(\cdot)$ is non-monotonic in f and (3.3) will admit multiple equilibrium solutions.

Considering that v_x is the precision of agents' signal, while v_z is the precision of the exogenous public signal, condition (3.5) means that a unique equilibrium exists whenever the precision of private signals is large relative to the accuracy of the public signal. Note that in the limit as $v_x \to \infty$ ($\sigma_x \to 0$) for given, v_z, $\left(v_z/\sqrt{v_x}\right)$ $\to 0, \sqrt{1+(v_z/v_x)} \to 1$ and $\sqrt{v_x + v_z} / v_x \to 0$, in (3.3) and (3.4). Consequently, the threshold equilibrium values converge to a unique limit $x^\blacktriangle = f^\blacktriangle = 1 - \frac{c}{\Delta s}$ for all z, and the size of attack q converges to 1 for all $f < (1 - c/\Delta S)$ and to 0 for all $f < (1 - c/\Delta S)$. This result reveals a sharp discontinuity, or strong nonlinearity, in the neighbourhood of $f^\blacktriangle = 1 - \frac{c}{\Delta s}$ when $\sigma_x \to 0$: a small variation in f suffices to trigger a large and sudden change in the size of attack and in the regime change. It also implies that a small departure from perfect information is sufficient to obtain a unique equilibrium.

However, the uniqueness result in the above game has been questioned by a number of authors who emphasise that public information is endogenously generated in most economic settings through markets and prices (see, e.g., Atkeson, 2001; Angeletos, Hellwig and Pavan, 2006, 2007; Angeletos and Werning, 2006). The reason is that if financial prices and other indicators of economic activity accurately reflect the state of fundamentals, agents can use this information, in addition to their private information, when deciding whether to attack the home currency or not. For example, private agents can use the market information provided by forward exchange rates and interest rate differentials to coordinate their decision to run against the home currency or not: if these asset prices indicate that a devaluation will occur, then everyone should attack; if, on the other hand, the asset prices reveal that a devaluation will not occur, then no one should attack. Thus, introducing endogenous sources of information into the basic framework of global games is crucial for the determinacy of equilibria. Interestingly, the most important result to emerge in this case is that the condition for uniqueness is reversed, as the idiosyncratic noise is now required to be large enough for uniqueness. Conversely, multiplicity is obtained when agents observe market signals with small enough noise. This result seems intuitive: if public information is highly accurate, then there is 'almost' common knowledge about the state of fundamentals and equilibrium multiplicity is again brought back into the model.

10.4 Currency regime choice with global capital markets

The currency collapses in a number of developed and emerging market economies in the last two decades, and the theoretical research demonstrating the inherent vulnerability to speculative attacks of government-controlled exchange rates, have fuelled a new interest in the issue of the appropriate foreign exchange regime for a country. A new approach emphasises the credibility and sustainability of policies and crisis prevention as primary criteria for assessing an exchange rate regime. This view, known as the bipolar view or two-corner hypothesis, states that there exist only two foreign exchange rate arrangements that are immune to crisis: (1) irrevocable fixed rates or 'hard' pegs, such as currency unions, currency boards or full dollarization; and (2) free floating. Intermediate regimes, including 'soft' pegs (such as conventional fixed peg, crawling peg, horizontal and crawling band) and managed floats, are fatally open to speculative attacks driven by policy inconsistencies or self-fulfilling expectations and thus are not viable in a world of highly integrated international capital markets (see, e.g., Eichengreen, 1994; Obstfeld and Rogoff, 1995; Summers, 2000; Fischer, 2001, 2008).

The bipolar view follows from the 'impossible trinity' principle or the trilemma, established by Mundell (1962, 1963) and Fleming (1962). The principle maintains that a country cannot achieve the following three goals simultaneously: exchange rate stability, capital market integration, and monetary autonomy. Any two of them, but not all, may be chosen. Thus, a country can retain monetary independence while allowing for free capital movement, but at the cost of exchange rate volatility; or it may have exchange rate stability and full capital market integration, but at the cost of losing monetary autonomy; or it can fix the exchange rate and retain monetary independence, but only by imposing controls on international capital flow, i.e. departing from full capital market integration. Since the global financial integration in the last two decades has strongly increased cross-border capital mobility, this necessarily reduces the policymaker's choice to surrender either exchange rate stability or monetary autonomy – that is, to choose between hard pegs and free floating, thereby 'hollowing out' the intermediate regime option.

The evidence on countries switching to corner solutions in the aftermath of the crises of the 1990s and early 2000s gave some initial support to this view and actively contributed to its favour in the international community and policy circles.[9] However, the new orthodoxy proved short lived, and from the late 1990s onwards it was challenged on several grounds.

First, some authors began to question the reported evidence for the two extreme solutions on the grounds that the country's actual exchange rate regime often diverges from the regime that it officially announces (see, e.g., Calvo and Reinhart, 2002; Bubula and Otker-Robe, 2002; Rogoff *et al.*, 2003; Reinhart and Rogoff, 2004; Shambaugh, 2004; Levy-Yeyati and Sturzenegger, 2003, 2005). The reasons for this dichotomy can be given as follows (see, e.g., Calvo and Reinhart, 2002; Levy-Yeyati and Sturzenegger, 2005, 2007; Genberg and Swoboda, 2005; Alesina and Wagner, 2006; von Hagen and Zhou, 2009). On the one hand, countries may display a 'fear of floating', by experiencing very low exchange rate variability (a *de facto* peg) while officially declaring a float. This behaviour could arise for several reasons, including lack of credibility, fear of inflation, currency mismatches and adverse balance sheet effects in case of devaluation (liability dollarisation). On the other hand, countries may display a 'fear of pegging', by announcing a pegged exchange rate while in practice implementing frequent changes in parity. Possible reasons include low quality institutions associated with poor economic management and political instability, potential trade-offs between competing policy objectives, and the option of retaining more flexibility to ensure the sustainability of the exchange rate regime when large real shocks occur.[10]

Second, other authors noticed that both hard pegs and floating regimes could be subject to market pressures (see, e.g. Williamson, 2000, 2002; Bubula and Otker-Robe, 2002; Goldstein, 2002; Esaka, 2010; Ghosh, Ostry and Tsangarides, 2010). Examples which question the hard end of the bipolar spectrum include the devaluation of the CFA frank under a currency union in 1994, the speculative pressure on Hong Kong SAR's currency board during the Asian crisis in 1998, and the collapse of the Argentina currency board in 2002; similarly, examples whch question the floating end of the spectrum include the pressure on the Canadian dollar in 1992, the Italian lira in 1995, the South African rand in 1998 and 2001, and the global crisis of 2008–09 with its epicentre in countries with floating regimes.

Third, it has been observed that, as yet, there is no clear evidence that intermediate regimes are vanishing (see, e.g., Frankel, 1999, 2004; Masson, 2001; Bubula and Otker-Robe, 2002; Rogoff *et al.*, 2003; McKinnon and Schnabl, 2004; Husain, Mody, and Rogoff, 2005; Bénassy-Quéré, Coeuré and Mignon, 2006; Fischer, 2008; Angkinand, Chiu and Willett, 2009; Esaka, 2010). In fact, the accumulated empirical evidence indicates that while in the past decade industrial countries moved toward the hard end of the spectrum with a combination of high exchange rate stability, financial openness and low monetary independence (mostly ascribed to the advent of the euro and some transitional economies), developing countries moved towards the intermediate solution with a combination of managed exchange rates, intermediate levels of monetary independence and financial integration.[11]

The key to understanding the country's choice of intermediate regimes is that they offer to combine the advantages of exchange rate stability, monetary autonomy, and different degrees of financial integration.[12] Analytical and empirical support for this view has been provided by the IMF Review on countries' choices of exchange rate regime (Ghosh, Ostry, and Tsangarides, 2010). It is pointed out that the main advantages of intermediate exchange rate arrangements are low nominal and real exchange rate volatility, greater trade integration, and low or minimal exchange rate overvaluation, all of which are associated with faster growth and better inflation performance. The study also reveals that the major flaws of intermediate regimes are: (i) greater vulnerability to currency and financial crises, although the risk for credit booms and growth crises are about as likely to occur under floating regimes as they are under the more rigid regimes; (ii) more rigid constraints upon the use of macroeconomic policies; and (iii) less timely external adjustments when current account deficits and the real exchange rate do not adjust as under a floating regime. A by-product of intermediate solutions in a world of increasing financial integration, however, is the massive increase in the demand for international reserves. In the last decade, particularly since 2003, reserve accumulations have accelerated dramatically, rising to 13 per cent of global GDP at the end of 2009 and doubling from their 2000 level. Most of this increase has taken place in Asian and Latin American emerging markets, where reserve holdings rose to 32 per cent of their GDP. The most spectacular changes occurred in China, which held, on average, over 23 per cent of global reserves in the years 2006 to 2008, and in Japan, which scored almost 14 per cent of the total.[13] The view that reserve hoarding behaviour reflects a self-insurance motive against the international liquidity shocks associated with capital markets globalisation has recently gained increasing support, basically on the evidence that countries with very large reserves proved better at dealing with the 2008–09 global financial crisis than those with smaller reserves. The rationale is that reserve accumulation provides a key tool for managing both domestic financial stability and the size of exchange rate adjustments in a world where short-term capital inflows turn out to be highly volatile, with frequent sudden stops and reversals.[14]

As argued by Aizenman (2011), the rising weight of financial factors in the demand for foreign reserves foreshadows a significant change to Mundell's trilemma. In fact, adding financial stability to the trilemma policy goals makes the trilemma framework evolve towards a policy quadrilemma, where concerns about asset price volatility, the stability of the domestic credit markets and the domestic banking sector, and external currency depreciation shift the demand for foreign reserves upwards in an attempt to self-insure against the risks of sudden reversal in capital flows and run independent macroeconomic policies.[15]

Notes

1 A more detailed analytical discussion of all these models, including the basic ones, is in Piersanti (2012).
2 Notable extensions of Krugman and Flood-Garber models in an inter-temporal optimising framework are, e.g., Obstfeld (1986a), Calvo (1987), Drazen and Helpman (1987), Bacchetta (1990), Claessens (1991), van Wijnbergen (1991).
3 This follows from the assumed constancy of both Y and i^* in equation (1.6).
4 Notice that the analysis would remain unchanged even if the lower bound on foreign reserves is assumed to be a negative finite number reflecting borrowing limits.
5 The general forward-looking solution to equation (1.10) is

$$S_t = \left(\frac{Y-\alpha i^* + \alpha}{\alpha}\right)^t k + \left(\frac{1}{Y-\alpha i^* + \alpha}\right)\sum_{j=t}^{\infty}\left(\frac{\alpha}{Y-\alpha i^* + \alpha}\right)^{j-t}\left(M_j^H + \overline{M}^F\right)$$

where

$$k \equiv S_0 - \left(\frac{1}{Y-\alpha i^* + \alpha}\right)\sum_{i=0}^{\infty}\left(\frac{\alpha}{Y-\alpha i^* + \alpha}\right)^i\left(M_j^H + \overline{M}^F\right).$$

Setting $k = 0$, which implies ruling out explosive bubbles, yields the above solution.
6 Policy rules with an escape clause were introduced by Flood and Isard (1989), and further analysed and extended by Lohmann (1990, 1992), and Persson and Tabellini (1990).
7 This opting-out cost can be interpreted in a number of ways: for example, it could reflect the loss of anti-inflation credibility and international reputation, voter dissension, or even removal from the office.
8 These values for f_L and f_H are found at the critical probabilities values, π_L and π_H, where the curves G_{f_L} and G_{f_H} are tangent to the 45° line, and are computed as

$$f_H = \alpha G\left[g^{-1}\left(\frac{1}{\alpha}\right)\right] - g^{-1}\left(\frac{1}{\alpha}\right)$$

and

$$f_L = \alpha G\left[-g^{-1}\left(\frac{1}{\alpha}\right)\right] + g^{-1}\left(\frac{1}{\alpha}\right)$$

9 See, e.g., Caramazza and Aziz (1998), Goldstein (1999), Mussa *et al.* (2000), Fischer (2001), Ghosh, Gulde and Wolf (2003).
10 Note that here we use the term "fear of pegging" for the cases where de facto regimes are more flexible than the de jure counterparts, as in Alesina and Wagner (2006) and von Hagen and Zhou (2009). On the other hand, Levy-Yeyati and Sturzenegger (2005), Genberg and Swoboda (2005) and others use the term "fear of pegging" to denote cases where countries in fact maintain relatively stable exchange rates but claim a more flexible regime due to a fear of being subjected to speculative attacks.
11 See, e.g., Rogoff *et al.* (2003), Bénassy-Quèrè, Coeurè and Mignon (2006), Fischer (2008), Aizenman, Chinn and Ito (2010).
12 The clearest statement on this is Jeffrey Frankel's (2004) insight that Mundell's impossible trinity does not imply that a country cannot have half stability and half independence, as a varying degree of capital mobility exists between the two extreme options of full capital controls and full financial integration. See also Willett (2003).
13 See Mateos y Lago, Duttagupta and Goyal (2009). See also Aizenman, Chinn and Ito (2008), Obstfeld, Shambaugh and Taylor (2010), Ghosh, Ostry and Tsangarides (2010).
14 See, e.g., Mendoza (2004), Aizenman and Lee (2007), Aizenman and Hutchison (2010), Obstfeld, Shambaugh and Taylor (2010).

15 See also Obstfeld, Shambaugh, and Taylor (2010), who link the reserve hoarding trend to three key factors associated with the fear of floating, the policies promoting the depth of domestic financial intermediation, and the financial integration with international financial markets.

References

Agénor, P.R. and Flood, R.P. (1994) 'Macroeconomic Policy, Speculative Attacks and Balance of Payments Crises', in F. Van der Ploeg, (ed.) *The Handbook of International Macroeconomics*, Basic Blackwell, Oxford.

Agénor, P.R. and Masson, P.R. (1999) 'Credibility, Reputation, and the Mexican Peso Crisis', *Journal of Money, Credit and Banking*, 31(1): 70–84.

Aghion, P., Bacchetta, P. and Banerjee, A. (2000) 'A Simple Model of Monetary Policy and Currency Crises', *European Economic Review*, 44(4–6): 728–38.

Aghion, P., Bacchetta, P. and Banerjee, A. (2001) 'Currency Crises and Monetary Policy in an Economy with Credit Constraints', *European Economic Review*, 45(7): 1121–50.

Aghion, P., Bacchetta, P. and Banerjee, A. (2004) 'A Corporate Balance-Sheet Approach to Currency Crises', *Journal of Economic Theory*, 119(1): 6–30.

Aizenman, J. (2011) 'The Impossible Trinity – from the Policy Trilemma to the Policy Quadrilemma', University of California, Santa Cruz, Economics Department.

Aizenman, J. and Hutchison, M.M. (2010) 'Exchange Market Pressure and Absorption by International Reserves: Emerging Market and Fear of Reserves Loss During the 2008–09 Crisis', *NBER Working Paper*, No. 16260, National Bureau of Economic Research, Cambridge, MA.

Aizenman, J. and Lee, J. (2007) 'International Reserves: Precautionary Versus Mercantilist Views, Theory and Evidence', *Open Economies Review*, 18(2): 191–214.

Aizenman, J., Chinn, M.D. and Ito, H. (2008) 'Assessing the Emerging Global Financial Architecture: Measuring the Trilemma's Configurations over Time', *NBER Working Paper Series*, No. 14533, National Bureau of Economic Research, Cambridge, MA.

Aizenman, J., Chinn, M.D. and Ito, H. (2010) 'The Emerging Global Financial Architecture: Tracing and Evaluating New Patterns of the Trilemma Configuration', *Journal of International Money and Finance*, 29(4): 615–41.

Alesina, A. and Wagner, A.F. (2006) 'Choosing (and Reneging on) Exchange Rate Regimes', *Journal of the European Economic Association*, 4(4): 770–99.

Angeletos, G.M. and Werning, I. (2006) 'Crises and Prices', *American Economic Review*, 96(5): 1720–36.

Angeletos, G.M., Hellwig, C. and Pavan, A. (2006) 'Signaling in a Global Game: Coordination and Policy Traps', *Journal of Political Economy*, 114(3): 452–84.

Angeletos, G.M., Hellwig, C. and Pavan, A. (2007) 'Dynamic Global Games of Regime Changes: Learning, Multiplicity, and the Timing of Attacks', *Econometrica*, 75(3): 711–56.

Angkinand, A., Chiu, E. and Willett, T.D. (2009) 'Testing the Unstable Middle and Two Corners Hypotheses about Exchange Rate Regimes', *Open Economy Review*, 20(1): 61–83.

Annicchiarico, B., Marini, G. and Piersanti, G. (2011) 'Budget Deficits and Exchange-Rate Crises', *International Economic Journal*, 25(2): 285–303.

Atkeson, A. (2001) 'Discussion of Morris and Shin's "Rethinking Multiple Equilibria in Macroeconomic Modelling"', in B.S. Bernanke and K.S. Rogoff (eds), *NBER Macroeconomics Annual 2000*, MIT Press, Cambridge MA.

Bacchetta, P. (1990) 'Temporary Capital Controls in a Balance-of-Payments Crisis', *Journal of International Money and Finance*, 9(3): 246–57.

Barro, R.J. and Gordon, D.B. (1983) 'A Positive Theory of Monetary Policy in a Natural Rate Model', *Journal of Political Economy*, 91(4): 589–610.

Bénassy-Quéré, A., Coeuré, B. and Mignon, V. (2006) 'On the Identification of De Facto Currency Pegs', *Journal of the Japanese and International Economies*, 20(1): 112–27.

Benigno, P. and Missale, A. (2004) 'High Public Debt in Currency Crises: Fundamentals versus Signalling Effects', *Journal of International Money and Finance*, 23(2): 165–88.

Bensaid, B. and Jeanne, O. (1997) 'The Instability of Fixed Exchange Rate Systems When Raising the Nominal Interest Rate Is Costly', *European Economic Review*, 41(8): 1461–78.

Berger, W. and Wagner, H. (2005), 'Interdependent Expectations and the Spread of Currency Crises', *IMF Staff Papers*, 52(1): 41–54.

Bernardo, A.E. and Welch, I. (2004) 'Liquidity and Financial Market Runs', *Quarterly Journal of Economics*, 119(1): 135–58.

Bertola, G. and Caballero, R.J. (1991) 'Sustainable Intervention Policies and Exchange Rate Dynamics', in P.R. Krugman and M. Miller (eds) *Exchange Rate Targets and Currency Bands*, Cambridge University Press, Cambridge.

Blackburn, K. (1988) 'Collapsing Exchange Rate Regimes and Exchange Rate Dynamics: Some Further Examples', *Journal of International Money and Finance*, 7(4): 373–85.

Blanco, H. and Garber, P.M. (1986) 'Recurrent Devaluation and Speculative Attacks on the Mexican Peso', *Journal of Political Economy*, 94(1): 148–66.

Bubula, A. and Otker-Robe, I. (2002) 'The Evolution of Exchange Rate Regimes Since 1990: Evidence from De Facto Policies', *IMF Working Paper*, 02/155, International Monetary Fund, Washington, DC.

Buiter, W.H. (1987) 'Borrowing to Defend the Exchange Rate and the Timing and Magnitude of Speculative Attacks', *Journal of International Economics*, 23(3–4): 221–39.

Buiter, W.H. and Grilli, V. (1991) 'Anomalous Speculative Attacks on Fixed Exchange Rate Regimes: Possible Resolution of the 'Gold Standard Paradox'', in P.R. Krugman and M. Miller (eds) *Exchange Rate Targets and Currency Bands*, Cambridge University Press, Cambridge.

Burnside, C. (2004) 'Currency Crises and Contingent Liabilities', *Journal of International Economics*, 62(1): 25–52.

Burnside, C., Eichenbaum, M. and Rebelo, S. (2001) 'Prospective Deficits and the Asian Currency Crisis', *Journal of Political Economy*, 109(6): 1155–97.

Burnside, C., Eichenbaum, M. and Rebelo, S. (2004) 'Government Guarantees and Self-Fulfilling Speculative Attacks', *Journal of Economic Theory*, 119(1): 31–63.

Caballero, R.J. and Krishnamurthy, A. (2001) 'International and Domestic Collateral Constraint in a Model of Emerging Market Crises', *Journal of Monetary Economics*, 48(3): 513–48.

Caballero, R.J. and Krishnamurthy, A. (2003) 'Excessive Dollar Debt: Financial Development and Underinsurance', *Journal of Finance*, 58(2): 867–93.

Caballero, R.J. and Krishnamurthy, A. (2004) 'Smoothing Sudden Stops', *Journal of Economic Theory*, 119(1): 104–27.

Calvo, G.A. (1987) 'Balance of Payments Crises in a Cash-in-Advance Economy', *Journal of Money, Credit and Banking*, 19(1): 19–32.

Calvo, G.A. (1998) 'Understanding the Russian Virus: With Special Reference to Latin America', Paper presented at the Deutsche Bank's conference on 'Emerging Markets: Can they be Crisis Free?', Washington, DC.

Calvo, G.A. (1999) 'Contagion in Emerging Markets: When Wall Street is a Carrier', reprinted in G.A. Calvo (2005) *Emerging Capital Markets in Turmoil. Bad Lack or Bad Policy?,* MIT Press, Cambridge MA.

Calvo, G.A. (2003) 'Explaining Sudden Stops, Growth Collapse, and BOP Crisis: The Case of Distortionary Output Taxes', *IMF Staff Papers*, 50: 1–20.

Calvo, G.A. and Mendoza, E.G. (2000) 'Rational Contagion and the Globalization of Securities Markets', *Journal of International Economics*, 51(1): 79–113.

Calvo, G.A. and Reinhart, C.M. (2002) 'Fear of Floating', *Quarterly Journal of Economics*, 117(2): 379–408.

Caramazza, F. and Aziz, J. (1998) 'Fixed or Flexible? Getting the Exchange Rate Right in the 1990s', *IMF Economic Issues*, 13, International Monetary Fund, Washington, DC.

Carlsson, H. and van Damme, E. (1993) 'Global Games and Equilibrium Selection', *Econometrica*, 61(5): 989–1018.

Céspedes, L.F., Chang, R. and Velasco, A. (2004) 'Balance Sheets and Exchange Rate Policy', *American Economic Review*, 94(4): 1183–93.

Chang, R. and Velasco, A. (2000a) 'Financial Fragility and the Exchange Rate Regime', *Journal of Economic Theory*, 92(1): 1–34.

Chang, R. and Velasco, A. (2000b) 'Banks, Debt Maturity and Financial Crisis', *Journal of International Economics*, 51(1): 169–94.

Chang, R. and Velasco, A. (2001) 'A model of Financial Crises in Emerging Markets', *Quarterly Journal of Economics*, 116(2): 489–517.

Chari, V.V. and Kehoe, P.J. (2003) 'Hot Money', *Journal of Political Economy*, 111(6): 1262–92.

Chari, V.V. and Kehoe, P.J. (2004) 'Financial Crises as Herds: Overturning the Critiques', *Journal of Economic Theory*, 119(1): 128–50.

Christiano, L.J., Gust, C. and Roldos, J.E. (2004) 'Monetary Policy in a Financial Crisis', *Journal of Economic Theory*, 119(1): 63–104.

Claessens, S. (1991) 'Balance of Payment Crises in an Optimal Portfolio Model', *European Economic Review*, 35(1): 81–101.

Cole, H.L. and Kehoe, T.J. (1996) 'A Self-Fulfilling Model of Mexico's 1994–1995 Debt Crisis', *Journal of International Economics*, 41(3–4): 309–30.

Cole, H.L. and Kehoe, T.J. (2000) 'Self-Fulfilling Debt Crisis', *Review of Economic Studies*, 67(1): 91–116.

Connolly, M.B. (1986) 'The Speculative Attack on the Peso and the Real Exchange Rate: Argentina, 1979–1981', *Journal of International Money and Finance*, 5(1): 117–30.

Connolly, M.B. and Taylor, D. (1984) 'The Exact Timing of the Collapse of an Exchange Rate Regime and Its Impact on the Relative Price of Traded Goods', *Journal of Money Credit and Banking*, 16(2): 194–207.

Corrado, L. and Holly, S. (2000) 'A Currency Crisis Model with a Misaligned Central Parity: A Stochastic Analysis', *Economic Letters*, 67(1): 61–8.

Corsetti, G., Pesenti, P. and Roubini, N. (1999) 'Paper Tigers? A Model of the Asian Crisis', *European Economic Review*, 43(7): 1211–36.

Corsetti, G., Pesenti, P., Roubini, N. and Tille, C. (2000) 'Competitive Devaluation: Toward a Welfare-Based Approach', *Journal of International Economics*, 51(1): 217–41.

Daekeun, P. and Sachs, J.D. (1996) 'The Timing of Exchange Regime Collapse under Capital Controls', *International Economic Journal*, 10(4): 123–41.

De Kock, G. and Grilli, V. (1993) 'Fiscal Policies and the Choice of Exchange Rate Regime', *Economic Journal*, 103(417): 347–58.

Delgado, F. and Dumas, B. (1993) 'Monetary Contracting Between Central Banks and the Design of Sustainable Exchange-Rate Zones', *Journal of International Economics*, 34(3–4): 201–24.

Dellas, H. and Stockman, A.C. (1993) 'Self-Fulfilling Expectations, Speculative Attacks, and Capital Controls', *Journal of Money, Credit, and Banking*, 25(4): 721–30.

Djajic, S. (1989) 'Dynamics of the Exchange Rate in Anticipation of Pegging', *Journal of International Money and Finance*, 8(4): 559–71.

Dornbusch, R. (1987) 'Collapsing Exchange Rate Regimes', *Journal of Development Economics*, 27(1–2): 71–83.

Drazen, A. (2000) 'Interest-Rate and Borrowing Defence against Speculative Attack', *Carnegie–Rochester Conference Series on Public Policy*, 53(1): 303–48.

Drazen, A. and Helpman, E. (1987) 'Stabilization with Exchange Rate Management', *Quarterly Journal of Economics*, 102(4): 835–56.

Drazen, A. and Masson, P.R. (1994) 'Credibility of Policies versus Credibility of Policymakers', *Quarterly Journal of Economics*, 109(3): 735–54.

Dumas, B. and Svensson, L.E.O. (1994) 'How Long do Unilateral Target Zones Last?', *Journal of International Economics*, 36(3–4): 467–81.

Eichengreen, B. (1994) *International Monetary Arrangements for the 21st Century*, Brookings Institution, Washington DC.

Esaka, T. (2010) 'Exchange Rate Regimes, Capital Controls, and Currency Crises: Does the Bipolar View Hold?', *Journal of International Markets, Institutions and Money*, 20(1): 91–108.

Fischer, S. (2001) 'Exchange Rate Regimes: Is the Bipolar View Correct?', *Journal of Economic Perspectives*, 15(2): 3–24.

Fischer, S. (2008) 'Mundell-Fleming Lecture: Exchange Rate Systems, Surveillance and Advice', *IMF Staff Papers*, 55(3): 367–83.

Fleming, M.J. (1962) 'Domestic Financial Policies under Fixed and under Floating Exchange Rates', *IMF Staff Papers*, 9(3): 369–80.

Flood, R.P. and Garber, P.M. (1984) 'Collapsing Exchange-Rate Regimes: Some Linear Example', *Journal of International Economics*, 17(1–2): 1–13.

Flood, R.P. and Garber, P.M. (1991) 'Linkages Between Speculative Attack and Target Zone Models of Exchange Rates', *Quarterly Journal of Economics*, 106(4): 1367–72.

Flood, R.P. and Hodrick, R.J. (1986) 'Real Aspects of Exchange Rate Regime Choice with Collapsing Fixed Rates', *Journal of International Economics*, 21(3–4): 215–32.

Flood, R.P. and Isard, P. (1989) 'Monetary Policy Strategies', *IMF Staff Papers*, 36(3): 612–32.

Flood, R.P. and Jeanne, O. (2005) 'An Interest Rate Defence of a Fixed Exchange Rate?', *Journal of International Economics*, 66(2): 471–84.

Flood, R.P. and Marion, N.P. (1999) 'Perspective on the Recent Currency Crisis Literature', *International Journal of Finance and Economics*, 4(1): 1–26.

Flood, R.P. and Marion, N.P. (2004) 'A Model of the Joint Distribution of Banking and Currency Crises', *Journal of International Money and Finance*, 23(6): 841–65.

Flood, R.P., Garber, P.M. and Kramer, C. (1996) 'Collapsing Exchange Rate Regimes: Another Linear Example', *Journal of International Economics*, 41(3–4): 223–34.

Forbes, K.J. (2001) 'Are Trade Links Important Determinants of Country Vulnerability to Crises?', *NBER Working Paper*, No. 8194, National Bureau of Economic Research, Cambridge, MA.

Forbes, K.J. (2004) 'The Asian Flu and Russian Virus: The International Transmission of Crises in Firm-Level Data', *Journal of International Economics*, 63(1): 59–92.

Frankel, J.A. (1999) 'No Single Currency Regime is Right for All Countries or At All Times', *NBER Working Paper*, No. 7338, National Bureau of Economic Research, Cambridge, MA.

Frankel, J.A. (2004) 'Lessons from Exchange Rate Regimes in Emerging Economies', in *Monetary and Financial Integration in East Asia: The Way Ahead*, ed. Asian Development Bank, Palgrave Macmillan Press, New York, vol. II.

Garber, P.M. and Spencer, M.G. (1995) 'Foreign Exchange Hedging and the Interest Rate Defense', *IMF Staff Papers*, 42(3): 490–516.

Genberg, H. and Swoboda, A.K. (2005) 'Exchange Rate Regimes: Does What Countries Say Matter?', *IMF Staff Papers*, 52 (Special Issue): 129–41.

Gerlach, S. and Smets, F. (1995) 'Contagious Speculative Attack', *European Journal of Political Economy*, 11(1): 45–63.

Ghosh, A.R., Gulde, A.M. and Wolf, H. (2003) *Exchange Rate Regimes: Choices and Consequences*, MIT Press, Cambridge MA.

Ghosh, A.R., Ostry, J.D. and Tsangarides, C. (2010) 'Exchange Rate Regimes and the Stability of the International Monetary System', *IMF Occasional Paper*, No. 270, International Monetary Fund, Washington, DC.

Glick, R. and Rose, A.K. (1999) 'Contagion and Trade: Why Are Currency Crises Regional', *Journal of International Money and Finance*, 18(4): 603–18.

Goldberg, L.S. (1991) 'Collapsing Exchange Rate Regimes: Shocks and Biases', *Journal of International Money and Finance*, 10(2): 252–63.

Goldberg, L.S. (1994) 'Predicting Exchange Rate Crises: Mexico Revisited', *Journal of International Economics*, 36(3–4): 413–30.

Goldfajn, I. and Valdés, R.O. (1997) 'Capital Flows and the Twin Crises: The Role of Liquidity', *IMF Working Paper*, No. 97/87, International Monetary Fund, Washington DC.

Goldstein, I. (2005) 'Strategic Complementarities and the Twin Crises', *Economic Journal*, 115(503): 368–90.

Goldstein, I. and Pauzner, A. (2005) 'Demand Deposit Contracts and the Probability of Bank Runs', *Journal of Finance*, 60(3): 1293–328.

Goldstein, M. (1999) *Safeguarding Prosperity in a Global Financial System: The Future International Financial Architecture,* Institute of International Economics, Washington, DC.

Goldstein, M. (2002) *Managed Floating Plus*, Institute for International Economics, No.66, Washington DC.

Grilli, V. (1986) 'Buying and Selling Attacks on Fixed Exchange Rate System', *Journal of International Economics*, 20(1–2): 143–56.

Guimarães, B. and Morris, S. (2007) 'Risk and Wealth in a Model of Self-Fulfilling Currency Attacks', *Journal of Monetary Economics*, 54(8): 2205–30.

Heinemann, F. and Illing, G. (2002) 'Speculative Attacks: Unique Equilibrium and Transparency', *Journal of International Economics*, 58(2): 429–50.

Husain, A.M., Mody, A. and Rogoff, K.S. (2005) 'Exchange Rate Regime Durability and Performance in Developing Versus Advanced Economies', *Journal of Monetary Economics*, 52(1): 35–64.

Isard, P. (1994) 'Realignments Expectations, Forward Rate Bias and Interventions in an Optimizing Model of Exchange Rate Adjustment', *IMF Staff Papers*, 41(3): 435–59.

Jeanne, O. (1997) 'Are Currency Crises Self-Fulfilling? A Test', *Journal of International Economics*, 43(3–4): 263–86.

Jeanne, O. (2000) 'Currency Crises: A Perspective on Recent Theoretical Developments', Special Papers in *International Economics*, No. 20, Princeton University Press, Princeton New Jersey.

Jeanne, O. and Masson, P.R. (2000) 'Currency Crises, Sunspots and Markov-Switching Regimes', *Journal of International Economics*, 50(2): 327–50.

Kaminsky, G.L. and Reinhart, C.M. (1999) 'The Twin Crises: the Causes of Banking and Balance-of-Payments Problems', *American Economic Review*, 89(3): 473–500.

Kaminsky, G.L., Lizondo, S. and Reinhart, C.M. (1998) 'Leading Indicators of Currency Crises', *IMF Staff Papers*, 45(1): 1–48.

Kodres, L. and Pritsker, M. (2002) 'A Rational Expectations Model of Financial Contagion', *Journal of Finance*, 57(2): 769–99.

Krugman, P.R. (1979) 'A Model of Balance-of-Payments Crises', *Journal of Money, Credit and Banking*, 11(3): 311–25.

Krugman, P.R. (1999) 'Balance Sheets, the Transfer Problem, and Financial Crises', in P. Isard, A. Razin and A. Rose, (eds) *International Finance and Financial Crises*, Kluwer Academic Publisher, Norwell Mass.

Krugman, P.R. and Rotemberg, J. (1991) 'Speculative Attacks on Target Zones', in P.R. Krugman and M. Miller (eds) *Exchange Rate Targets and Currency Bands*, Cambridge University Press, Cambridge.

Kydland, F.E. and Prescott, E.C. (1977) 'Rules Rather Than Discretion: The Inconsistency of Optimal Plans', *Journal of Political Economy*, 85(3): 473–92.

Lahiri, A. and Végh, C.A. (2003) 'Delaying the Inevitable: Interest Rate Defense and Balance of Payment Crises', *Journal of Political Economy*, 111(2): 404–24.

Levy-Yeyati, E. and Sturzenegger, F. (2003) 'To Float or to Fix: Evidence on the Impact of Exchange Rate Regimes on Growth', *American Economic Review*, 93(4): 1173–93.

Levy-Yeyati, E. and Sturzenegger, F. (2005) 'Classifying Exchange Rate Regimes: Deeds vs. Words', *European Economic Review*, 49(6): 1603–35.

Levy-Yeyati, E. and Sturzenegger, F. (2007) 'Fear of Appreciation', *World Bank Policy Research Working Paper*, No. 4387, The World Bank, Washington, DC.

Lohmann, S. (1990) 'Monetary Policy Strategies – A Correction', *IMF Staff Papers*, 37(2): 440–45.

Lohmann, S. (1992) 'Optimal Commitment in Monetary Policy: Credibility versus Flexibility', *American Economic Review*, 82(1): 273–86.

Loisel, O. and Martin, P. (2001) 'Coordination, Cooperation, Contagion and Currency Crises', *Journal of International Economics*, 53(2): 399–419.

McKinnon, R. and Schnabl, G. (2004) 'The East Asian Dollar Standard, Fear of Floating, and Original Sin', *Review of Development Economics*, 8(3): 331–60.

Masson, P.R. (1995) 'Gaining and Losing ERM Credibility: the Case of the United Kingdom', *Economic Journal*, 105(430): 571–82.

Masson, P.R. (1999) 'Contagion: Macroeconomic Models with Multiple Equilibria', *Journal of International Money and Finance*, 18(4): 587–602.

Masson, P.R. (2001) 'Exchange Rate Regime Transitions', *Journal of Development Economics*, 64(2): 571–86.

Mateos y Lago, I., Duttagupta, R. and Goyal, R. (2009) 'The Debate on the International Monetary System', *IMF Staff Position Note*, No. 09/26, International Monetary Fund, Washington, DC.

Mendoza, R.U. (2004) 'International Reserve-Holding in the Developing World: Self Insurance in a Crisis-Prone Era?', *Emerging Market Review*, 5(1): 61–82.

Metz, C.E. (2002) 'Private and Public Information in Self-Fulfilling Currency Crises', *Journal of Economics*, 76(1): 65–85.

Miller, V. (1999) 'The Timing and Size of Bank-Financed Speculative Attacks', *Journal of International Money and Finance*, 18(3): 459–70.

Miller, V. (2003) 'Bank Runs and Currency Peg Credibility', *Journal of International Money and Finance*, 22(3): 385–92.

Morris, S. and Shin, H.S. (1998) 'Unique Equilibrium in a Model of Self-Fulfilling Currency Attacks', *American Economic Review*, 88(3): 587–97.

Morris, S. and Shin, H.S. (1999) 'A Theory of the Onset of Currency Attacks', in P.R. Agénor, M. Miller, D. Vines and A. Weber (eds) *The Asian Financial Crisis: Causes, Contagion and Consequences*, Cambridge University Press, Cambridge UK.

Morris, S. and Shin, H.S. (2001) 'Rethinking Multiple Equilibria in Macroeconomic Modelling', in B.S. Bernanke and K.S. Rogoff (eds) *NBER Macroeconomics Annual 2000*, MIT Press, Cambridge MA.

Morris, S. and Shin, H.S. (2003) 'Global Games: Theory and Application', in M. Dewatripont, L.P. Hansen and S.J. Turnovsky (eds) *Advances in Economics and Econometrics* (Proceedings of the Eighth World Congress of the Econometric Society), Cambridge University Press, Cambridge UK.

Morris, S. and Shin, H.S. (2004) 'Coordination Risk and the Price of Debt', *European Economic Review*, 48(1): 133–53.

Mundell, R.A. (1962) 'The Appropriate Use of Monetary and Fiscal Policy for Internal and External Stability', *IMF Staff Papers*, 9(1): 70–77.

Mundell, R.A. (1963) 'Capital Mobility and Stabilization Policy under Fixed and Flexible Exchange Rates', *Canadian Journal of Economics and Political Science*, 29(4): 475–85.

Mussa, M., Masson, P.R., Swoboda, A., Jadresic, E., Mauro, P. and Berg, A. (2000) 'Exchange Rate Regimes in an Increasingly Integrated World Economy', *IMF Occasional Paper*, No. 193, International Monetary Fund, Washington, DC.

Neut, A. and Velasco, A. (2004) 'Asset Prices and Self-Fulfilling Macroeconomic Pessimism', IMF Conference in honour of G.A. Calvo, International Monetary Fund, Washington, DC, April 15–16.

Obstfeld, M. (1984) 'Balance-of-Payments Crises and Devaluation', *Journal of Money, Credit and Banking*, 16(2): 208–17.

Obstfeld, M. (1986a) 'Speculative Attack and the External Constraint in a Maximizing Model of the Balance of Payments', *Canadian Journal of Economics*, 19(1): 1–22.

Obstfeld, M. (1986b) 'Rational and Self-Fulfilling Balance of Payments Crises', *American Economic Review*, 76(1): 72–81.

Obstfeld, M. (1994) 'The Logic of Currency Crises', *Cahiers Économiques et Monétaires*, Banque de France, 43:189–213.

Obstfeld, M. (1996) 'Models of Currency Crises with Self-Fulfilling Features', *European Economic Review*, 40(3–5): 1037–47.

Obstfeld, M. (1997) 'Destabilizing Effects of Exchange-Rate Escape Clauses', *Journal of International Economics*, 43(1–2): 61–77.

Obstfeld, M. and Rogoff, K.S. (1995) 'The Mirage of Fixed Exchange Rates', *Journal of Economic Perspectives*, 9(4): 73–96.

Obstfeld, M., Shambaugh, J.C. and Taylor, A.M. (2010) 'Financial Stability, the Trilemma, and International Reserves', *American Economic Journal: Macroeconomics*, 2(2): 57–94.

Osakwe, P.N. and Schembri, L.L. (2002) 'Real Effects of Collapsing Exchange Rate Regimes: An Application to Mexico', *Journal of International Economics*, 57(2): 299–325.

Ozkan, G.F. and Sutherland, A. (1995) 'Policy Measures to Avoid a Currency Crisis', *Economic Journal*, 105(429): 510–519.

Ozkan, G.F. and Sutherland, A. (1998) 'A Currency Crisis Model with an Optimizing Policymaker', *Journal of International Economics*, 44(2): 339–64.

Persson, T. and Tabellini, G. (1990) *Macroeconomic Policy, Credibility and Politics*, Harwood Academic, Chur, Switzerland.

Piersanti, G. (2012) *The Macroeconomic Theory of Exchange Rate Crises*, Oxford University Press, Oxford.

Radelet, S. and Sachs, J.D. (1998) 'The East Asian Financial Crisis: Diagnosis, Remedies, Prospects', *Brookings Papers on Economic Activity*, 29(1): 1–74.

Reinhart, C. and Rogoff, K.S. (2004) 'A Modern History of Exchange Rate Regimes: A Reinterpretation', *Quarterly Journal of Economics*, 119(1): 1–48.

Rochet, J.C. and Vives, X. (2004) 'Coordination Failures and the Lender of Last Resort: Was Bagehot Right After All?', *Journal of the European Economic Association*, 2(6): 1116–47.

Rogoff, K.S., Husain, A.M., Mody, A., Brooks, R. and Oomes, N. (2003) 'Evolution and Performance of Exchange Rate Regimes', *IMF Working Paper*, No 03/243, International Monetary Fund, Washington, DC.

Sachs, J.D., Tornell, A. and Velasco, A. (1996) 'The Mexican Peso Crisis: Sudden Death or Death Foretold', *Journal of International Economics*, 41(3–4): 265–83.

Salant, S.W. and Henderson, D.W. (1978) 'Market Anticipations of Government Policies and the Price of Gold', *Journal of Political Economy*, 86(4): 627–48.

Savastano, M.A. (1992) 'Collapse of a Crawling Peg Regime in the Presence of a Government Budget Constraint', *IMF Staff Papers*, 39(1): 79–100.

Schneider, M. and Tornell, A. (2004) 'Balance Sheet Effects, Bailout Guarantees and Financial Crises', *Review of Economic Studies*, 71(07): 883–913.

Shambaugh, J.C. (2004) 'The Effect of Fixed Exchange Rates on Monetary Policy', *Quarterly Journal of Economics*, 119(1): 301–52.

Summers, L.H. (2000) 'International Financial Crises: Causes, Prevention, and Cures', *American Economic Review, Papers and Proceedings*, 90(2): 1–16.

van Wijnbergen, S. (1991) 'Fiscal Deficits, Exchange Rates Crises and Inflation', *Review of Economic Studies*, 58(1): 81–92.

Velasco, A. (1996) 'Fixed Exchange Rates: Credibility, Flexibility and Multiplicity', *European Economic Review*, 40(3–5): 1023–35.

von Hagen, J. and Zhou, J. (2009) 'Fear of Floating and Fear of Pegging: A Simultaneous Choice Model of De Jure and De Facto Exchange Rate Regimes in Developing Countries', *Open Economies Review*, 20(3): 293–315.

Werner, A.M. (1995) 'Exchange Rates Targets Zones, Realignments, and the Interest Rate Differential: Theory and Evidence', *Journal of International Economics*, 39(3–4): 353–67.

Willett, T. (2003) 'Fear of Floating Needn't Imply Fixed Rates: An OCA Approach to the Operation of Stable Intermediate Currency Regimes', *Open Economies Review*, 14(1): 71–91.

Williamson, J. (2000) *Exchange Rate Regimes for Emerging Markets: Reviving the Intermediate Options. Policy Analyses in International Economics*, 60, Institute for International Economics, Washington, DC.

Williamson, J. (2002) 'The Evolution of Thought on Intermediate Exchange Rate Regimes', *Annals of the American Academy of Political and Social Science*, 579(1): 73–86.

Willman, A. (1988) 'The Collapse of the Fixed Exchange Rate Regime with Sticky Wages and Imperfect Substitutability between Domestic and Foreign Bonds', *European Economic Review*, 32(9): 1817–38.

Wyplosz, C. (1986) 'Capital Controls and Balance of Payments Crises', *Journal of International Money and Finance*, 5(2): 167–79.

11 Firms' internationalisation, innovation strategies and product quality

Piergiuseppe Morone and Francesco Renna

11.1 Introduction

Traditional international trade models rest upon the assumption of firms' symmetry within a particular industry and in a particular country. This assumption implies that they all face the same demand functions and produce under the same technologies. As a result, all firms produce the same goods, set the same price and supply the same quantities. Thus, if there are exports, all firms in the industry are exporters (Basile, 2001, p. 1185). However, a growing body of work has extensively documented, both theoretically and empirically, the considerable inter-firm heterogeneity in export performance among firms in the same industry within the same country. This variation stems from various sources of firm-level heterogeneity, which relate primarily to technological advantages, as well as to other firm-specific characteristics (Basile, 2001, p. 1186).

In this chapter we address the relationship between internationalisation, innovation and quality upgrading for a sample of Italian small and medium sized enterprises (SMEs) within the theoretical framework of innovation theory and 'new' new trade theory. Using firm-level data, we develop an econometric model in order to assess the impact of product innovation, quality upgrading (measured both by the ISO9000 certification and investments in product quality) and cost saving technical processes on firms' exports, under the assumption of firms' heterogeneity. Export behaviour is defined in a dual way: both as the probability for a firm to export, and as the export turnover of exporting firms. Building upon existing methods, we include in our empirical model various other firm characteristics (skilled labour force, labour productivity and firm age) as well as industry and geographical dummies.

Our core finding suggests that product innovation and quality upgrading are essential drivers of exports, whereas cost-saving activities are never statistically significant. This, in turn, would suggest that the best performing firms coped with the growing international competitiveness directing their efforts towards knowledge-intensive activities, a result which is indeed in line with a knowledge-based view of modern economies.

The remainder of this chapter is structured as follows: section 2 provides a brief literature review; section 3 describes the empirical methodology and the data-set; section 4 presents the econometric results; and section 5 concludes.

11.2 Innovation, quality and SMEs export performances

11.2.1 Theoretical background

The relationship between innovation, quality differentiation and exports has attracted increasing attention over the last two decades. Twenty years ago, Grossman and Helpman (1991) proposed a growth model wherein innovation is conceived as a tool to enhance both horizontal differentiation (i.e. producing new varieties of goods) and vertical differentiation (i.e. developing higher quality goods). The authors explain that firms face strong competitive pressure, 'as any single innovation enjoys a limited run at the technological frontier' (1991, p. 43), because of continual technical improvements by rival firms. 'Almost every product exists on a *quality ladder*, with variants below, that may already have become obsolete, and others above, that have yet to be discovered' (1991, p. 43). In this vein, innovation-oriented activities enable firms to manufacture state of the art products and thereby move up the quality ladder vis-à-vis competitors. Subsequently, Grossman and Helpman (1995) extended this model of quality ladders to address trade issues and showed how countries holding the technological lead in a specific product will dominate the world market in that product.

Although highly relevant, this seminal model fell short of explaining differences within-countries and within-industries in exporting performances across firms. The question at the forefront is why some firms perform well in terms of exports, whereas other firms (operating in the same country and in the same industry) are forced to exit (or are unable to enter) international markets? On this point, the empirical evidence is unambiguous, showing that exporting is concentrated:

> A minority of the firms that export make up the bulk of exporting activity measured on a value basis. [. . .] in most [. . .] countries, the largest 5 per cent of exporting firms account for more than three-quarters of total exports.
>
> (WTO, 2008)

This observation challenges both the old and the new trade theories and calls for new theoretical explanations. Various authors (Eaton and Kortum, 2002; Bernard *et al.*, 2003; Melitz, 2003) have addressed this point, building upon the heterogeneity of firms. Melitz (2003) constructed a model in which only a few highly productive firms engage in exporting. The underlying idea is that firms have a heterogeneous level of productivity and only highly productive firms are able to make sufficient profits to cover the large fixed costs required for exporting. In this regard, being present on international markets becomes a task closely related to productivity levels.

As noted by Boermans and Roelfsema (2012), in Melitz's original model he assumes that productivity differences across firms are exogenously assigned by a draw from a known productivity distribution. Therefore, individual firm productivity levels are revealed (or have already materialised) to the firm's management before the internationalisation decision is taken – this implying that in the steady state the causality runs from productivity to international activities.[1]

However, an alternative hypothesis would be that inspired managerial decisions could influence productivity. Melitz and Constantini (2008) present a model in which firms anticipate trade liberalisation by making investments to upgrade the production process through innovation. The reason is that the higher expected profits from internationalisation are contingent upon higher productivity. Using the same logic, Bustos (2011) shows that, in particular, firms with medium productivity levels have an interest in investing in productivity, since these firms are at the fringe and are most probably hit by the selection processes following trade liberalisation (Boermans and Roelfsema, 2012).

This dual interpretation of the phenomenon gave rise to two competing explanations, which can be synthetically summarised as the 'learning by exporting' hypothesis and the 'self-selection' hypothesis. In a nutshell, in the first case firms already operating in international markets would benefit (and learn) from operating in highly competitive international markets by gaining from knowledge spillovers; in the second case, only those firms which are already comparatively more productive would self-select into international markets, and hence higher productivity (and the associated knowledge endowment) is conceived as a precondition to hit and survive in international markets.

In both cases, however, a common idea of 'new' new trade theory models is that heterogeneous performances in international markets are crucially dependent upon the specific characteristics of a firm. Along with productivity, firms vary in several other aspects, often closely connected. For instance, firms operating in the same industry preserve significant differences in size (with larger firms being more inclined to export), in attitude towards innovation activities, and in the vertical differentiation of their products. The importance of such firm-specific characteristics in explaining differentials in trade flows has been widely acknowledged in the recent theoretical literature, and these 'Melitz-type models' constitute the theoretical foundations for empirical research based on firm-level data, such as the one proposed here.

11.2.2 The case study

The empirical investigation presented hereafter focuses on Italian firms and, specifically, on Italian small and medium sized enterprises. The chosen case study is particularly relevant since, as recognised by early studies on the Italian specialisation model (Modiano, 1982 and 1984; Amendola, 1991), Italian firms have traditionally competed by climbing the quality ladder, and hence have produced high-value goods (i.e. the so-called *Made in Italy* products). Following this line of reasoning, several recent studies (Basile *et al.*, 2006; Bugamelli, 2007; Lamieri and Lanza, 2006; Borin and Lamieri, 2007) have showed that Italian exports gained momentum in the first years of the twenty-first century mainly via quality upgrading.[2] Vertical differentiation has also been achieved through innovative efforts, which proved to be very effective in sustaining international competitiveness (Morone and Testa, 2008). In an earlier paper, Sterlacchini (1999) found that the intensity of innovative activities (expenditures on design and trial production and the technological level of capital stock), other than R&D, had a positive and significant impact upon Italian SMEs' probability of exporting and on their export shares. Basile (2001) corroborated this finding, showing how the export intensity of innovating firms is systematically higher than that of non-innovating firms.

The decision to focus our attention on SMEs is not a trivial one. Small firms are the backbone of the Italian economy; their importance as sources of employment and value-added, as well as their contribution to economic growth, is well acknowledged by scholars and analysts (European Commission, 2008).[3] Export orientation has been shown to enhance the probability of survival for SMEs; it is associated with increased productivity and competitiveness (Berry *et al.*, 2002; Bagchi-Sen, 1999) and helps small firms to grow fast (Levy *et al.*, 1999; Samiee and Walters, 1990). For such reasons, assessing the determinants of the export propensity of SMEs will cast light upon the overall economic conditions and competitiveness of the country.

11.3 Model of export propensity and data

11.3.1 The estimation model

In order to investigate how a firm's decision to export is related to its specific characteristics, innovation activities and quality upgrading, we define $EE_{i,2003}$ as 1 if a firm is exporting in the year 2003 and 0 otherwise, and regress it on a vector (X_i) of firm-specific characteristics, product innovation measured as the introduction of new products (I_i), and a vector of quality measures (Q_i). Thus, we define the probability of exporting as $pr(EE_{i,2003} = 1) = \Phi(Z_i)$ where $Z_i = \beta_0 + \beta_1' X_i + \beta_2 I_t + \beta_3' Q_i$ and $\Phi(Z)$ is the standardised cumulative normal distribution. Maximum likelihood analysis is used to obtain estimates of the parameters. Subsequently, we estimate a linear model, restricted to exporting firms, where the independent variables are as above and the dependent variable is defined as the logarithm of turnover imputable to exports ($\ln ET_{i,2003}$). While we recognise that the '*learning by exporting*' literature suggests that innovation itself may be a result, rather than the engine, of exporting, no attempt will be made in this study to correct for the reverse causality. As such, the results presented in this study should be interpreted as the correlation between innovation and export, and not as the effects of innovation upon export.

11.3.2 Database and variables' definition

The data were retrieved from the 9th wave of the Capitalia survey on Italian manufacturing firms with more than ten employees, covering the period 2001–2003 (Capitalia, 2005). The survey covers more than 4,000 manufacturing firms, which were selected using a sampling method stratified by geographical area, industry and firm size.[4] The survey collects firm-level information on turnover level, structural characteristics, labour force, inter-firms relationship, financial structure, attitudes toward foreign markets and various measures of product quality and firms' innovation. Using this data-set, we first define our binary dependent variable ($EE_{i,2003}$) used in the probit model, which is coded 1 if the firm exported during the year 2003, and 0 otherwise. Subsequently, we defined the logarithm of exports' turnover ($\ln ET_{i,2003}$) used in the second estimation model as the share of total turnover imputable directly to exporting activities.

With regard to independent variables, we define 14 sectoral dummies and choose the *food, beverage and tobacco* sector as the reference omitted category; this enables us to capture the export performance of the other sectors with respect to the reference category. We also consider the following: firm's age, measured as the years since establishment; skilled labour force, measured as the number of employees with a university degree; firm's productivity (calculated as the turnover per worker); and three geographical dummies (accounting for firms localised in the North-East, North-West and centre of Italy, using South as the reference omitted category).

With regard to innovation and quality, the database provides valuable information on the following: (1) investment in product innovation; (2) investment in product quality; (3) ISO9000 certification for Quality Level structure; (4) and investment aimed at curbing production costs.

Investment in product innovation is a dummy variable coded 1 if the firm's main investment objective over the period 2001–2003 aimed at the production of new products. This measure presents both advantages and disadvantages. The main advantage is that, in principle, it should capture any innovation introduced by any firm and, hence, should not be affected by differences which exist across sectors as well as across size classes (such as those observed in the propensity to patent). Nonetheless, our measure of innovation suffers from two shortcomings: first, it relies only on the perception of the firms' managers in answering the questionnaire; and second, it does not discriminate between innovations that are new to the firm or to the market.[5]

Quality upgrading is captured by two dummies: ISO9000 quality certification (yes = 1, no = 0) and investments in product quality (1 if the firm has invested, 0 otherwise). Following Pekovic and Galia (2009) we consider ISO9000 certification as a core Quality Level structure. The ISO9000 standards were created in order to facilitate a mutual understanding of quality management system requirements in national and international trade. The notion that ISO9000 is a generic standard implies that the same standards can be applied to any organisation, regardless of company size, product line, and economic sector. Companies that adopt the ISO9000 standard can be assured that their quality programs are built on a firm foundation of state of the art quality practices (Pekovic and Galia, 2009). Strictly speaking, the ISO9000 system is based on management practices intended to integrate quality concerns in to the daily management of organisations. In practice, the implementation of this standard requires profound strategic changes, which concern the whole organisation and involve changes in the fundamental behaviour and applied routine of employees (Pekovic and Galia, 2009, p. 830). In general, ISO9000 certification can deliver business advantages for a firm and, in particular, Douglas *et al.* (2003) contend that ISO9000 certification leads to improvements in quality.

Our second quality measure refers to investments undertaken to improve existing products' quality over the period 2001–2003. As for product innovation, the advantage of measuring product quality using investments undertaken to improve quality is that it accounts for any effort undertaken by a firm's management to improve product quality; however, doubts remain regarding the effectiveness of such efforts. As we believe, both our quality proxies attempt to provide a direct measurement of quality. In this regard, our quality measures differ from other measures considered in previous studies, which typically use the price index of traded goods as a proxy for product quality.[6] This commonly used indirect indicator has some critical shortcomings: for instance, it is based upon the hypothesis that price variations are exclusively linked to qualitative variations of the product in question.[7] Tightly related to this point is our decision to include investments in cost-saving activities in the regression model. As a matter of fact, this variable represents an alternative to quality upgrading, as it would entail a price curbing strategy undertaken by firm *i*.

The descriptive statistics of the variables included in this study are reported in Table 11.1.

It is conceivable that the export strategy adopted by Italian firms depends upon the market they serve. For example, innovations aimed at reducing the cost of production could be the key factor in exporting to low cost countries such as China, while investing in product quality may be the key factor in exporting to more developed economies. In Table 11.2 we test whether there are differences in the volume of turnover toward major export zones associated with each business practice (ISO9000 certification, new product development, quality improvement, or cost-cutting innovation).

We do not find that the effects of the quality, innovation, or cost reduction affect the export turnover differently by region of export. IS09000 certification and new product development are positively related to higher export turnover, no matter where the final export destination is. Investments in product quality are less important in increasing the turnover from exports to China. Technology aiming at curbing production costs does not seem to play a role in export turnover, not even when exports are direct toward China. Overall, these results suggest that Italian exports are strongly influenced by product quality.

11.4 Results and interpretation

In this section we will present the results of our empirical analysis. We will first report the results obtained from running the probit model, and then we will present the results obtained from running the log-linear model of the export-related turnover for exporting firms.

As seen in Table 11.3, which reports the coefficients of the probit model specification discussed in section 3.1, most of our independent variables are

Table 11.1 Summary statistics

Descriptive statistics of the variables used in the study	*Linear estimations*	
	Coeff.	*P >| t |*
Export in 2003	0.7472	0.4347
Business strategies		
Product quality investments	0.5335	0.4990
ISO9000	0.5238	0.4995
Product innovation investments	0.2122	0.4089
Costs saving investments	0.1862	0.3894
Control variables		
Age	30.3728	18.7536
Skilled labour force	6.2695	22.9807
Labour productivity	12.1674	0.6806
Geographical dummies – South is the reference group		
North-East	0.3213	0.4671
North-West	0.3519	0.4777
Centre	0.1762	0.3810
Sectors – food, beverages, and tobacco is the reference group		
Textile and apparel	0.1189	0.3237
Leather and related products	0.0434	0.2037
Wood and furniture	0.0286	0.1668
Paper and printing	0.0554	0.2287
Coke and refined petroleum products	0.0070	0.0832
Chemicals and chemical products	0.0527	0.2234
Rubber and plastics products	0.0534	0.2249
Non-metallic mineral products	0.0616	0.2404
Metal products	0.1707	0.3763
Mechanical products	0.1289	0.3352
Electrical equipment	0.0751	0.2636
Motor vehicles, trailers and semi-trailers	0.0217	0.1457
Other sectors	0.0693	0.2540
N.	2583	

Source: author's elaborations on Capitalia database

statistically significant and correctly signed. For easy interpretation of the results, we also report the marginal effect of the variables. We observe that a firm's age positively affects future export decisions, suggesting that older firms have a higher probability of being exporters. Specifically, we note for each year of increase in a firm's age, there is a corresponding increase in the probability of being an exporter equal to 0.1 percentage points. Also, a skilled labour force positively affects the probability of exporting: a unit increase in the number of employees with a university degree is associated with a 0.4 percentage increase in the probability of

Table 11.2 Mean export turnover by business strategy

Mean export turnover (in euros) by business strategy						
	ISO9000			*Product quality investments*		
	Yes	*No*	*Difference*	*Yes*	*No*	*Difference*
Export turnover	2.50E+09	6.47E+08	1.85e+09***	1.72E+09	9.42E+08	7.73e+08***
Export turnover UE15	2.76E+09	9.93E+08	1.85e+09***	2.47E+09	1.38E+09	1.10e+09***
Export turnover China	7.58E+07	3.19E+07	4.39e+07**	5.70E+07	5.70E+07	3260.647
	Product innovation investments			*Costs saving investments*		
	Yes	*No*	*Difference*	*Yes*	*No*	*Difference*
Export turnover	2.19E+09	1.11E+09	1.08e+09***	1.54E+09	1.99E+09	4.55E+08
Export turnover UE15	2.60E+09	1.76E+09	8.44e+08***	2.15E+09	1.94E+09	2.17E+08
Export turnover China	9.03E+07	4.68E+07	4.35e+07**	7.36E+07	5.19E+07	2.17E+07

*, **, and *** indicate significance of the difference in the means at the 10, 5 and 1 per cent level, respectively (1 tail test).

Source: author's elaborations on Capitalia database

Table 11.3 Probit analysis

Dependent variable: Firms' exporting status, 2003				
	Probit estimations		*Marginal effects*	
	Coeff.	*P >| t |*	*Coeff.*	*P >| t |*
Control variables				
Age	0.005	0.002	0.001	0.001
Skilled labour force	0.015	0.000	0.004	0.000
Labour productivity	0.277	0.000	0.077	0.000
Geographical dummies				
North-East	0.346	0.010	0.096	0.010
North-West	0.306	0.049	0.085	0.054
Centre	0.082	0.628	0.023	0.629
Sectors				
Textile and wearing apparel	0.711	0.002	0.209	0.006
Leather and related products	1.258	0.000	0.300	0.000
Wood and furniture	0.227	0.318	0.076	0.331
Paper and printing	–0.183	0.425	–0.066	0.422
Coke and refined petroleum products	–0.948	0.000	–0.340	0.000
Chemicals and chemical products	0.130	0.635	0.045	0.636
Rubber and plastics products	0.659	0.027	0.197	0.033
Non-metallic mineral products	–0.382	0.031	–0.140	0.028
Metal products	0.188	0.479	0.064	0.485
Mechanical products	0.862	0.001	0.240	0.003
Electrical equipment	0.430	0.127	0.138	0.138
Motor vehicles, trailers and semi-trailers	0.241	0.520	0.081	0.518
Other sectors	0.833	0.003	0.235	0.005
Quality, innovation and costs saving				
Product quality investments	0.082	0.036	0.023	0.040
ISO9000	0.181	0.044	0.051	0.045
Product innovation investments	0.294	0.000	0.082	0.000
Costs saving investments	0.055	0.478	0.015	0.481
Constant	–3.651	0.000		
N.	0.2583			
Pseudo R-squared	0.1252			

Source: author's elaborations on Capitalia database

exporting in 2003. Finally, and most significantly, labour productivity considerably affects a firm's attitude towards foreign markets: as shown in Table 11.3, a unit increase in labour productivity produces an increase in the probability of being exporter equal to 7.7 percentage points.

These findings are in agreement with most of the existing empirical literature. First, one of the most robust and most important results found across countries in this literature is that exporting firms are more productive than non-exporting firms. Bernard *et al.* (2007) estimate that US exporting plants are more productive than non-exporting plants by 14 per cent for value-added and 3 per cent for total factor productivity. Similarly, Mayer and Ottaviano (2007) estimate that French exporters exhibit a 15 per cent higher total factor productivity than non-exporters and a 31 per cent higher labour productivity. Second, data suggests that US and European firms that export are more capital-intensive and skill-intensive with respect to their choice of inputs than firms that do not export (WTO, 2008). Third, another related result is that older plants are more likely to have more intensive export activity (e.g. Johanson and Vahlne, 1990; Bernard and Jensen, 2004).

Sectoral marginal effects show, with few exceptions, a higher probability of being engaged in future exports for firms operating in traditional sectors. Specifically, *Textiles and apparel* and *Leather and related products* are among the sectors with which the highest probability of being exporters is associated. A third sector associated with a high probability of exporting is *Mechanical products*, a finding often understood in the literature as the effect of a progressive shift from the production of typical Italian final goods to the production of machinery and equipment for making them (De Arcangelis and Ferri, 2005). Overall, these findings suggest that Italian firms still have a substantial comparative advantage in traditional sectors.

The analysis of marginal effects also reveals the presence of strong geographical effects. Ceteris paribus, firms located in the North-East and North-West are respectively 9.6 and 8.5 percentage points more likely to engage in export activities when compared with firms located in the South of the country.

Quality variables are both statistically significant and have a positive impact upon exports. All other things being equal, firms complying with the ISO9000 quality standard are, on average, about 5 percentage points more likely to export than firms without ISO9000 certification. Similarly, firms that invested in product quality upgrading increase, ceteris paribus, their probability of exporting by more than 2 percentage points. Investments in product innovation are also associated with a sizable impact upon a firm's exporting probability. In fact, firms that invested in product innovation were, ceteris paribus, 8.2 percentage points more likely to be an exporter in 2003. Interestingly, cost-saving investments do not significantly impact upon export probability. This is a non-trivial result, as it suggests that price competition is not effective for Italian SMEs aiming to penetrate foreign markets. Conversely, the relevance of product innovations, coupled with the positive impact of our quality measures, suggest that climbing the quality ladder is a valuable strategy for promoting exports.

We will now investigate the impact of quality and innovation upon exports, restricting our analysis to exporting firms and assessing their impact upon export turnover growth.

The results reported in Table 11.4 confirm our earlier findings. Again, age, skills and labour productivity are significant and are positively correlated with the

Table 11.4 Linear model (restricted to exporting firms)

Dependent variable: Firms' export turnover (in log), 2003	*Linear estimations*	
	Coeff.	*P* >| *t* |
Control variables		
Age	0.004	0.004
Skilled labour force	0.020	0.000
Labour productivity	0.817	0.000
Geographical dummies		
North-East	0.160	0.097
North-West	0.295	0.001
Centre	0.082	0.483
Sectors		
Textile and wearing apparel	1.094	0.000
Leather and related products	1.445	0.000
Wood and furniture	0.452	0.039
Paper and printing	0.379	0.058
Coke and refined petroleum products	–0.557	0.392
Chemicals and chemical products	0.609	0.025
Rubber and plastics products	0.657	0.016
Non-metallic mineral products	1.058	0.000
Metal products	0.701	0.002
Mechanical products	1.169	0.000
Electrical equipment	0.798	0.001
Motor vehicles, trailers and semi-trailers	1.185	0.000
Other sectors	0.816	0.001
Quality, innovation and costs saving		
Product quality investments	0.218	0.000
ISO9000	0.703	0.000
Product innovation investments	0.248	0.003
Costs saving investments	0.003	0.972
Constant	7.720	0.000
N.	1939	
R-squared	0.277	

Source: author's elaborations on Capitalia database

share of turnover imputable to exporting activities. Being located in the South of the country is also detrimental for exporting firms' turnover, while the traditional sectors are again among the most dynamic ones.

With regard to innovation and quality, one can often observe a strong (and highly significant) relationship between ISO9000 certification and export turnover, which is coupled with a significant and sizable effect for investments in product quality and product innovation. Furthermore, while the marginal effect of product innovation was much bigger than the marginal effect of product quality in the estimation of the propensity to export, we cannot reject the hypothesis that the coefficients on the two variables are identical in the equation of the turnover from export. Again, the coefficient of our price competition proxy variable turns out to be not significant.

11.5 Conclusions

In this chapter, we have investigated the impact of quality and innovation upon the export performances of Italian SMEs operating in the manufacturing sector. Following the existing literature, we defined a model where, along with a set of control variables, export is determined by product innovation activities and quality differentiation undertaken by small and medium enterprises.

The empirical investigation is rooted in innovation theory and 'new' new trade theory and, specifically, it is based on the assumption of firms' heterogeneity. The novelty of our assessment rests upon the measurement of quality, which was twofold: on the one hand we used a dummy variable that captures investment efforts in product quality improvements; on the other hand, we considered compliance with the ISO9000 standard as a measure of quality upgrading.

Running a probit model, we found that product innovation and quality strategies are important drivers of the propensity to export, whereas price competition (proxied by cost saving investments) turned out to be never significant for Italian SMEs operating in manufacturing sectors. A similar result emerges from the analysis of the turnover from exports. Unfortunately, the possibility of reverse causality due to the '*learning by exporting*' effect cast some doubts upon whether the effect estimated in this study represents the true causal effect. However, since reverse causality, if present and not corrected for, would lead to an overstatement of the true effect, we can deduce that cost-saving investments do not play a key role in the export strategy of Italian firms.

This is indeed an interesting result, as it would show that the best performing firms meet the new challenges posed by international markets in a less habitual way: not by chiefly aspiring to obtain competitiveness through cost reduction, but by generating entrepreneurial rents through enhanced knowledge creation, as was reflected in product innovation and quality upgrading. As observed by Maskell (1999), this should be interpreted as a gradual shift towards a knowledge-based economy, which, in turn, is itself an outcome of the ongoing process of internationalisation.

Notes

1 A large body of empirical studies for developed economies provide evidence in support of this hypothesis (Bernard and Jensen, 2004; Bernard *et al.*, 2011; Wagner, 2007, 2011).

2 While the volume of exported commodities remained unchanged, the overall value of exports grew; this was interpreted as a rise in the quality of exported commodities matched with an increase in the unit value of goods sold abroad. Note here that higher prices might simply reflect a strategy of translating higher costs into higher prices abroad, at the expense of lower market share and a less sustainable competitive position in the long run. We shall return to this point in the following section, where we shall propose a direct quality measure, using a qualitative variable rather than the traditional price index of traded goods.

3 There are approximately 65 SMEs per 1000 inhabitants in Italy, substantially above the EU-27 average, which is approximately 40. In line with this, the relative importance of SMEs for the Italian economy by far exceeds the EU average, as illustrated by a considerably above-EU-average share of persons employed and value-added accounted for by SMEs. It should be noted that this elevated importance is mainly due to micro-enterprises, while medium-sized enterprises are, in fact, underrepresented vis-a-vis the EU average.

4 Precisely, the 9th wave of Capitalia contains information on 4,289 firms.

5 As discussed elsewhere (Imbriani *et al.*, 2012), both of these problems refer to what Smith (2005) has labelled the 'fundamental definitional issue' of what should be considered 'new':

> [D]oes an innovation have to contain a basic new principle that has never been used in the world before, or does it only need to be new to a firm? Does an innovation have to incorporate a radically novel idea, or only an incremental change? In general, what kinds of novelty count as an innovation?'.
>
> (2005, p. 149)

6 Often, the average unit price of exported goods is used as a quality proxy. This price is calculated as the ratio between the export value and the export volume. However, it should be borne in mind that this index does not represent a pure price index.

7 For example, if a man's t-shirt produced in Japan costs about four times as much as a similar t-shirt made in Thailand, then it is immediately assumed that the t-shirt made in Japan is of better quality than the one made in Thailand. Nonetheless, it is known that price differences can be the result not only of a difference in quality, but also of different price structures (Faruq, 2006). Recently, other more complex indicators were put forward to overcome these problems (see, for instance, Martinez-Zarzoso and Suarez-Burguet, 2000); however, the average unit price remains the index most commonly used when applying this type of empirical analysis.

References

Amendola, A. (1991) 'Competitività e Specializzazione delle Esportazioni Italiane: Dinamica della Domanda Mondiale e Struttura dei Mercati di Esportazione', in C. Imbriani (ed.) *Commercio estero, Competitività e Specializzazione dell'Italia* – Consiglio nazionale delle ricerche – Progetto finanziato Struttura ed evoluzione dell'economia Italiana, Franco Angeli.

Bagchi-Sen, S. (1999) 'The Small and Medium Sized Exporters' Problems: An Empirical Analysis of Canadian Manufacturers', *Regional Studies*, 33(3): 231–45.

Basile, R. (2001) 'Export Behaviour of Italian Manufacturing Firms over the Nineties: the Role of Innovation', *Research Policy*, 30(8): 1185–1201.

Basile, R., de Nardis, S. and Girardi, A. (2006) 'Pricing to Market of Italian Exporting Firms', *ISAE Working Papers* 70, ISAE – Institute for Studies and Economic Analyses – (Rome).

Bernard, A.B. and Jensen, J.B. (2004) 'Why Some Firms Export', *Review of Economic Studies*, 86(2): 561–69.

Bernard, A.B., Eaton, J., Jensen, B. and Kortum, S. (2003) 'Plants and Productivity in International Trade', *American Economic Review*, 93(4): 1268–90.

Bernard, A.B., Jensen, J.B., Redding, S.J. and Schott, P.K. (2007) 'Firms in International Trade', *Journal of Economic Perspectives*, 21(3): 105–30.

Bernard, A.B., Jensen, J.B., Redding, S.J. and Schott, P.K. (2011) 'The Empirics of Firm Heterogeneity and International Trade', *NBER Working Paper*, No. 17627.

Berry, A., Rodriguez, E. and Sandee, H. (2002) 'Firm and Group Dynamics in the Small and Medium Enterprise Sector in Indonesia', *Small Business Economics*, 18(1–3): 141–61.

Boermans, M.A. and Roelfsema, H. (2012) 'The Effects of Internationalization on Innovation: Firm-Level Evidence for Transition Economies', Tjalling C. Koopmans Research Institute, Discussion Paper Series No. 12–04, Utrecht University.

Borin, A. and Lamieri, M. (2007) 'Misurare La Qualità Dei Beni Nel Commercio Internazionale', *Working Paper della fondazione Manlio Masi*, No. 1/2007.

Bugamelli, M. (2007) 'Export Prices, Product Quality and Firms' Characteristics: An Analysis on a Sample of Italian firms', Temi di discussione (Economic working papers) 634, Bank of Italy, Economic Research Department.

Bustos, P. (2011) 'Trade Liberalization, Exports, and Technology Upgrading: Evidence on the Impact of MERCOSUR on Argentinean Firms', *American Economic Review*, 101(1): 304–40.

Capitalia (2005) *Indagine Sulle Imprese Manifatturiere. Nono Rapporto Sull'industria Italiana E Sulla Politica Industriale*, Rome.

De Arcangelis, G. and Ferri, G. (2005) 'The Specialization of the Districts: From Typical Final Goods to Machinery for Making Them?', in Banca d'Italia, proceedings of the conference: Economie locali, modelli di agglomerazione e apertura internazionale. Nuove ricerche della Banca d'Italia sullo sviluppo territoriale.

Douglas, A., Coleman, S. and Oddy, R. (2003) 'The case for ISO 9000', *The TQM Magazine*, 15(5): 316–24.

Eaton, J. and Kortum, S. (2002) 'Technology, Geography, and Trade', *Econometrica*, 70(5): 1741–79.

European Commission (2008) 'SBA FACT SHEET ITALY', Available online at: http://ec.europa.eu/enterprise/policies/sme/files/craft/sme_perf_review/doc_08/spr08_fact_sheet_it_en.pdf (accessed 27 January 2011).

Faruq, H. (2006) 'New Evidence on Product Quality and Trade', *Caepr Working Papers* 2006–019, Center for Applied Economics and Policy Research, Economics Department, Indiana University Bloomington.

Grossman, G. and Helpman, E. (1991) *Innovation and Growth in the Global Economy*, MIT Press, Cambridge MA.

Grossman, G. and Helpman, E. (1995) 'Technology and Trade', in G. Grossman and K. Rogoff (eds) *Handbook of International Economics*, Vol. 3, North-Holland Publishers.

Imbriani, C., Morone, P. and Testa, G. (2012) 'Innovation, Quality and Exports: The case of Italian SMEs', mimeo—Sapienza University of Roma.

Johanson, J. and Vahlne, J.E. (1990) 'The Mechanism of Internationalization', *International Marketing Review*, 7(4): 11–24.

Lamieri, M. and Lanza, A. (2006) 'La Ripresa Passa Attraverso la Qualità: Una Analisi del Potere di Mercato Delle Imprese Italiane. Imprese e Territorio', *Rivista di analisi economica*, 1(1): 22–9.

Levy, B., Berry, A. and Nugent, J.B. (1999) *Fulfilling the Export Potential of Small and Medium Firms*, Boston, MA: Kluwer Academic Publishers.

Martinez-Zarzoso, I. and Suarez-Burguet, C. (2000) 'The Determinants of Trade Performance: Influence of R&D on Export Flows', *Applied Economics*, 32(15): 1939–46.

Maskell, P. (1999) 'Competitiveness, Localised Learning and Regional Development', in R. Treptow, T. Hastings, P. Maskell and J. Musil (eds) *Economic and Cultural Transitions Towards a Learning City. The Case of Jena*, OECD, Friedrich-Schiller-Universität. OECD Publishing.

Mayer, T. and Ottaviano, G.I.P. (2007) 'The Happy Few: The Internationalisation of European Firms', Bruegel-CEPR EFIM 2007 Report, Brussels: Bruegel Blueprint.

Melitz, M.J. (2003) 'The Impact of Trade on Intra-Industry Reallocations and Aggregate Industry Productivity', *Econometrica*, 71(6): 1695–1725.

Melitz, M.J. and Costantini, J.A. (2008) 'The Dynamics of Firm Level Adjustment to Trade Liberalization', in E. Helpman, D. Marin and T. Verdier (eds) *The Organization of Firms in a Global Economy*, Harvard University Press, Massachusetts.

Modiano, P. (1982) 'Competitività e Collocazione Internazionale delle Esportazioni Italiane: il Problema dei Prodotti Tradizionali', *Economia e Politica Industriale*, 33.

Modiano, P. (1984) 'La Collocazione Internazionale dell'Industria Italiana: un Tentativo di Interpretazione di Alcune Tendenze Recenti', *Economia Italiana*, 3.

Morone, P. and Testa, G. (2008) 'Firms Growth, Size and Innovation', *Economics of Innovation and New Technology*, 17(4): 311–29.

Pekovic, S. and Galia, G. (2009) 'From Quality to Innovation: Evidence from Two French Employer Surveys', *Technovation*, 29(12): 829–42.

Samiee, S. and Walters, P.G.P. (1990) 'Influence of Firm Size on Export Planning and Performance', *Journal of Business Research*, 20(3): 235–48.

Smith, K.H. (2005) 'Measuring Innovation', in J. Fagerberg, D. Mowery and R. Nelson, (eds) *The Oxford Handbook of Innovation*, Oxford University Press, New York, US.

Sterlacchini, A. (1999) 'Do Innovative Activities Matter to Small firms in Non-R&D Intensive Industries? An Application to Export Performance', *Research Policy*, 28(8): 819–32.

Wagner, J. (2007) 'Exports and Productivity: A Survey of the Evidence from Firm Level Data', *The World Economy*, 30(1): 60–82.

Wagner, J. (2011) 'International Trade and Firm Performance: A Survey of Empirical Studies since 2006', *IZA Discussion Paper*, 5916.

WTO (2008) *World Trade Report 2008: Trade in a Globalizing World*, Geneva: World Trade Organization.

12 Do foreign ownership and technological environment affect firm survival?

Evidence on Italian manufacturing and services firms

Filippo Reganati, Anna M. Ferragina and Rosanna Pittiglio

12.1 Introduction

Although foreign multinational enterprises (FMNEs) generally possess some specific characteristics – e.g. they are generally larger in size, more productive, and more capital- and research-intensive, which reduces their likelihood of closure compared to domestic non-multinational firms (DOMFs) – it is commonly argued that FMNEs face a 'liability of foreignness', i.e. they are more likely to exit the market than purely domestic firms (Rodrick, 1997). Theoretically, the expected relationship between foreign ownership and firm survival is ambiguous. The 'footloose' character of FMNEs may be explained by their position within an international production network; thus, they can easily relocate production between countries in response to adverse shocks in the host country. Using optimal portfolio theory, Flamm (1984) showed that US multinationals rapidly adjust their operations in response to changes in host country environments based upon particular country risks. The exit propensity may also depend on the nature of FDI involved: if FDI is horizontal, multinationals are mainly motivated by market-seeking determinants and are thus less likely to be influenced by changes in production costs in host countries. Conversely, vertical foreign investment, which is primarily driven by cost-saving forces, is more sensitive in reacting to changes in production costs and may be more likely to cease as a result of sudden shocks (Inui, Kneller, Matsuura and McGowan, 2009). However, the 'rooted' character of FMNEs with respect to DOMFs may be explained by a result which has emerged from the finance literature, which analyses the effect of sunk entry costs on firm exit (Dixit and Pindyck, 1994): the larger the amount of irrecoverable costs, the greater the value of waiting before making an exit decision. Thus, if the sunk costs of investing abroad are higher than those for establishing a purely domestic plant in the host country, foreign affiliates are less likely to exit.[1]

After controlling for several firm and industry differences, a large amount of empirical research has found that FMNEs are more footloose than DOMFs

(Colombo and Delmastro, 2000, for Italy; Bernard and Sjöholm, 2003, for Indonesia; Görg and Strobl, 2003b, for Ireland; Girma and Görg, 2004, for the UK; Van Beveren, 2007, for Belgium; Alvarez and Görg, 2009, for Chile; Inui *et al.*, 2009, for Japan; Bandick, 2010, for Sweden), whereas other studies have found that foreign-owned firms have the same survival chances as DOMFs (Mata and Portugal, 2002, for Portugal; Özler and Taymaz, 2007, for Turkey).

Recently, by investigating the survival of foreign subsidiaries in Denmark over the period 1895–2001, Kronborg and Thomsen (2009) have shown that FMNEs have a higher survival probability than domestic enterprises; however, the foreign survival advantage appears to have been eroded by globalisation. But within the literature, one issue that has been somewhat overlooked is that of which factors may moderate or strengthen the footloose behaviour of FMNEs. In particular, it could be argued that, given the strong concentration of FMNEs in industries with higher R&D intensity and greater technological complexity, the technological intensity of an industry might moderate their footloose behaviour.[2] This question is particularly crucial if we consider that within the existing literature the relationship between the level of technological intensity in a sector and the survival probability of firms active in that sector is not clear-cut (Audretsch, 1995). On the one hand, being active in a highly innovative sector enables firms to create new products, stimulating growth and thereby their likelihood of survival. There is compelling evidence that product and process innovation are important for survival, because even incumbent firms must continuously innovate in order to mitigate the threat of disruption by new technologies (Christensen, 1997). On the other hand, it has been argued that the risk of exit may be higher for firms in high-tech sectors because of the uncertainty associated with innovation (Ericson and Pakes, 1995).

Our study aims to add some value to the existing literature and fill this research gap by trying to answer the following research questions: How does foreign ownership affect the survival of Italian firms? To what extent are FMNEs located in Italy more 'footloose' than DOMFs? If and to what extent does the level of technological intensity in a sector differently affect the survival dynamics of FMNEs versus DOMFs? To this end, we apply survival methods and estimate a Cox Proportional Hazards model (CPHM hereafter) (Cox, 1972) in which we look for the effect of foreign ownership on firm survival, controlling for firm- and industry-specific covariates.

Our study contributes to the empirical literature on firm survival in several ways. First, our empirical analysis is based on a very large data-set with a much wider coverage of the firm universe than has been considered in previous empirical research on firm survival in Italy. Second, unlike most of the empirical literature, our sample of firms is not restricted to manufacturing sectors, but also covers the services sector. Third, by decomposing firm activities into different technological classes, we shed some light on the impact that technological environment may have upon Italian firms' survival according to their ownership status.

The plan of the chapter is as follows. Section 2 describes the data and provides some descriptive statistics related to FMNEs and DOMFs disaggregated by manufacturing and services and by technological classes. Section 3 presents the model used in the survival analysis. Section 4 provides the results of the CPHM. Finally, Section 5 summarises and concludes.

12.2 Data and description

The empirical analysis included in this chapter has been conducted using firm-level data from the AIDA database (Analisi Informatizzata Delle Aziende), a commercial data-set provided by Bureau Van Dijk which contains extensive firm-level information on Italian firms. In particular, AIDA provides detailed yearly information on a wide set of economic and financial variables, such as sales, costs and number of employees, value added, tangible fixed assets, start-up year, sector of activity, and legal and ownership status. From the ownership section of the data, we constructed a foreign ownership indicator for a firm. A firm is considered to be foreign if it is majority owned, wholly owned, or the main known shareholder is foreign. In a sense, this variable states that a foreign firm is entitled to control or to manage the Italian located affiliate, since it has the largest ownership share. Firm exit is defined as the termination of a firm's activities. Information about this is provided by a 'legal status' variable in the AIDA data-set. It indicates whether a firm is active or inactive. An inactive firm is defined as a firm that is in liquidation, dissolved, or in receivership. By omitting all observations for which the necessary data are incomplete, we have obtained an unbalanced panel of approximately one million observations covering the years 2004 to 2008.[3]

The advantage of using this data-set is twofold. First, it is highly representative of the entire universe of corporate companies (e.g. for 2007, our sample covers approximately 87 per cent of total employees declared by the Italian National Institute of Statistics – ISTAT, 2008). Second, our data-set accurately reflects the actual size distribution of firms in the Italian economy characterised, as is well known, by a large number of small- and medium-sized enterprises.[4]

Before formally testing for the relationship between ownership and firm survival, we provide some descriptive statistics based upon our data. Tables 12.1–12.3 compare the distribution of Italian firms by ownership status, sector of activity, and size, with the latter being measured by the number of employees.

According to the figures, DOMFs represent the largest percentage of Italian firms (more than 99 per cent), which are primarily smaller firms (i.e. those firms

Table 12.1 Distribution of Italian firms by size and ownership status (percentages, sample average)

	FMNEs	*DOMFs*	*Total*
Total sample	0.60	99.40	100.00
Size 1–49	61.08	94.59	94.39
Size 50–249	27.30	4.73	4.87
Size > 250	11.62	0.68	0.74

Source: author's elaborations on AIDA database.

Table 12.2 Distribution of Italian manufacturing firms by size and ownership status (percentages, sample average)

	FMNEs	*DOMFs*	*Total*
Manufacturing sector	**0.76**	**99.24**	**100.00**
Size 1–49	45.95	91.15	**90.80**
Size 50–249	37.48	8.02	**8.24**
Size > 250	16.57	0.83	**0.95**
High and medium-high technology manufacturing sectors	**1.61**	**98.39**	**100.00**
Size 1–49	45.63	90.00	**89.29**
Size 50–249	36.43	8.83	**9.27**
Size > 250	17.94	1.18	**1.45**
Low and medium-low technology manufacturing sectors	**0.42**	**99.58**	**100.00**
Size 1–49	46.45	91.60	**91.41**
Size 50–249	39.09	7.70	**7.83**
Size > 250	14.46	0.70	**0.75**

Source: author's elaborations on AIDA database.

with less than 50 employees account for approximately 95 per cent of total DOMFs). The number of FMNEs is substantially smaller than that of indigenous firms, resulting in a considerably larger average size for foreign firms. Tables 12.2 and 12.3 show the manufacturing and service sectors according to the level of technological intensity.[5] Specifically, following the OECD taxonomy, we aggregate manufacturing (service) sectors into two groupings: (i) High and

Table 12.3 Distribution of Italian service firms by size and ownership status (percentages, sample average)

	FMNEs	*DOMFs*	*Total*
Services	**0.52**	**99.48**	**100.00**
Size 1–49	71.03	96.15	96.02
Size 50–249	20.60	3.24	3.34
Size > 250	8.37	0.60	0.64
Knowledge-intensive services	**0.51**	**99.49**	**100.00**
Size 1–49	69.42	95.37	95.24
Size 50–249	21.58	3.82	3.92
Size >250	9.00	0.80	0.84
Less knowledge-intensive services	**0.53**	**99.47**	**100.00**
Size 1–49	72.02	96.64	96.51
Size 50–249	20.00	2.88	2.97
Size > 250	7.98	0.48	0.52

Source: author's elaborations on AIDA database.

Table 12.4 Average exit rate by ownership status and sector (percentages)

	FMNEs	*DOMFs*	*Total*
Manufacturing Sectors	**5.83**	**4.75**	**4.74**
High and medium-high technology manufacturing sectors	4.80	4.54	4.58
Low and medium-low technology manufacturing sectors	7.50	4.74	4.81
Services	**5.82**	**6.50**	**6.50**
Knowledge-intensive services	6.84	6.72	6.72
Less knowledge-intensive services	5.39	6.12	6.11
Total sample	**5.77**	**6.00**	**6.00**

Source: author's elaborations on AIDA database.

medium-high technology (knowledge-intensive) industries; and (ii) Low and medium-low technology (less knowledge-intensive).

At the subsample level, we observe a pattern of distribution of firms similar to the one observed at the aggregated level. More specifically, both the high and low technology sectors firms are prevalenty small-sized.

In contrast, for the knowledge-intensive and less knowledge-intensive services, firms with 1–49 employees tend to prevail, independently of the ownership status (FMNEs).

Table 12.4 reports the average exit rates of firms (measured by the number of exiting firms relative to the total number of firms); these rates are reported both as a total and according to ownership status.

The tables suggest that, compared with the total sample and the service sector, the percentage of exit is larger in DOMFs, with a rate of 6 and 6.50 per cent, respectively. In the manufacturing sectors, the highest exit rate (5.83 per cent) is registered for FMNEs. Moreover, when we consider the different levels of technology, we observe the expected higher exit rates for firms with low technology, independently of the ownership status. However, the results for the services sectors are surprising, as the firms active in less knowledge-intensive sectors show an exit rate lower than the others.

12.3 Empirical methodology

The empirical analysis is carried out by using survival methods. These methods allow us to control for both the occurrence of an event (i.e. whether a firm exits) and the timing of an event (that is, when the exit takes place). Therefore, these methods take into account the evolution of the exit risk and its determinants over time.

The central concept in survival analysis is the hazard function [$h(t)$] which is defined as the probability that a firm will exit the market at moment t given that it has survived until this period t and conditional on a vector of covariates X_{it}, which may include both time-varying and time-constant variables:

$$h(t, X_{it}) = \lim_{dt \to 0} \frac{Pr(t \leq T < 1 + dt \mid T \geq 1, X_{it})}{dt} \quad (12.1)$$

where T is a non-negative continuous random variable (duration). Specifically, we undertake a multivariate analysis in order to evaluate the effect of foreign ownership on the risk of exit, controlling simultaneously for the effect of the other variables considered. Thus, we estimate the following semi-parametric CPHM:

$$h\,(t, X_{it}) = h_0\,(t)\exp\,(\beta' X) \quad (12.2)$$

where $h_0\,(t)$ stands for the (unspecified) baseline hazard when all of the covariates are set to zero; that is, the hazard function obtained for $\exp\,(\beta' X) = 1$, and β' is a corresponding vector of regression coefficients to be estimated. The logarithm expression of equation (12.2) gives us a linear model that can be estimated by the maximum likelihood method:

$$\ln h(t) = \ln h_0\,(t) + \beta' X_{it}. \quad (12.3)$$

An interesting feature of the model that motivates its name is that the effect of a covariate operates in additive fashion on $h_0\,(t)$ so that a unit change in a covariate leads to a proportional effect on the hazard rate. Also, to interpret the estimates, we should note that hazard ratios lower (higher) than 1 imply that the hazard rate decreases (increases) and the corresponding probability of survival increases (decreases), with all other things being equal.

12.3.1 Variables

Following the prevalent literature on firm survival, we include the following variables in our model:[6]

a) Firm-specific variables

- SIZE is given by the number of employees. Several factors suggest a negative relationship between firm size and the probability of exit. First, the output levels of larger firms are more likely to be close to their industries' minimum efficient scales. Second, compared with small firms, large firms may have a higher chance of survival due to their easier access to capital markets and their greater capacity to recruit qualified workers.
- AGE is defined as the difference between year t and the official year of a firm's incorporation. Because older firms are more likely to possess characteristics that have previously assisted them in preventing exit, we expect that these firms will have a lower chance of exit.

- PROD is labour productivity, i.e. the net value added per employee. Because several theoretical models (Jovanovic, 1982; Hopenhayn, 1992) have predicted that the exit of firms is largely motivated by productivity differences at the firm level, we expect that the survival rate of firms will be higher for more productive firms.
- KL represents a firm's capital intensity and is given by the fixed assets over total employment. Theoretically, we would expect more capital-intensive firms to be more likely to survive. In fact, firms with higher capital–labour ratios may have a lower ratio of variable to fixed costs. Given the basic shutdown rule that a firm will remain in operation as long as it can cover variable costs, firms with low variable-cost production techniques may be more likely to withstand negative shocks than high variable-cost production producers (Doms, Dunne, and Roberts, 1995). In addition, if capital is firm-specific, it may involve sunk costs and become a barrier to exit (Dixit, 1989).
- PROFIT is the operating margin for total sales. Higher profits reduce exiting, while lower profits stimulate the decision to exit; thus, we expect that the operating profit ratio will positively affect the likelihood of survival.
- INW is the foreign multinational ownership dummy that takes a value of 1 if firm i is foreign-owned and 0 otherwise.

b) Industry-specific variables

- HERF is the Herfindhal concentration ratio at the industry level (3-digit Ateco). Theoretically, the expectation of market concentration on survival is not straightforward. A higher market concentration may lead to higher price-cost margins in the industry, which, ceteris paribus, should increase a firm's probability of survival. However, firms in highly concentrated markets may be subject to fiercely aggressive behaviour by rivals, and this behaviour may reduce the chances of survival.[7]
- MES is the minimum efficient scale for each industry. Following Comanor and Wilson (1967), we measured the MES by the average value of shipments of the largest establishments producing more than 50 per cent of industry shipments divided by industry shipments (3-digit Ateco). Theoretically, the expectation of the effect of the MES on firm survival is also ambiguous. Industries with a larger MES are expected to have higher price-cost margins and thus a higher probability of firm survival. However, the MES should exert a positive influence on the hazard rate because the output of new firms is typically less than the MES level (Audretsch, 1991).
- FP is a proxy for foreign penetration and is defined as the share of employment in FMNEs in sector j at time t. A positive relation between this variable and the hazard rate suggests that a greater presence of MNEs is associated with a decrease in the probability of exit for DOMFs and therefore might capture the presence of spillover effects from FDI. Conversely, a negative correlation with the hazard rate means that a greater presence of MNEs is associated with a decrease in the probability of exit and might suggest the presence of competition/displacement effects on firm survival.

Finally, time dummies are included to capture business cycle effects, and industry dummies are included to control for industry-specific characteristics which may affect the survival rates.[8]

12.4 Estimation results

12.4.1 Estimation results for manufacturing and services

In this section we present the regression results of the CPHM of equation (12.2), as applied to our sample of Italian firms over the period from 2004 to 2008. The estimations are stratified by 2-digit Ateco industrial classifications, which allow for equal coefficients of the covariates across strata (industries) but baseline hazards that are unique for each stratum (industry). For each regression, we report the hazard rates and associated robust standard errors, which are adjusted for clustering at the firm level.

The Wald test provides satisfactory support for our model specification.

Table 12.5 shows the results of the model applied both to all sectors (column 1), and separately for manufacturing (column 2) and services (column 3). In the estimates for the total economy, we observe that the coefficient on INW is higher than 1 and statistically significant at the 1 per cent level. More specifically, Italian firms with foreign-firm participation have an approximately 40 per cent increased risk of failure with respect to the DOMFs. This result seems to support the hypothesis that FMNEs are more 'footloose' than national firms. Corresponding to several other studies of the manufacturing sector (e.g., Audretsch and Mahmood, 1995; Colombo and Delmastro, 2000; Bernard and Sjöholm, 2003; Girma and Görg, 2004; Van Beveren, 2007; Esteve Pérez and Mañez Castillejo, 2008), the 'footloose' hypothesis for FMNEs is also confirmed when we divide the entire sample according to the macro-sector of activity in which firms are involved (i.e. manufacturing and services). In particular, the hazard ratios for the coefficients on INW indicate that the hazard rates of foreign firms are approximately 1.6 and 1.3 times higher for FMNEs than for national firms, depending on their sector of activity. In summary, although there was no clear a priori indication about the conditional correlation between multinational ownership and exit patterns, our findings allow us to draw the conclusion that DOMFs are more firmly rooted in the local economy, while FMNEs are more likely to exit.

Let us now turn to the other firm- and industry-specific characteristics. Looking at Table 12.5, we observe that the results are generally in accordance with our expectations and with the results of previous studies. In particular, a robust and expected finding across the three different samples is that older and more productive firms are found to have a lower risk of exit than younger and less productive firms. This result is in line with the predictions of the Industrial Organisation selection models literature (Jovanovic, 1982; Hopenhayn, 1992) and is highly standard in the empirical evidence on firm survival (Dunne, Roberts, and Samuelson, 1988; Mata and Portugal, 2002; Esteve Pérez and Mañez Castillejo, 2008; among others). Firm size has a positive effect on firm survival only in the service sectors (i.e. small firms face a higher hazard of exit compared with large

Table 12.5 Estimation results: Cox Proportional Hazard Model

	All sectors (Ateco 15–99)	*Manufacturing (Ateco 15–37)*	*Services (Ateco 40–99)*
Firm variables			
INW	1.4034	1.5780	1.3441
	(0.0612)a	(0.1144)a	(0.0735)a
AGE	0.9950	0.9950	0.9947
	(0.0005)a	(0.0009)a	(0.0005)a
SIZE	0.9998	0.9999	0.9996
	(0.0000)a	(0.0000)	(0.0001)a
PROD	0.9860	0.9428	0.9868
	(0.0023)a	(0.0127)a	(0.0023)a
KL	0.9812	1.0929	0.9815
	(0.0079)b	(0.0329)a	(0.0081)c
PROFIT	0.9965	0.9960	0.9968
	(0.0007)a	(0.0014)a	(0.0008)b
Industry variables			
HERF	1.0000	1.0000	1.0000
	(0.0000)a	(0.0000)a	(0.0000)a
MES	1.0295	1.0255	1.0328
	(0.0048)a	(0.0110)b	(0.0054)a
FP	0.7670	0.9743	0.7148
	(0.0698)a	(0.1981)	(0.0739)a
Industry dummies	*yes*	*yes*	*yes*
Time dummies	*yes*	*yes*	*yes*
No. of obs	*1,031,172*	*281,341*	*725,740*
No. of subjects	*426,488*	*110,217*	*306,762*
No. of failures	*76,045*	*7,483*	*56,538*
Wald's test (χ^2)	*289.26a*	*120.92a*	*220.55a*

a,b,c indicate statistical significance at the 1, 5 and 10 per cent levels.
Robust standard errors, adjusted for clustering at the firm level, are in parentheses.

firms). Higher firm profits correspond to a lower risk of failure. Similar to the results of several previous studies (Esteve Pérez and Mañez Castillejo, 2008; Özler and Taymaz, 2007), the positive effect of this variable may be attributed to the relationship of higher profits with higher efficiency and/or higher degree of market power.

Regarding the industry-specific covariates, our results show that firms in industries with a higher level of minimum efficient scale have a higher probability of exit. Corresponding to the studies by Audretsch and Mahmood (1995) and

Görg and Strobl (2003b), the negative effect of this variable may be explained as follows: firms may face more obstacles in achieving an efficient production scale and may suffer a cost disadvantage *vis-à-vis* the most efficient firms in the market. Our results also indicate that higher levels of industry concentration correspond to an increased probability of firm exit; this result suggests that firms in highly concentrated sectors are subject to stronger competition, which increases their likelihood of exit.[9]

Finally, we have found that the share of MNEs in the sector has a positive effect on firm survival only in the services sector, within which the exit risk is decreased by more than 28 per cent. These effects may be present because DOMFs are more imitative and thus more able to absorb knowledge spillovers from foreign firms. In manufacturing, our findings of no significant impact for the foreign presence differ from those of both Görg and Strobl (2004) and Burke, Görg, and Hanley (2008). However, our results are consistent with the literature on FDI spillovers in Italy, which has found a weak presence of spillovers from FDI in manufacturing (Imbriani and Reganati, 2002; Castellani and Zanfei, 2007), but evidence of a significant effect in the services sector (Pittiglio, Reganati, and Sica, 2008).

12.4.2 Estimation results for sectors with different technological intensity

Tables 12.6 and 12.7 present estimates for both manufacturing and services by separating firm activities into different technological classes. The estimates reveal that being controlled by a FMNE increases the hazard rate of a firm regardless of the technological level of the sector of activity within which it operates. Thus, within both static and dynamic industries, foreign ownership exerts a strongly negative influence upon the survival of firms by increasing the chance of exit; this result suggests that the behaviour and strategies of MNEs differ from those of DOMFs. In the low and medium-low technology manufacturing sectors, the chances of exit for FMNEs increases by approximately 89 per cent with respect to non-MNEs; in the high and medium-high technology manufacturing industries, the chance of exit increases by approximately 45 per cent. The same result is obtained in the services sector: FMNEs belonging to the less knowledge-intensive services have a 34 per cent higher risk of exiting compared with DOMFs belonging to the same sector, and exceed by 43 per cent the exit risk of DOMFs within knowledge-intensive services.

Some further remarks should be made. We observe significantly higher hazard ratios for FMNEs in the low and medium-low technology manufacturing sectors, compared with the high and medium-high technology manufacturing sectors. This seems to support the hypothesis that while the exit behaviour of foreign firms in less technology-intensive sectors is sensitive to cost opportunities, in more technology-intensive sectors, ceteris paribus, the behaviour of foreign firms is subject to the important sunk costs of establishing production, which determine more persistent investments. However, we observe the opposite result in the services sectors. FMNEs in less knowledge-intensive services exhibit lower exit

Table 12.6 Estimation results by technological intensity in manufacturing: Cox Proportional Hazard Model

	Low and medium-low technology	*High and medium-high technology*
Firm variables		
INW	1.8962	1.4485
	(0.2049)a	(0.1437)a
AGE	0.9958	0.9928
	(0.0011)a	(0.0018)a
SIZE	0.9996	1.0000
	(0.0003)	(0.0001)
PROD	0.9512	0.8401
	(0.0156)a	(0.0208)a
KL	1.1101	1.0933
	(0.0446)a	(0.0476)b
PROFIT	0.9667	1.0001
	(0.0089)a	(0.0016)
Industry variables		
HERF	1.0001	1.0001
	(0.0000)c	(0.0000)a
MES	1.0222	1.0266
	(0.0130)c	(0.0217)
FP	0.8113	1.0025
	(0.2948)	(0.2479)
Industry dummies	*yes*	*yes*
Time dummies	*yes*	*yes*
No. of obs.	*201,054*	*80,287*
No. of subjects	*79,008*	*31,258*
No. of failures	*12,653*	*4,830*
Wald's test (χ^2)	*98.03a*	*93.05a*

a, b, c indicate statistical significance at the 1, 5 and 10 per cent levels.
Bootstrap standard errors, adjusted for clustering at the firm level, are in parentheses.

rates compared with FMNEs engaged in knowledge-intensive services. This result suggests that specific features of services may play an important role. In fact, there is a range of less knowledge-intensive services that are more likely to be non-tradable and thus may be supplied to local markets by foreign and DOMFs only through location in those markets. Therefore, it is not surprising that activities in these sectors are less volatile than those in services with lower knowledge intensity.

Turning to the other firm- and industry-specific characteristics, we observe that the results are generally in accordance with our previous results at the more aggregate level. In particular, older and more productive firms are found to have a lower risk of exit than younger and less productive firms. Firm size has a positive effect upon firm survival only in the service sectors. Regardless of the technological level of the sector, capital-intensive manufacturing firms show a higher rate of exit (between 9 and 11 per cent) than less capital-intensive firms.[10]

In contrast, capital intensity has a positive effect upon the survival of firms belonging to less knowledge-intensive services sectors. Higher firm profits correspond to a lower risk of failure only for firms belonging to low and medium-low technology industries in the manufacturing sector and to less knowledge-intensive services.

With regard to the industry-specific covariates, our results show that firms in industries with a higher minimum efficient scale have a higher probability of exit

Table 12.7 Estimation results by knowledge intensity in the services: Cox Proportional Hazard Model

	Less knowledge-intensive services	*Knowledge-intensive services*
Firm variables		
INW	1.3417	1.4338
	(0.1019)a	(0.1194)a
AGE	0.9966	0.9898
	(0.0008)a	(0.0010)a
SIZE	0.9996	0.9997
	(0.0001)b	(0.0001)a
PROD	0.9872	0.9750
	(0.0023)a	(0.0078)a
KL	0.9107	0.9878
	(0.0431)b	(0.0093)
PROFIT	0.9966	1.0002
	(0.0008)a	(0.0030)
Industry variables		
HERF	1.0001	1.0002
	(0.0000)a	(0.0000)
MES	1.0306	1.0132
	(0.0094)a	(0.0086)
FP	0.4870	0.9645
	(0.0733)a	(0.1603)
Industry dummies	*yes*	*yes*
Time dummies	*yes*	*yes*
No. of obs.	*358,911*	*224,112*
No. of subjects	*148,870*	*96,384*
No. of failures	*26,310*	*17,894*
Wald's test (χ^2)	*116.00a*	*139.84a*

a,b,c indicate statistical significance at the 1, 5 and 10 per cent levels.
Bootstrap standard errors, adjusted for clustering at the firm level, are in parentheses.

in the manufacturing and services sectors of lower technological intensity. Higher levels of industry concentration correspond to an increased probability of firm exit across all of the subsamples, with the exception of knowledge-intensive services.

Finally, we have found that the share of MNEs in the sector has a positive effect upon firm survival in less knowledge-intensive services, in which the exit risk decreases by more than 50 per cent. These effects may be present because DOMFs are more imitative and thus more able to absorb the knowledge spillovers from foreign firms.

12.5 Conclusions

This chapter has empirically investigated the relationship between foreign ownership and firm survival in Italy. In order to achieve this goal, we have conducted an analysis for the period from 2004 to 2008 based on CPHM. This study has allowed us to determine whether foreign firms are more likely to exit the market than DOMFs, and the extent to which the exit probability of foreign firms depends upon the technological environment in which they operate.

The analysis for this period reveals that manufacturing and service firms owned by FMNEs are more likely to exit the market than national firms. These results stand, even when other firm- and industry-specific variables are controlled, thus supporting the idea that FMNEs are inherently footloose.

This chapter has also investigated how foreign ownership affects the survival prospects of Italian manufacturing and service firms according to the technological intensity of the sector in which they operate, and has highlighted some differences between static and dynamic industries. In particular, our findings reveal that foreign ownership generally exerts a negative influence upon firm survival in both static and dynamic industries. However, there is evidence of a stronger negative influence upon the survival of firms operating in low and medium-low technology manufacturing industries. The opposite effect occurs in the services sector, in which FMNEs belonging to high knowledge-intensive services have greater exit risks than foreign firms belonging to less knowledge-intensive services. The much higher hazard ratios for FMNEs in low and medium-low technology manufacturing sectors, compared with the ratios for high and medium-high technology manufacturing sectors, seem to support the hypothesis that the exit behaviour of foreign firms is indeed related to the role of cost opportunities, which are generally relevant in less technology-intensive sectors, and to the role of the sunk costs of establishing production, which are typically higher in less traditional sectors, ceteris paribus. Finally, we find that the presence of MNEs has a positive effect upon the survival of those firms that operate only in the less knowledge-intensive service sector. This effect may exist because DOMFs are more imitative in these sectors and are thus more able to absorb the knowledge spillovers from foreign firms.

Overall, our study enriches our understanding of the determinants of firms' survival in Italy and provides insight on a number of policy issues, especially those related to the risk of adopting ownership-specific incentive policies to account for the very different sensitiveness of manufacturing and services activities in terms of survival, and the very heterogeneous results when a more disaggregated analysis is conducted.

Acknowledgements

We owe special thanks to Carla Carlucci for her support with the database construction and to Umberto De Marco for his assistance with the AIDA data collection. Any errors or omissions remain the responsibility of the authors.

Appendix A

High and medium-high technology manufacturing sectors	**Knowledge-intensive services**
24 Manufacture of chemicals and chemical products 25 Manufacture of rubber and plastic products 29 to 35 Manufacture of machinery and equipment n.e.c. Manufacture of electrical and optical equipment Manufacture of motor vehicles, trailers and semi-trailers Manufacture of other transport equipment	60 Land transport; transport via pipelines 61 Water transport 62 Air transport 64 Post and telecommunications 65 to 67 Financial intermediation 70 to 74 Real estate, renting and business activities 80 Education 85 Health and social work 92 Recreational, cultural and sporting activities
Low and medium-low technology manufacturing sectors	**Less knowledge-intensive services**
15 to 22 Manufacture of food products, beverages and tobacco; textiles and textile products; leather and leather products; wood and wood products; pulp, paper and paper products; publishing and printing 23 Manufacture of coke, refined petroleum products and nuclear fuel 26 to 28 Minerals, basic metals and fabricated metal products; other non-metallic mineral products 36 Furniture	50 to 52 Wholesale and retail trade 55 Hotels and restaurants 63 Supporting and auxiliary transport activities; activities of travel agencies 75 Public administration and defence; compulsory social security 90 Sewage and refuse disposal, sanitation and similar activities 93 Other service activities

Notes

1 However, the literature is controversial on this point. Some authors posit that multinationals should face higher sunk costs when establishing a new firm because new firms are typically more skill-intensive and capital-intensive than incumbent firms. Other authors claim that multinationals, such as multi-unit enterprises, are likely

to benefit from lower sunk costs in terminating a plant's operations due to the greater efficiency of their internal factor markets in re-deploying the production equipment and labour force of the closed plant (Baden-Fuller, 1989).

2 In one recent study on Portugal, Varum, Rocha, and Da Silva (2010), have shown that foreign firms operating in more technology-intensive industries face lower hazards during crises.

3 We focus on this period because, since 2004, AIDA has shown wider coverage of corporate enterprises operating in both the manufacturing and services sectors.

4 Approximately 95 per cent of the firms present in our database have less than 50 employees, compared with the official statistics of 98.5 per cent in 2006 (ISTAT, 2008).

5 For a detailed list of these sectors along with their ATECO codes, see Appendix A.

6 All financial variables were converted into constant 2,000 euros using three-digit industry price level deflators.

7 Görg and Strobl (2004) found a significantly positive effect of concentration on hazard rates at the firm level, while Mata and Portugal (1994) found a negative though insignificant effect of concentration on exit patterns.

8 Simple correlation coefficients were also calculated between the variables to assess whether multicollinearity was present. Correlations between the independent variables were generally low. Results are available upon request.

9 The empirical evidence of the effect of market concentration on firm survival is mixed. Görg and Strobl (2003a) found a positive effect, while Mata and Portugal (1994) and Strotmann (2007) found a negative effect.

10 Similar results were found by Bandick (2010) for Sweden, and by Özler and Taymaz (2007) for Turkey.

References

Alvarez, R. and Görg, H. (2009) 'Multinationals and plant exit: evidence from Chile', *International Review of Economics and Finance*, 18(1): 45–51.

Audretsch, D. (1991) 'New firm survival and the technological regime', *Review of Economics and Statistics*, 73(4): 520–26.

Audretsch, D. (1995) *Innovation and industry evolution*, Cambridge, Mass: MIT Press.

Audretsch, D. and Mahmood, T. (1995) 'New-firm survival: new results using a hazard function', *Review of Economics and Statistics*, 77(1): 97–103.

Baden-Fuller, C. (1989) 'Exit from declining industries and the case of steel castings', *The Economic Journal*, 99(398): 949–61.

Bandick, R. (2010) 'Multinationals and plant survival', *Review of World Economics*, 146(4): 609–34.

Bernard, A.B. and Sjöholm, F. (2003) 'Foreign owners and plant survival', NBER Working Paper, 10039.

Burke, A., Görg, H. and Hanley, A. (2008) 'The impact of foreign direct investment on new firm survival in the UK: evidence for static versus dynamic industries', *Small Business Economics*, 31(2): 395–407.

Castellani, D. and Zanfei, A. (2007) 'Multinational companies and productivity spillovers: is there a specification error?', *Applied Economics Letters*, 14(14): 1047–51.

Christensen, C.M. (1997) 'Patterns in the evolution of product competition', *European Management Journal*, 15(2): 117–27.

Colombo, M. and Delmastro, M. (2000) 'A note on the relation between size, ownership status and plant's closure: sunk costs vs. strategic size liability', *Economic Letter*, 69(3): 421–27.

Comanor, W. and Wilson, T. (1967) 'Advertising, market structure, and performance', *Review of Economics and Statistics*, 9(4): 423–40.

Cox, D.R. (1972) 'Regression models and life-tables', *Journal of the Royal Statistical Society*, Series B, 34(2): 187–220.

Dixit, A. (1989) 'Entry and exit decisions under uncertainty', *Journal of Political Economy*, 97(2): 620–38.

Dixit, A. and Pindyck, R. (1994) *Investment under uncertainty*, Princeton: Princeton University Press.

Doms, M., Dunne, T. and Roberts, M. (1995) 'The role of technology use in the survival and growth of manufacturing plants', *International Journal of Industrial Organization*, 13(4): 523–42.

Dunne, T., Roberts, M.J. and Samuelson, L. (1988) 'Patterns of firm entry and exit in US manufacturing industries', *Rand Journal of Economics*, 19(4): 495–515.

Ericson, R. and Pakes, A. (1995) 'Markov-Perfect industry dynamics: a framework for empirical work', *Review of Economic Studies*, 62(1): 53–82.

Esteve Pérez, S. and Mañez Castillejo, J. (2008) 'The resource-based theory of the firm and firm survival', *Small Business Economics*, 30(3): 231–49.

Flamm, K. (1984) 'The volatility of offshore investment', *Journal of Development Economics*, 16(3): 231–48.

Girma, S. and Görg, H. (2004) 'Blessing or curse? Domestic plants survival and employment prospects after foreign acquisition', *Applied Economics Quarterly*, 50(1): 89–110.

Görg, H. and Strobl, E. (2003a) "Footloose' Multinationals?', *Manchester School*, 71(1): 1–19.

Görg, H. and Strobl, E. (2003b) 'Multinational companies, technology spillovers and plant survival', *Scandinavian Journal of Economics*, 105(4): 581–95.

Görg, H. and Strobl, E. (2004) 'Foreign direct investment and local economic development: beyond productivity spillovers', *GEP The Leverhulme Centre for Research on Globalisation and Economic Policy*, University of Nottingham.

Hopenhayn, H.A. (1992) 'Entry, exit, and firm dynamics in long run equilibrium', *Econometrica*, 60(5): 1127–50.

Imbriani, C. and Reganati, F. (2002) 'Do multinational enterprises affect domestic firms' productivity?', *Studi Economici*, 78: 5–18.

Inui, T., Kneller, R., Matsuura, T. and McGowan, D. (2009) 'Globalization, productivity and plant exit – evidence from Japan', *E.S.R.I.*, Cabinet Office and Nihon University.

ISTAT (2008) *Statistiche in breve. La demografia d'impresa. Anni 2001–2006*, Roma: Istat.

Jovanovic, B. (1982) 'Selection and the evolution of industry', *Econometrica*, 50(3): 649–70.

Kronborg, D. and Thomsen, S. (2009) 'Foreign ownership and long-term survival', *Strategic Management Journal*, 30(2): 207–219.

Mata, J. and Portugal, P. (1994) 'Life duration of new firms', *Journal of Industrial Economics*, 42(3): 227–46.

Mata, J. and Portugal, P. (2002) 'The survival of new domestic and foreign-owned firms', *Strategic Management Journal*, 23(4): 323–43.

Özler, S. and Taymaz, E. (2007) 'Foreign ownership, competition and survival dynamics', *Review of Industrial Organization*, 31(1): 23–42.

Pittiglio, R., Reganati, F. and Sica, E. (2008) 'Horizontal and vertical spillovers from FDI in the Italian productive system', *Quaderni DSEMS* 08–2008, University of Foggia.

Rodrick, D. (1997) *Has globalization gone too far?*, Washington, DC: Institute for International Economics.

Strotmann, H. (2007) 'Entrepreneurial survival', *Small Business Economics*, 28(1): 87–104.

Van Beveren, I. (2007) 'Footloose multinationals in Belgium?', *Review of World Economics*, 143(3): 483–507.

Varum, C.A., Rocha, V.C. and Da Silva, H. (2010) 'Foreign ownership and firm hazard during crises: The moderating role of industry's technological intensity', *The Empirical Economics Letters*, 9(12): 1131–37.

Index

Locators shown in *italics* refer to figures and tables.

additionality, input and output of LAC innovation impacts, 182–9, *186*, *187*
agents role within social networks, 39–40
Agénor, P., 219–20, 225
agglomeration, of economies: definition and features, 91–2; *see also applications*, e.g. services, company
Agrawal, A., 41
Aguirregabiria, V., 164
AIDA database, 259
Aizenman, J., 232
Aldrich, H., 82
Alonso-Borrego, C., 164
Anderson, A., 62
Annicchiarico, B., 220
Antonietti, R., 95
Antonucci, T., 159
approaches, resource-based analysis of cluster company differences, 28, *29*
arrangements, trade, significance for trading systems, 201
Aschhoff, B., 184
Autor, D., 162

Bacchetta, P., 219–20
Basile, R., 245
Batjargal, B., 65
Becattini, G., 26
beliefs, economic, salience as feature of implementation cycles, 142–4
Bell, G., 30
Bell, M., 21
Benigno, P., 225
Bensaid, B., 225
Berman, E., 162
Bernard, A., 251
Bertola, G., 220
Betts, J., 163
Bhagwati, J., 201
Bhave, M., 63
Biggiero, L., 23
BIP (Innovation Programme for Bremen), 118
Blackburn, K., 220
Blanchflower, D., 160
Blanco, H., 220
Boermans, M., 243
Bogliacino, F., 161
Boschma, R., 18, 21, 23, 25
boundaries, regional significance for Bremen, 119, *119*
Bremen (case study): history and structural characteristics, 117–19; internationalisation of technologies in, 125; research results on technological competences in, 125–7, *126*; salience of inventor networks and classes, 123–5, *124*; technological competences of, 120–23, *121*, *122*, *123*; topography and commuting within, 119, *119*
Breschi, S., 5, 18, 19, 21, 41
Broekel, T., 21, 23
Buiter, W., 220
Burger, M., 93
Burgess, S., 160
Burnside, C., 220
businesses *see* companies
Bustos, P., 244

Caballero, R., 220
Cable, D., 62
Cainelli, G., 95
Calvo, G., 225
Campbell, K., 68

Cantner, U., 5–6, 25
Cantwell, J., 113
capital, impact of internationalisation of flows of, 209–10, *211*
capital, markets: currency regime choice in global, 230–1; models of currency crises in, 215–26, *224*; uniqueness vs. multiplicity in crisis models of, 226–30
capital, social: approach to cluster company differences, 30, *31–2*; comparison of use in team and solo start-ups, 71–4, *72–4*, 77–9, *78*; effect on creation of companies, 62–5, 74–7, *75–6*, 79–81, *80*; importance in creation of companies, 60–62; research on use in companies, 65–85, *68*, *69*, *70*, *72–4*, *75–6*, *78*, *80*, *81*
Carlsson, H., 226
Casavola, P., 163
case studies: relationship of product innovation and quality, 243–5; research on Italian FMNEs and DOMFs, 257–69, *259*, *260*, *261*, *267*, *269*; research on propensity of SMEs, 245–53, *248*, *249*, *250*, *252*; technological competences and RINs, 117–27, *119*, *121*, *122*, *123*, *126*
Cassi, L., 41, 46, 47
change, technological, and impact of employment compensation, 155–8; *see also* skills based technology change
Chari, V., 225
choices, foreign regime, salience for currency crisis prevention, 230–2
Christiano, L., 225
citations, patent: role in knowledge flow creation, 43–4, *44*, 53–5, *54*; salience for social networks and knowledge flow, 40–42, *41*
classes, technology; salience for Bremen technological competence, 123–5, *124*
Cloodt, M., 115
clusters, company: centrality of for innovation, 17–19; importance of knowledge for, 17–19; role and importance of heterogeneity of, 25–32, *27*, *29*, *31–2*; *see also* networks, innovation; networks, knowledge
clusters, knowledge, salience as locus of information, 5
Coad, A., 161
codified knowledge, 3–4
Cole, H., 225
Combes, P., 93
communities, knowledge, importance of social networks in, 40
commuting, passenger; significance in Bremen region, 119, *119*
companies; importance of competitiveness and internationalisation, 1–2, *3*; use of knowledge, 4–5; *see also* clusters, company; sectors, company; *see also elements e.g.* creation, company; cycles, implementation; innovation, company; performance, company; *see also type e.g.* small and medium enterprises
companies, services of: classification of high and medium technology, *270*; data on DOMF and FMNE, 259–61, *259*, *260*, *261*; geography of regional specialisation in, 96-7, *97*; Hirschmann linkages as applied to development of, 94–6, 99, 101; impact of DOMF and FMNE on, 264–9, *267*, *268*; impact of urbanisation and knowledge density, 92–4, 99; research on specialisation in, 98–103, *100*
'comparative advantage' theory (Ricardo), 196–7
compensation, role of technology change in mechanisms of, 155–8
'compensation theory' (Marx), 156
competences, technological: case study, 117–27, *119*, *121*, *122*, *123*, *124*, *126*; research on profile of regional, 114–17; salience of regions for development of, 113–14
competitiveness: importance of long-term for company, 1–2, *3*; *see also factors impacting e.g.* innovation; internationalisation
conclusions, research: export propensity of SMEs, 253; foreign ownership and company survival, 269–70; measuring social capital in company start-ups, 81–5; measuring social capital in company start-ups, 81–5; of regional technological profiles, 125–7; specialisation in company services, 102–3
Connolly, M., 220
Constantini, J., 244
Cooke, P., 30
coordination, need for, salience as policy rationale for innovation, 175–6
Corrado, L., 220
Corredoira, R., 30
Corsetti, G., 220

Cowan, R., 39
Cox Proportional Hazards model (CPHM), 258–69, *259*, *260*, *261*, *265*, *267*, *268*
creation, company: comparison of social capital in team and solo start-ups, 71–4, *72–4*, 77–9, *78*; effects of social capital on, 62–5, 74–7, *75–6*, 79–81, *80*; research on social capital and, 65–85, *68*, *69*, *70*, *72–4*, *75–6*, *78*, *80*, *81*; use of social capital in, 60–62
creation, knowledge: and learning processes, 2–5; model of, 42–55, *44*, *48*, *50*, *52*, *54*
crises, currency: fundamentals approach to, 215–21; policy choice in prevention of, 230–2; self-fulfilling based approach to, 221–6, *224*; uniqueness vs. multiplicity in models of, 226–30
crises, economic, significance for trading systems, 201–2
currency: fundamentals approach to crises of, 215–21; policy choice in crisis prevention of, 230–2; self-fulfilling based approach to crises of, 221–6, *224*; uniqueness vs. multiplicity in crisis models of, 226–30
cycles, implementation: characteristics and assessment of model of, 144–50; features and role of expectations, 142–4

Daekeun, P., 220
data envelopment analysis (DEA), 28, *29*
data, research, on characteristics of FMNEs and DOMFs, 259–61, *259*, *260*, *261*
David, P., 176, 184
Davidsson, P., 62–3
DEA (data envelopment analysis), 28, *29*
deindustrialisation, significance for trading systems, 197–8, *198*
De Kock, G., 225
Delegate for Technology of Bremer Senate, 118
Delgado, F., 220
Dellas, H., 219–20
development, regional: salience of innovation and, 112–14; significance of RTA for, 115
development, sectoral, salience of regions for, 113–14
diffusion, knowledge, model of, 42–55, *44*, *48*, *50*, *52*, *54*
Djajic, S., 219
domestic non-multinational firms (DOMFs): comparison with FMNEs, 257–8; impact on company survival, 266–9, *267*, *268*; impact on manufacturing and services, 264–6; methodology models of survival of, 261–4; research data and description of, 259–61, *259*, *260*, *261*
Doms, M., 161, 163
Dornbusch, R., 219, 220
Douglas, A., 247
Drazen, A., 225
Dumas, B., 220
Dunne, M., 163
Duranton, G., 93

economics: features and significance of New Economy, 131–42, *133*, *134*, *135*, *136*, *137*, *138*, *140*, *141*, *142*; relationship of knowledge and information, 2–5; salience of regions in knowledge-based, 112–13; use of LKS, 18; *see also factors affecting e.g.* competences, technological; internationalisation; *see also particular expectations e.g.* beliefs, economic; expectations, economic
economies: definition and features of agglomeration of, 91–2; significance for trading systems, 201–2; *see also applications e.g.* services, company
Edwards, S., 212
Eichenbaum, M., 220
Elfring, T., 63
Elhorst, J., 99
employees and employment: impact of product innovation, 161–5; relationship with innovation, 155–61; significance for trading systems, 200–1
enterprises, small and medium: innovation, quality and export performance, 243–5; research on export propensity of, 245–53, *248*, *249*, *250*, *252*
Entorf, H., 160
entrepreneurship, solo *see* start-ups, solo
entrepreneurship, team *see* start-ups, team
environments, standards for, significance for trading systems, 200–1
environments, technological: impact on company survival, 266–9, *267*, *268*; research data on DOMF and FMNE, 259–61, *259*, *260*, *261*
European Labour Force Surveys, 94
European Patent Office (EPO), 119

Evangelista, R., 159
exchange, foreign: models of currency crises in, 215–26, *224*; policy choice in crisis prevention of, 230–2; uniqueness vs. multiplicity in crisis of, 226–30
expectations, economic, salience as feature of implementation cycles, 142–4
exports: research on propensity of SME, 245–53, *248*, *249*, *250*, *252*; theories and case study of SME, 243–5
externalities, urbanisation, impact on company services, 92–6

failures, institutional, salience as policy rationale for innovation, 175–6
Falk, M., 164
Flamm, K., 257
Flood, R., 218, 219–20
Florin, J., 62
flows, capital, impact of internationalisation of, 209–10, *211*
flows, knowledge: model of creation of, 42–55, *44*, *48*, *50*, *52*, *54*; salience of patent citations for, 40–42, *41*; use of SNA for analysing, 19
FMNEs *see* foreign multinational enterprises
Folta, T., 28
Foray, D., 3
Forbes, R., 220
foreign multinational enterprises (FMNEs): comparison with DOMFs, 257–8; impact on company survival, 266–9, *267*, *268*; impact on manufacturing and services, 264–6; methodology models of survival of, 261–4; research data and description of, 259–61, *259*, *260*, *261*
Francois, J., 95

Garber, P., 218, 219, 220
geography, of company service regional specialisation, 96–7, *97*
Gera, S., 163
Gerlach, S., 220
Giuliani, E., 21, 23, 25, 30
Glick, R., 220
globalisation: impact on capital markets currency regime, 230–2; significance for trading systems, 198–200, *199*; *see also* internationalisation
Goldberg, L., 220
good ('public good'), salience as policy rationale for innovation, 174
Gordon, 140–41
Goshal, S., 60, 61
Goux, D., 164
Graf, H., 5–6, 25
Granovetter, M., 63
Greenan, N., 160
Griliches, Z., 184
Grilli, V., 220, 225
Grossman, G., 243
growth, impact of internationalisation of, 210–12
Guellec, D., 160
Guerrieri, P., 95
Gust, G., 225

Hagedoorn, J., 115
Hall, B., 161, 184
Hall, R., 140
Harrison, R., 161
Haskel, J., 163
Hayton, J., 65
Heden, Y., 163
Helpman, E., 243
Henderson, R., 40–41
heterogeneity, role in performance of cluster companies, 25–32, *27*, *29*, *31–2*
Hirschmann, A., 94–6, 98, 99, 101, 102
Hobijn, B., 131–2
Hodrick, R., 220
Holly, S., 220
Honig, B., 62
Hulsink, W., 63

incentives, tax, role in LAC innovation policy implementation, 178–80
industries *see* companies
information, characteristics and relationship of economics, 2–5
information, asymmetric, salience as policy rationale for innovation, 174–5
initial public offerings (IPOs), features and significance of New Economy, 132–9, *134*, *135*, *136*, *137*, *138*
innovation: centrality of company clusters and social networks, 17–19; impact of internationalisation of, 210–12; link with internationalisation, 9–11; policies of Bremen, 117–19; salience for regional development, 112–14; strengths and link with knowledge, 5–7; *see also* networks, innovation; performance, innovation
innovation, company: characteristics of LAC performance, 172–3, *173*;

implementation in LAC of policies of, 176–80, *177*; rationales for policy intervention involving, 174–6; results of LAC evaluations of, 180–89, *186–7*
innovation, product: impact on employees, 161–5; relationship with employment, 155–61; research on developing model of, 245–53, *248*, *249*, *250*, *252*; theories and case study of relations with quality, 243–5
Innovation Programme for Bremen (BIP), 118
Innovision 2010, 118
inputs, additionality of, of study of LAC innovation impact, 182–5, *186*
institutions, failure of, salience as policy rationale for innovation, 175–6
internationalisation: importance for companies, 1–2, *3*; link with innovation, 9–11; measurement within regions, 116–17; process, degree and effects of, 206–12, *208*, *209*, *210*, *211*; salience of regions for, 113–14; *see also* globalisation; *see also locus of e.g.* Bremen
International Patent Classification (IPC), 120
interventions, policy *see* policies, governmental
interviews, features of in social capital research, 65–6
inventors, salience of networks of for Bremen technological competence, 123–5, *124*
IPC (International Patent Classification), 120
IPOs (initial public offerings), features and significance of New Economy, 132–9, *134*, *135*, *136*, *137*, *138*
ISP (Special Investment Programme), 118
Italy (case study): comparison of DOMF and FMNEs in, 257–8; DOMF and FMNE impact on manufacturing and services, 264–6; impact of DOMF and FMNE technological intensity, 266–9, *267*, *268*; models of DOMF and FMNE survival in, 261–4; relationship between product innovation and quality in, 243–5; research data and description of FMNEs and DOMFs in, 259–61, *259*, *260*, *261*; research on propensity of SMEs in, 245–53, *248*, *249*, *250*, *252*

Jaffe, A., 40–41
Jeanne, O., 219, 225
Jenkinson, T., 132, 133
Jonard, N., 39
Jorgenson, D., 140
Jovanovic, B., 131–2

Kehoe, P., 225
Kehoe, T., 225
Keynes, J., 142
Klette, T., 184
knowledge: and learning processes, 2–5; impact of density of on company services, 92–4, 99; importance for companies, 1–2, *3*; importance for industrial clusters, 17–19; strengths and link with innovation, 5–7; theoretical and empirical literature on networks and, 39–42, *41*; *see also* flows, knowledge; localised knowledge spillovers; networks, knowledge; *see also factors impacting e.g.* economies; innovation; internationalisation
knowledge, local, salience for LKS measurement, 19–25, *20*, *22*, *24*
Kodres, L., 225
Kong, 159
Kox, H., 94
Kramer, C., 219
Kronborg, D., 258
Krugman, P., 220
Krugman-Flood-Garber model of currency crises, 219
Kruse, G., 30

labour and workers: impact of product innovation, 161–5; relationship with innovation, 155–61; significance for trading systems, 200–1
Lahiri, A., 219
Latin American countries (LACs): features and performance of innovation in, 172–3, *173*; implementation of innovation policies in, 176–80, *177*; results of innovation evaluations in, 180–89, *186–7*
Layard, R., 159
Lazerson, M., 26
Lazzarini, S., 30

learning: relationship with knowledge creation, 2–5; role in knowledge flow creation, 44–7
Leduc, S., 131, 139
Lee, J., 201
Lemma, 148–9, 150
LeSage, J., 99, 101
Li, X., 28
Lissoni, F., 5, 18, 19, 21, 41
Ljungqvist, A., 132, 133
localised knowledge spillovers (LKS): advances in measurement of, 19–25, *20*, *22*, *24*; use by economists, 18
Loisel, O., 220
Lorenzoni, G., 26
Luque, A., 164

McCann, B., 28
McDermott, G., 30
McEvily, B., 30
Machin, S., 160, 163, 164
Mairesse, J., 164
manufacturing, company: classification of high and medium technology, *270*; impact of DOMF and FMNE on, 264–9, *267*, *268*; research data on DOMF and FMNE, 259–61, *259*, *260*, *261*
Marini, G., 220
Marion, N., 220
markets, capital: currency regime choice in global, 230–1; models of currency crises in, 215–26, *224*; uniqueness vs. multiplicity in crisis models of, 226–30
Marsden, P., 68
Marshall, A., 17, 91–3, 94
Martin, P., 220
Martinez-Fernández, M., 26, 28
Marx, K., 156
Masson, P., 225
Maurin, E., 164
Mayer, T., 251
mechanisms, compensation, role of technology change in triggering, 155–8
Meliciani, V., 95
Melitz, M., 243, 244
Mendoza, E., 225
Mesquita, L., 30
methodologies, research: measuring social capital in company start-ups, 65–71, *68*, *69*, *70*; of export propensity study, 245–7; of study of LAC innovation impact, 180–82; of survival of FMNEs and DOMFs, 261–4; on regional technological profiles, 114–17
migration, impact of internationalisation of, 208–9, *210*
Miller, V., 220
Missale, A., 225
models and theories: application of SDM for company services, 98–9, *100*; creation and diffusion of knowledge flows, 42–55, *44*, *48*, *50*, *52*, *54*; export propensity of SME, 245–53, *248*, *249*, *250*, *252*; fundamentals approach to currency crises, 215–26, *224*; of 'intrinsic' implementation cycles, 144–50; relationship between product innovation and quality, 243–5; research methodology of DOMF and FMNE survival, 261–4; uniqueness vs. multiplicity in currency crisis, 226–30
Moen, J., 184
Molinia-Morales, F., 26, 28
Morone, P., 4
Morrison-Paul, C., 163
multiplicity, vs. uniqueness in currency crisis models, 226–30

Nahapiet, J., 60, 61
Nefussi, B., 103
networks, innovation: measurement within regions, 115–16; salience for LKS measurement, 19–25, *20*, *22*, *24*; salience for regional development, 114
networks, inventor, salience for Bremen technological competences, 123–5, *124*
networks, knowledge, salience for LKS measurement, 19–25, *20*, *22*, *24*
networks, social: centrality for innovation of clusters of, 17–19; structure and role in knowledge flow creation, 47–9, *48*; theoretical and empirical literature on knowledge and, 39–42, *41*
New Economy, features and significance of productivity and initial public offerings, 131–42, *133*, *134*, *135*, *136*, *137*, *138*, *140*, *141*, *142*, 150
Nickell, F., 159

Obstfeld, M., 219, 220, 225
Organisation for Economic Cooperation and Development (OECD), 202–3
Osakwe, P., 220
Ottaviano, G., 251
outputs, additionality of, of LAC innovation impacts, 185, 187–9, *187*

outsourcing, significance for trading systems, 198–200, *199*

Pace, R., 99, 101
Pack, H., 212
Park, I., 201
Parra-Requena, G., 21
Pástor, L., 132, 133, 134
patents: role in knowledge flow creation, 43–4, *44*, 50–52, *52*; role in regional technological profile research, 114–17; role in study of Bremen technological competences, 120–27, *121*, *122*, *123* *124*, *126*; salience for social networks and knowledge flow, 40–42, *41*
performance, company: characteristics of innovation in LAC, 172–3, *173*; effect of social capital on, 74–7, *75–6*, 79–81, *80*; rationales for policies controlling, 174–6; results of LAC evaluations of, 180–89, *186–7*
performance, export: research on propensity of, 245–53, *248*, *249*, *250*, *252*; theories and case study of SME, 243–5
performance, innovation, role in performance of cluster companies, 25–32, *27*, *29*, *31–2*
Pesenti, P., 220
Pianta, M., 159, 160
Piersanti, G., 220
Piva, M., 161, 163, 164
Ploder, M., 6
Polanyi, M., 3–4
policies, governmental: choices of in currency crisis prevention, 230–2; implementation in LAC of innovation, 176–80, *177*; rationales for company innovation, 174–6; results of LAC evaluations of, 180–89, *186–7*
policies, trade: attacks on and implications for trading systems, 196–202, *198*, *199*; choices of in currency crisis prevention, 230–2; prevalence of world trade restrictions, 202–6, *203*, *204*, *205*, *206*
Polmeier, W., 160
Popper, K., 11–12
Porter, M., 1
PPP (purchasing power parity), 216
prevention, crisis, policy choices in currency, 230–2
Pritsker, M., 225
productivity economic, features and significance of New Economy, 131–42, *133*, *134*, *135*, *136*, *137*, *138*, *140*, *141*, *142*, 150
products, quality of: research on developing model of, 245–53, *248*, *249*, *250*, *252*; theories and case study of relations with innovation, 243–5
profiles, technological, research on regional, 114–17
prophecies, self-fulfilling, role as element of implementation cycles, 142–4
protectionism, significance for trading systems, 201–2
'public good', salience as policy rationale for innovation, 174
Puga, D., 93
purchasing power parity (PPP), 216

quality, product: research on developing model of, 245–53, *248*, *249*, *250*, *252*; theories and case study of relations with innovation, 243–5

Rabellotti, R., 26
Rao, R., 161
Raspe, O., 93, 94
Rebelo, S., 220
regimes, currency, salience of choice for control of, 230–2
regional innovation networks (RINs), 112–14
regions: research on technological profiles of, 114–17; salience for innovation and technological competences, 112–14; specialisation of in company services, 94–103, *97*, *100*; *see also particular e.g.* Bremen
Reid, N., 6
relations, social, centrality for innovation of clusters of, 17–19
Renzulli, L., 82
research: and description of DOMFs and FNMEs, 259–61, *259*, *260*, *261*; measuring social capital in company start-ups, 65–85, *68*, *69*, *70*, *72–4*, *75–6*, *78*, *80*, *81*; methodology of export propensity, 245–7; of regional technological profiles, 114–17; results of export propensity of SMEs, 247–53, 248–53, *248*, *249*, *250*, *252*; specialisation in company services, 98–103, *100*; *see also* methodologies, research; results, research; *see also type e.g.* case studies

restrictions, trade, world prevalence, 202–6, *203*, *204*, *205*, *206*
results, research: measuring social capital in company start-ups, 71–81, *72–4*, *75–6*, *78*, *80*; of export propensity case study, 247–53, *248*, *249*, *250*, *252*; of impact of DOMF and FMNE manufacturing and services, 264–6; of impact of technologies on FMNEs and DOMFs 266–9, *267*, *268*; of study of LAC company innovation impact, 182
Revealed Technological Advantage (RTA), 115, 120–22
Ricardo, D., 196
RINs (regional innovation networks), 112–14
Ritter, J., 150
rivalry, knowledge, role in social networks, 39–40
Roberts, K., 85
Roelfsema, H., 243
Roldos, J., 225
Rose, A., 220
Rotemberg, J., 220
Roubini, N., 220
RTA (Revealed Technological Advantage), 115, 120–22
Rubalcaba, L., 94

Sachs, J., 220, 225
Salvatore, D., 210–11
Sammarra, A., 23
Samuelson, P., 200
Samuelsson, M., 62–3
Savastano, M., 219, 220
Savona, M., 159
SBTC (skills based technology change) hypothesis, 161–5
Schembri, L., 220
Schmitz, G. 26
Schumpeter J.A., 1
Schwellnus, C., 103
SDM (Spatial Durbin Model), 98–9, *100*
sectors, company, salience of regions for, 113–14
Seim, K., 164
services, company *see* companies, services of
Shane, S., 62
Shin, K., 20
Shleifer, A., 130, 131, 139, 141, 143, 144–9
Sichel, 141
Siegel, D., 163
Sill, K., 131, 139
Simonetti, R., 159
simulations, numerical, significance in knowledge flow creation, 49–55, *50*, *52*, *54*
Sinclair, P., 159
Singh, J., 41
skills based technology change (SBTC) hypothesis, 161–5
small and medium enterprises (SMEs): innovation, quality and export performance, 243–5; research on export propensity of, 245–53, *248*, *249*, *250*, *252*
Smets, F., 220
Smith, A., 1
SNA (Social Network Analysis), 19
Snijders, T., 67
Social Network Analysis (SNA), 19
Spatial Durbin Model (SDM), 98–9, *100*
Special Investment Programme (ISP), 118
specialisation, regional; geography of in company services, 96–7, *97*; importance of Hirschmann linkages, 94–6; research on company service, 98–103, *100*
Spencer, M., 219
spillovers, salience as policy rationale for innovation, 174
spillovers, localised knowledge spillovers: advances in measurement of, 19–25, *20*, *22*, *24*; use by economists, 18
Staber, 19
standards, environmental, significance for trading systems, 200–1
start-ups, company: comparison of social capital in team- and solo-, 71–4, *72–4*, 77–9, *78*; effects of social capital, 62–5, 74–7, *75–6*, 79–81, *80*; importance of social capital in, 60–62; research on social capital and, 65–85, *68*, *69*, *70*, *72–4*, *75–6*, *78*, *80*, *81*
start-ups, solo: effects of social capital for, 62–5, 74–7, *75–6*, 79; importance of social capital in, 60–62; research on social capital and, 65–85, *68*, *69*, *70*, *72–4*, *75–6*, *78*, *80*, *81*
start-ups, team: comparison with solo start-ups in use of social capital, 71–4, *72–4*; effects of social capital for, 62–5, 74–7, *75–6*, 80–81, *80*; importance of social capital in, 60–62; research on social capital and, 65–85, *68*, *69*, *70*, *72–4*, *75–6*, *78*, *80*, *81*
Steiner, M., 6

Steinmueller, E., 175
Stiglitz, J., 174
Stiroh, K., 140
Stockman, A., 219–20
Strogatz, S., 48–9
structures, social, features and change in Bremen, 117–19
survival, company: impact of DOMF and FMNE technological intensity, 266–9, *267*, *268*; methodology models of DOMF and FMNE, 261–4; research data and description of, 259–61, *259*, *260*, *261*; research methodology on DOMF and FMNE, 261–4; *see also* manufacturing, company; services, company
Svensson, L., 220
systems, trading *see* trade and trading

tacit knowledge, 3–4
tariffs and non-tariffs, restrictions of, significance for trade restrictions, 202–6, *203*, *204*, *205*, *206*
Taylor, D., 220
Taylor, R., 4
technologies: and impact of employment compensation, 155–8; impact on company survival, 266–9, *267*, *268*; internationalisation of in Bremen, 125; research data on DOMF and FMNE, 259–61, *259*, *260*, *261*; research on profile of regional, 114–17; *see also* competences, technological
Ter Wal, A., 18, 25
theories and models *see* models and theories
Thomsen, S., 258
Tomas-Miquel, J., 28
Toole, A., 184
Tornell, A., 225
trade and trading: attacks on and implications, 196–202, *198*, *199*; impact of internationalisation of, 208, *208*, *209*; significance of arrangements for, 201; *see also* restrictions, trade; *see also impacts e.g.* globalisation; internationalisation
Trajtenberg, M., 40–41

uncovered interest parity (UIP), 216
uniqueness vs. multiplicity in currency crisis models of, 226–30
urbanisation: impact of externalities of on company services, 92–4; research on company service, 98–103, *100*
Uzzi, P., 63

van Damme, E., 226
Van der Gaag, M., 67
Van Oort, F., 93, 94
Van Reenen, J., 160, 164
Van Wijnbergen, S., 220
Végh, C., 219
Velasco, A., 225
ventures, solo and team: comparison of social capital use in, 71–4, *72–4*, 77–9, *78*; effect of social capital on performance of, 74–7, *75–6*, 79–81, *80*; research on success of, 65–85, *68*, *69*, *70*, *72–4*, *75–6*, *78*, *80*, *81*
Veronesi, P., 132, 133, 134
Vivarelli, M., 159, 160, 161, 163
Von Hippel, E., 46–7

Wadhwani, S., 160
Watts, D., 48–9
Weick, K., 85
Weiss, A., 174
Werner, A., 220
West, G., 84–5
Whittington, K., 30
Willman, A., 220
Woerz, J., 95
workers and labour: impact of product innovation, 161–5; relationship with innovation, 155–61; significance for trading systems, 200–1
World Trade Organization (WTO), 201, 243
Wyplosz, C., 219–20

Zaheer, A., 30
Zahra, S., 65
Zhang, S., 28
Zirulia, L., 46, 47